Gunjan Jeswani, Ajita Khichariya (Eds.)
Micro- and Nano-emulsion Technologies

Also of interest

Ethics in Nanotechnology.
Emerging Technologies Aspects
Marcel Van de Voorde, Gunjan Jeswani (Eds.), 2021
ISBN 978-3-11-070181-4, e-ISBN 978-3-11-070188-3

Ethics in Nanotechnology.
Social Sciences and Philosophical Aspects
Marcel Van de Voorde, Gunjan Jeswani (Eds.), 2021
ISBN 978-3-11-071984-0, e-ISBN 978-3-11-071993-2

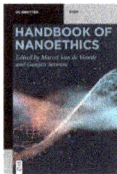

Handbook of Nanoethics
Jeswani, van de Voorde (Eds.), 2021
ISBN 978-3-11-066923-7, e-ISBN (PDF) 978-3-11-066928-2,
e-ISBN (EPUB) 978-3-11-066947-3

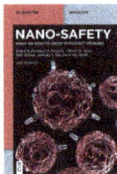

Nano-Safety.
What We Need to Know to Protect Workers
Fazarro, Sayes, Trybula, Tate, Henks (Eds.), 2023
ISBN 978-3-11-078182-3, e-ISBN (PDF) 978-3-11-078183-0,
e-ISBN (EPUB) 978-3-11-078195-3

Hydrogels.
Antimicrobial Characteristics, Tissue Engineering, Drug Delivery Vehicle
Khan, Rauf, Xu (Eds.), 2024
ISBN 978-3-11-133349-6, e-ISBN (PDF) 978-3-11-133408-0,
e-ISBN (EPUB) 978-3-11-133422-6

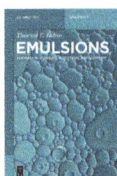

Emulsions.
Formation, Stability, Industrial Applications
Tadros, 2016
ISBN 978-3-11-045217-4, e-ISBN (PDF) 978-3-11-045224-2,
e-ISBN (EPUB) 978-3-11-045226-6

Micro- and Nano-emulsion Technologies

—

Pharmaceutical Design, Pharmaceutical Development, and Drug Delivery

Edited by
Gunjan Jeswani and Ajita Khichariya

DE GRUYTER

Authors

Dr. Gunjan Jeswani
School of Pharmacy
Shri Shankaracharya Professional University
Bhilai 490020
Chhattisgarh
India

Dr. Ajita Khichariya
Kamla Institute of Pharmaceutical Sciences
Shri Shankaracharya Professional University
Bhilai 490020
Chhattisgarh
India

ISBN 978-3-11-159362-3
e-ISBN (PDF) 978-3-11-159365-4
e-ISBN (EPUB) 978-3-11-159383-8

Library of Congress Control Number: 2025942293

Bibliographic information published by the Deutsche Nationalbibliothek
The Deutsche Nationalbibliothek lists this publication in the Deutsche Nationalbibliografie;
detailed bibliographic data are available on the Internet at http://dnb.dnb.de.

© 2026 Walter de Gruyter GmbH, Berlin/Boston, Genthiner Straße 13, 10785 Berlin
Cover image: Yevhenii Khil/iStock/Getty Images Plus
Typesetting: Integra Software Services Pvt. Ltd.

www.degruyterbrill.com
Questions about General Product Safety Regulation:
productsafety@degruyterbrill.com

Foreword

The realm of pharmaceutical sciences is witnessing a paradigm shift in drug delivery technologies, where precision, efficacy, and patient-centric approaches are at the forefront of innovation. In this context, micro- and nanoemulsions have emerged as revolutionary platforms, bridging the gap between traditional drug formulations and modern nanotechnology-driven therapeutics.

Their ability to encapsulate hydrophobic drugs, improve bioavailability, and ensure controlled or sustained release makes them ideal carriers for oral, transdermal, transnasal, ophthalmic, and parenteral drug delivery systems. Hence, they are positioned as pivotal components in advanced pharmaceutical formulations. These emulsions reduce systemic toxicity and side effects, thereby improving patient compliance. Their nanoscale structure facilitates deep tissue penetration, making them particularly effective for oncology, neurological disorders, dermatology, and vaccine delivery. Furthermore, micro/nanoemulsions are being integrated into cosmeceuticals and wound healing formulations, enhancing therapeutic efficacy through superior skin permeation and bioactive delivery.

With continued advancements in formulation science, characterization techniques, and regulatory frameworks, micro/nanoemulsions are set to redefine pharmaceutical innovation, paving the way for next-generation drug delivery systems that are more effective, safer, and patient-friendly.

This book is a meticulously compiled resource that captures the essence of these cutting-edge drug delivery systems. The editors have successfully focused on one of the most promising drug delivery systems and have comprehensively covered all the important areas in this field. This book will serve as a valuable guide for researchers, formulators, and industry professionals, helping them advance research, optimize formulations, and aid in manufacturing innovative pharmaceutical products.

With contributions from leading experts, this book is a testament to the collective efforts of researchers in advancing the frontiers of micro/nanoemulsion technologies. As we stand at the intersection of pharmaceutical nanotechnology and therapeutic innovation, this compilation will undoubtedly inspire future research, drive technological advancements, and contribute to improving patient care through superior drug delivery mechanisms.

Dr. Deependra Singh
Chairman, Education Regulations Committee,
Pharmacy Council of India

https://doi.org/10.1515/9783111593654-202

Foreword

The appeal of microemulsions and nanoemulsions lies in their versatility and effectiveness. As industries push the boundaries of what's possible, the demand for cutting-edge drug delivery systems and high-performance cosmetic formulations has grown exponentially. These technologies offer solutions that were once considered unattainable – efficient transdermal delivery, increased stability of bioactive compounds, and targeted release mechanisms that promise to revolutionize therapeutic and cosmetic sciences.

Notable examples from the market underscore the transformative impact of these technologies. In the pharmaceutical sector, products like sandimmun Neoral, a microemulsion formulation of cyclosporine, have significantly improved the oral bioavailability of the drug, ensuring more consistent therapeutic effects for transplant patients.

This book is a timely and comprehensive exploration of these advancements. It brings together a wealth of knowledge on the formulation, characterization, and applications of microemulsions and nanoemulsions. Each chapter is thoughtfully crafted, guiding readers through the complexities of emulsion science. The discussions extend to practical challenges and emerging trends, making this book an invaluable resource for researchers, practitioners, and industry leaders.

Notably, this volume goes beyond the basics. It delves into specialized applications, such as the role of nanoemulsions in transnasal drug delivery, their efficacy in anticancer treatments, and their promising applications in wound healing and cosmeceuticals. The inclusion of clinical studies and real-world examples further enhances the book's relevance, bridging the gap between theoretical research and practical implementation.

As we move forward, the importance of these technologies in developing effective and targeted drug delivery systems cannot be overstated. This book not only captures the current state of the art but also serves as a catalyst for future research. It is a must-read for anyone committed to advancing the fields of pharmaceutical development and formulation science.

Ajit Singh
Chairman, ACG Group
Email: ajit.singh@acg-world.com

https://doi.org/10.1515/9783111593654-203

Contents

List of contributing authors

Mohammad Adnan Raza
Department of Pharmaceutics
Rungta College of Pharmaceutical Sciences
and Research
Bhilai 490024
Chhattisgarh
India

Anjila Firdous
Department of Pharmaceutics
Rungta College of Pharmaceutical Sciences
and Research
Bhilai 490024
Chhattisgarh
India

Tejas B. Patil
School of Pharmacy and Technology
Management, NMIMS
Shirpur, Dhule 425405
Maharashtra
India

Nousheen Khatoon
Department of Pharmacology
Rungta College of Pharmaceutical Sciences
and Research
Bhilai 490024
Chhattisgarh
India

Ayushmaan Roy
Faculty of Pharmacy
Kalinga University
Naya Raipur
Chhattisgarh
India
And
Department of Pharmaceutics
Rungta College of Pharmaceutical Sciences
and Research
Bhilai 490024
Chhattisgarh
India

Vijayalakshmi Ghosh
Department of Biotechnology
GD Rungta College of Science and Technology
Bhilai 490024
Chhattisgarh
India

Ajazuddin
Department of Pharmaceutics
Rungta College of Pharmaceutical Sciences
and Research
Bhilai 490024
Chhattisgarh
India
And
Rungta College of Engineering and Technology
Bhilai 490024
Chhattisgarh
India

Ashish Pandey
School of Pharmacy
Shri Shankaracharya Professional University
Durg, Bhilai
Chhattisgarh
India
ashispharma@gmail.com

Shweta Ramkar
Kamla Institute of Pharmaceutical Sciences
Shree Shankaracharya Professional University
Durg, Bhilai
Chhattisgarh
India
shwetaramkar@gmail.com

Preeti K. Suresh
University Institute of Pharmacy
Pandit Ravishankar Shukla University
Raipur
Chhattisgarh
India
preetisureshuiop@gmail.com

https://doi.org/10.1515/9783111593654-205

Mitali Singh
Sahu Onkar Saran School of Pharmacy
Faculty of Pharmacy
IFTM University
Moradabad 244102
Uttar Pradesh
India

Mukesh Kumar Singh
Sahu Onkar Saran School of Pharmacy
IFTM University
Moradabad 244102
Uttar Pradesh
India

Arun Kumar Mishra
Sahu Onkar Saran School of Pharmacy
Faculty of Pharmacy
IFTM University
Moradabad 244102
Uttar Pradesh
India

Sanjay Kumar Gupta
Department of Pharmacology
Rungta College of Pharmaceutical Sciences
and Research
Durg, Bhilai 490024
Chhattisgarh
India

Aakash Gupta
Department of Pharmacology
Rungta College of Pharmaceutical Sciences
and Research
Durg, Bhilai 490024
Chhattisgarh
India

Debarshi Kar Mahapatra
Department of Pharmaceutical Chemistry
Shri Rawatpura Sarkar Institute of Pharmacy
Durg, Kumhari 490042
Chhattisgarh
India

Alka Sahu
Department of Pharmaceutics
Rungta College of Pharmaceutical Sciences
and Research
Durg, Bhilai 490024
Chhattisgarh
India

Sumit Sahu
Department of Pharmaceutics
Rungta College of Pharmaceutical Sciences
and Research
Durg, Bhilai490024
Chhattisgarh
India

Priya Komre
Department of Pharmacology
Rungta College of Pharmaceutical Sciences
and Research
Durg, Bhilai 490024
Chhattisgarh
India

Yamini Sahu
Department of Pharmacology
Rungta College of Pharmaceutical Sciences
and Research
Durg, Bhilai 490024
Chhattisgarh
India

Amrita Thakur
School of Pharmacy
Vishwakarma University
Survey No. 2, 3, 4, Kondhwa Main Rd
Laxmi Nagar
Betal Nagar, Kondhwa
Pune 411048
Maharashtra
India

Gunjan Jeswani
School of Pharmacy
Shri Shankaracharya Professional University
Bhilai 490020
Chhattisgarh
India

Hend I. Shahin
Department of Pharmaceutical Sciences
Wegmans School of Pharmacy
St. John Fisher University
3690 East Ave
Rochester, NY 14618
USA

Lipika Chablani
Department of Pharmaceutical Sciences
Wegmans School of Pharmacy
St. John Fisher University
3690 East Ave
Rochester, NY 14618
USA

Swarnali Das Paul
Shri Shankaracharya Professional University
Bhilai
Chhattisgarh
India

Disha Kesharwani
Columbia Institute of Pharmacy
Raipur
Chhattisgarh
India

Monika Bhairam
Columbia Institute of Pharmacy
Raipur
Chhattisgarh
India

Gunjan Kalyani
Columbia Institute of Pharmacy
Raipur
Chhattisgarh
India

Sandhya Mishra
Columbia Institute of Pharmacy
Raipur
Chhattisgarh
India

Shrikant Dargude
Department of Pharmaceutical Sciences
School of Health Sciences and Technology
Dr. Vishwanath Karad MIT World Peace
University
Kothrud, Pune
Maharashtra
India

Ajita Khichariya
Kamla Institute of Pharmaceutical Sciences
Shri Shankaracharya Professional University
Bhilai 490020
Chhattisgarh
India

Anuruddha Chabukswar
Department of Pharmaceutical Sciences
School of Health Sciences and Technology
Dr. Vishwanath Karad MIT World Peace
University
Kothrud, Pune
Maharashtra
India

Satish Polshettiwar
Department of Pharmaceutical Sciences
School of Health Sciences and Technology
Dr. Vishwanath Karad MIT World Peace
University
Kothrud, Pune
Maharashtra
India

Swati Jagdale
Department of Pharmaceutical Sciences
School of Health Sciences and Technology
Dr. Vishwanath Karad, MIT World Peace
University
Kothrud, Pune
Maharashtra
India

Gunjan Jeswani and Ajita Khichariya

Introduction

Micro- and nanoemulsions have emerged as revolutionary systems in pharmaceutical science due to their exceptional ability to enhance the solubility, stability, and bioavailability of therapeutic agents. These colloidal dispersions, characterized by their nanoscale droplet size and thermodynamic stability, have opened new avenues for effective drug delivery across various routes, including oral, transdermal, and parenteral. The development of micro- and nanoemulsion systems integrates principles of nanotechnology with pharmaceutical design, enabling precise control over drug release profiles and targeted delivery. This book aims to provide a comprehensive understanding of micro- and nanoemulsion technologies, focusing on their formulation, characterization, and diverse applications in medicine, consumer products, and industry.

Evolution of emulsion, microemulsion, and nanoemulsion

Origin of emulsions: natural emulsiogenesis

The earliest emulsions formed naturally on the Earth about 3–4 billion years ago with the emergence of waterborne microbes. These microorganisms produced organic hydrocarbon molecules that aggregated into liquid oil droplets dispersed in water, creating the first natural oil-in-water emulsions. Over time, photosynthetic phytoplankton, bacteria, algae, and fungi significantly increased hydrocarbon production, leading to large crude oil deposits around 300 million years ago. This period saw the formation of both oil-in-water and water-in-oil emulsions naturally within porous reservoir rocks.

Early human use and definition

The term "emulsion" originates from the Latin word *emulgere* meaning "to milk out," inspired by the milky-white dispersions secreted by plants approximately 300–400

Gunjan Jeswani, School of Pharmacy, Shri Shankaracharya Professional University, Bhilai 490020, Chhattisgarh, India
Ajita Khichariya, Kamla Institute of Pharmaceutical Sciences, Shri Shankaracharya Professional University, Bhilai 490020, Chhattisgarh, India

https://doi.org/10.1515/9783111593654-001

million years ago. Emulsions, by definition, consist of droplets of one immiscible liquid phase dispersed in another, stabilized by surfactants or amphiphilic agents. The invention of the homogenizer by Gaulin in 1900 marked a significant advancement in emulsion technology, enabling the production of submicron droplets through high-pressure homogenization.

Evolution of microemulsions

The term "microemulsion" was first introduced by J.H. Schulman in 1959 to describe systems that form spontaneously due to a dramatic reduction in interfacial tension. Unlike traditional emulsions, microemulsions are thermodynamically stable systems with nanoscale droplets typically ranging from 10 to 100 nm. They consist of multiple components, such as water, oil, surfactants, and sometimes cosurfactants, which self-assemble into various structures like spherical micelles or bicontinuous phases.

Microemulsions differ fundamentally from conventional emulsions in that they do not require external energy for formation but rely on spontaneous emulsification driven by entropy and low interfacial tension. The introduction of the phase inversion temperature (PIT) method further expanded the scope of microemulsions by enabling their formation through temperature-induced changes in surfactant solubility.

Emergence of nanoemulsions

The concept of nanoemulsions emerged in the late twentieth century, gaining significant traction due to advancements in nanotechnology. Unlike microemulsions, nanoemulsions are kinetically stable but thermodynamically unstable systems with droplet sizes ranging from 20 to 200 nm. They require external energy inputs, such as high-pressure homogenization or ultrasonication for formation. Nanoemulsions are particularly valued in pharmaceuticals for their ability to enhance drug bioavailability and stability.

The first peer-reviewed article mentioning nanoemulsions was published in 1996 by Calvo et al., although the term was inconsistently used with "submicron emulsions." The rapid growth in publications on nanoemulsions around 2010 reflects their rising importance, surpassing microemulsions in research focus due to their versatility in drug delivery, cosmetics, and food industries.

Process of emulsification

The emulsification process for micro- and nanoemulsions involves creating stable dispersions of immiscible liquids, such as oil and water, with the help of surfactants. This process can be broadly categorized into high-energy and low-energy methods, each offering distinct advantages based on the desired application and properties of the emulsion.

High-energy methods

High-energy methods rely on external mechanical forces to break down larger droplets into nanoscale droplets. These techniques include:

- **High-pressure homogenization:** Involves forcing the oil-water mixture through a narrow orifice at high pressure, causing intense shear forces that break down droplets into nanometric sizes. This method is particularly effective for producing kinetically stable nanoemulsions.
- **Ultrasonication:** Utilizes ultrasonic waves to generate cavitation forces that fragment oil droplets into smaller sizes. This technique is efficient for producing both nanoemulsions and microemulsions with a uniform size distribution.
- **Microfluidization:** A specialized form of homogenization in which the mixture is passed through microchannels, creating high shear and turbulence, resulting in smaller and uniform droplets.

These methods are characterized by their ability to produce nanoemulsions that remain kinetically stable due to their small droplet size, which prevents coalescence and flocculation.

Low-energy methods

Low-energy methods take advantage of the intrinsic physicochemical properties of the emulsion components, utilizing spontaneous emulsification mechanisms without external forces. Key techniques include:

- **PIT:** Involves adjusting the temperature to invert the phases (oil-in-water to water-in-oil or vice versa). As the temperature crosses a critical point, the spontaneous formation of small droplets occurs.
- **Spontaneous emulsification:** This method involves mixing the oil phase containing a lipophilic surfactant with the aqueous phase. As the surfactant diffuses, it creates interfacial turbulence, leading to the spontaneous formation of nano-sized droplets.
- **Self-emulsifying drug delivery systems:** These are mixtures of oil, surfactant, and drug that spontaneously form nano- or microemulsions upon contact with aqueous fluids, enhancing drug solubilization and bioavailability.

The low-energy methods are particularly useful for systems where maintaining thermodynamic stability is crucial, such as in microemulsions that form spontaneously under equilibrium conditions.

Patented technology

Ascendia's **EmulSol**® technology creates stable, optically clear nanoemulsions without relying on organic solvents and using minimal cosurfactants, thanks to a high-pressure homogenization process. By carefully selecting specific long-chain triglycerides and combining them with an ionizable surfactant, Ascendia has eliminated the need for organic solvents in its formulation approach. Although EmulSol formulations are prepared through a conventional homogenization process, the proprietary blend of oils and surfactants results in a physically stable suspension of oil droplets in the water phase, making the formulation safer for administration. Removing solvents reduces injection site irritation, making the technology particularly suitable for pediatric applications, while minimizing surfactants enhances both safety and chemical stability. Ascendia has applied this innovative technology to develop its lead pipeline product, ASD-002, a novel injectable form of the antithrombotic drug **clopidogrel**.

Key differences in emulsification of micro- and nanoemulsions

- **Thermodynamic/kinetic stability**
    - Emulsions are kinetically stable.
    - Microemulsions are thermodynamically stable, forming spontaneously without external energy, but they are sensitive to changes in temperature and dilution.
    - Nanoemulsions are kinetically stable, requiring external energy for their formation but remain stable under temperature and dilution changes.
- **Droplet size and uniformity**
    - Emulsions have larger droplet sizes (1–100 µm).
    - Microemulsions generally form smaller and more uniform droplets (10–100 nm) with a clear or translucent appearance.
    - Nanoemulsions can have a wider range of droplet sizes (20–500 nm) and typically appear milky.
    - Miniemulsions offer smaller droplets (100–500 nm).
- **Surfactant requirements**
    - Microemulsions require a high concentration of surfactants and sometimes cosurfactants to stabilize the interface; therefore, they are thermodynamically stable and help prevent Ostwald ripening.
    - Nanoemulsions can be stabilized with lower surfactant concentrations, making them preferable for certain pharmaceutical applications.
    - Miniemulsions offer improved stability due to the presence of cosurfactants.

The choice of emulsification process is thus dictated by the intended application, desired stability, and physicochemical characteristics of the active ingredients.

Key breakthrough applications

In medicine

Micro- and nanoemulsions have demonstrated immense potential in medical applications, particularly in enhancing the therapeutic efficacy of poorly water-soluble drugs. One of the breakthrough applications of microemulsions in medicine is demonstrated by the cyclosporin microemulsion formulation (Neoral®), which has significantly enhanced the pharmacokinetic profile of cyclosporin compared to its standard oil-based formulation. The microemulsion system allows for more consistent and rapid absorption, higher peak blood concentrations, and reduced intra- and interpatient variability. These advantages translate to improved therapeutic outcomes in renal and liver transplantation by minimizing the risk of graft rejection and optimizing immunosuppression. Additionally, the microemulsion formulation is less affected by food and bile, making it particularly beneficial for patients with hepatic dysfunction or biliary diversion.

Similarly, breakthrough applications of nanoemulsions in medicine are exemplified by Diprivan®, a commercially available nanoemulsion formulation of propofol, which is widely used as a general anesthetic. The nanoemulsion system of Diprivan® significantly enhances the solubility and bioavailability of propofol, allowing for rapid onset and short duration of action, which are critical for procedures requiring quick induction and recovery from anesthesia. The formulation is stabilized using egg lecithin as a surfactant, which helps in maintaining the kinetic stability of the nano-sized oil droplets and prevents coalescence. Additionally, the small droplet size in the nanoemulsion facilitates a larger surface area for absorption, leading to more predictable pharmacokinetics and reduced pain on injection compared to conventional macroemulsions. This advancement demonstrates the potential of nanoemulsions to improve the safety, efficacy, and patient compliance of injectable drugs, making Diprivan® a landmark example in the clinical application of nanoemulsion technology.

In consumer products

The utilization of micro- and nanoemulsions extends beyond pharmaceuticals to consumer products, including cosmetics, personal care, and nutraceuticals. Their ability to enhance the solubility of active ingredients and provide a smooth, non-greasy feel has led to their widespread adoption in skincare formulations, sunscreens, and hair care products. This section explores the role of emulsions in achieving product stability, improved sensory properties, and enhanced efficacy.

In industrial sector

In the industrial sector, micro- and nanoemulsions are employed for their superior dispersion capabilities, enhanced mass transfer, and stability. Applications span from agrochemicals, where they improve pesticide delivery and reduce environmental impact, to the oil and gas industry, where they aid in enhanced oil recovery and corrosion inhibition. The section examines case studies highlighting the economic and operational benefits of these systems in various industrial processes.

Challenges

Microemulsions and nanoemulsions, despite their promising applications, face several challenges in their development and commercialization. One major hurdle is terminological confusion due to the overlap in droplet size ranges and the inconsistent use of terms like "low-energy emulsification" for both systems, making it difficult to clearly differentiate them. Additionally, achieving and maintaining long-term stability is complex, as nanoemulsions are thermodynamically unstable and prone to coalescence, creaming, and Ostwald ripening over time. Microemulsions, while thermodynamically stable, require high surfactant concentrations that can raise toxicity concerns, especially in pharmaceutical applications. Furthermore, scaling up production while maintaining narrow droplet size distributions and consistent performance remains a technical challenge. Advanced techniques like microfluidization and ultrasonication have improved size control, but these methods can be energy-intensive and may not always be feasible for large-scale manufacturing. Overcoming these challenges is crucial for harnessing the full potential of micro- and nanoemulsions across diverse industries.

Conclusion

The evolution of emulsions, microemulsions, and nanoemulsions represents a remarkable journey from naturally occurring systems to sophisticated delivery platforms with diverse applications across pharmaceuticals, food, and cosmetics. Their small droplet size and ability to disperse hydrophobic components make them particularly valuable for enhancing solubility, stability, and targeted delivery. As research advances, these systems hold immense promise for high-impact applications such as immunotherapy and targeted drug delivery. While many preclinical studies are still in their early stages, a deeper understanding of the intricate interactions between emulsions and biological systems – from cells to tissues and organs – will be pivotal in accelerating their translation into real-world clinical and commercial applications. The future of micro- and nanoemulsions is undoubtedly bright, with ongoing innovations paving the way for groundbreaking advancements in medicine and beyond.

Mohammad Adnan Raza, Anjila Firdous, Tejas B. Patil,
Nousheen Khatoon, Ayushmaan Roy, Vijayalakshmi Ghosh,
Sanjay kumar Gupta, and Ajazuddin*

Chapter 1
Techniques for characterizing micro/nanoemulsions in pharmaceutical formulations

Abstract: Emulsions are colloidally stable dispersions of two immiscible liquids stabilized by surfactants and cosurfactants, exhibiting droplet sizes ranging from 2 to 100 nm. Nanoemulsion and microemulsion formulations are commonly used for drug delivery. They can be formulated using a variety of high- and low-energy processes. This chapter reviews different techniques for characterizing these nano- and microemulsions, discussing the significance of optimum formulation for nanodroplet systems, emphasizing droplet size, solubilization, colloidal stability, and optical and rheological characteristics. Methods such as dynamic light scattering (DLS) and laser diffraction are employed to ascertain droplet size and polydispersity index (size distribution) – critical parameters affecting the emulsion's stability and performance. DLS is also used to assess zeta potential, that is, to analyze surface charge and stability of emulsions. Electron mi-

---

Mohammad Adnan Raza and **Anjila Firdous** contributed equally.

Acknowledgment: The authors acknowledge the e-library of Rungta College of Pharmaceutical Sciences and Research, Kohka, Kurud Road, Bhilai, Chhattisgarh, India, for providing the necessary literature to compile the work. The authors also wish to acknowledge DST-FIST project Level-0 (SR/FST/College-2018-431-C) for providing financial assistance.

*Corresponding author: Ajazuddin,** Product Development Laboratory, Department of Pharmaceutics, Rungta College of Pharmaceutical Sciences and Research, Kohka, Kurud Road, Bhilai, Chhattisgarh, India; Research and Development, Rungta College of Engineering and Technology, Kohka, Kurud Road, Bhilai, Chhattisgarh, India, e-mail: write2ajaz@gmail.com
Mohammad Adnan Raza, Anjila Firdous, Department of Pharmaceutics, Rungta College of Pharmaceutical Sciences and Research, Bhilai 490024, India
Tejas B. Patil, School of Pharmacy and Technology Management NMIMS, Shirpur, Dhule 425405, India
Nousheen Khatoon, Department of Pharmacology, Rungta College of Pharmaceutical Sciences and Research, Bhilai 490024, India
Ayushmaan Roy, Department of Pharmaceutics, Rungta College of Pharmaceutical Sciences and Research, Bhilai 490024, India; Faculty of Pharmacy, Kalinga University, Naya Raipur, India
Vijayalakshmi Ghosh, Department of Biotechnology, GD Rungta College of Science and Technology, Bhilai 490024, Chhattisgarh, India
Sanjay kumar Gupta, Department of Pharmacology, Rungta College of Pharmaceutical Sciences and Research, Bhilai 490024, Durg, Chhattisgarh, India, e-mail: sanjay.gupta0311@gmail.com

https://doi.org/10.1515/9783111593654-002

croscopic techniques (e.g., transmission electron microscopy and scanning electron microscopy) are used to visualize the morphology and structural details at the nanoscale. The characterization of nano- and microemulsions is conducted by differential scanning calorimetry to assess the physical state of the drug inside the formulations and to investigate potential interactions between the drug and other constituents of the nano- and microemulsions. This chapter also addresses the importance of rheological studies in understanding the flow behavior and viscoelastic properties of emulsions, as well as spectroscopic techniques as Fourier-transform infrared spectroscopy and nuclear magnetic resonance to investigate the molecular interactions and composition of the emulsions. The application of these characterization techniques in optimizing formulation processes and ensuring the quality of pharmaceutical micro/nanoemulsions is also discussed. The in vitro drug release, vitro permeation, stability and thermodynamic stability, shelf life, dispersibility, viscosity, surface tension, refractive index, percentage transmittance, pH, and osmolarity of nanoemulsions are all further examined.

Keywords: Nanoemulsion, microemulsion, dynamic light scattering, zeta potential, transmission electron microscopy, rheology

1.1 Introduction to micro/nanoemulsions in pharmaceutical formulations

Micro- and nanoemulsions have been an interest among the pharmaceutical researchers for drug delivery application owing to their distinct physicochemical characteristics and high loading efficacy leading to better bioavailability of drugs [1, 2]. Nanoemulsions differ from conventional emulsions due to the in situ droplet dimensions being within submicron and nanoscales (typically 10–100 nm for nanoemulsions, >200 nm and <500 nm). Consequently, they exhibit exceptional stability apart from low optical transparency [3]. Their properties make them very flexible carriers over a wide range of active pharmaceutical ingredients (APIs), especially those with low water solubility leading to enhanced better solubilizing, stabilization, and controlled release behavior [4, 5].

In pharmaceuticals, micro- and nanoemulsions are investigated as carriers for different drug types (e.g., hydrophilic, lipophilic, or amphiphilic compounds) [6]. These emulsions have very small droplet sizes that increase the surface area and exposure to biological membranes resulting in improved absorption and bioavailability. These features have the potential of a variety of oral, topical, and injectable drug applications within their framework [7]. While microemulsions and nanoemulsions are both useful in pharmaceuticals, they have contrasting stability, preparation methods of the formulation components, and surfactant demands.

Microemulsions are thermodynamically stable, clear, or translucid dispersions of oil in water, that form spontaneously but where the droplet size is much smaller than those found in conventional emulsion; whereas nanoemulsions are kinetically stabile transparent colloidal dispersion (compared to traditional macro- and microemulsion systems) [8]. Most of the microemulsions have droplet sizes in the range from 10 to 100 nm but their size can go up to thousands of times this limit, which makes it difficult for them to pass through all body sites. In contrast, nanoemulsions have a relatively more pure droplet size within the range of 0.5–100 nm [9]. Both types of emulsions are typically transparent or opalescent, but nanoemulsions tend to be clearer. This confers aesthetic superiority in formulations where appearance assumes significant importance [10].

Stability is the prime difference between micro- and nanoemulsions. Being thermodynamically stable, it means that microemulsions form spontaneously and will not separate over time. Nanoemulsions are only kinetically stable as they require energy input for the formation, which would prolong their stability, although only temporarily and not permanently [11]. Microemulsions are spontaneously formed through a high concentration of surfactant and cosurfactant; hence, they are quite easy to prepare. Nanoemulsions, however, require high energy preparation methods such as ultrasonication or high-pressure homogenization, more commonly at lower surfactant concentrations than microemulsions, which may reduce the risk for toxicity [10, 12]. Oil and water are arranged in microdomains within a continuous phase to create specific structured phase behaviors in microemulsions. Nanoemulsion represents a more uniform and evenly distributed phase without forming coexisting phases, which may be more advantageous for the stability and activity of certain pharmaceutical applications [13].

Microemulsions have vast applications in pharmaceuticals, food, and cosmetics and are beneficial for the controlled and targeted delivery of active ingredients. Nanoemulsions are highly valuable for drug delivery applications, enhancing bioavailability, and enabling controlled release, especially in relation to low solubility drugs [14, 15]. Additionally, nanoemulsions have an improved ability to cross biological barriers, making them promising for advanced drug delivery. While microemulsions benefit from spontaneous formation, they require high surfactant concentrations, which can pose risks of toxicity or irritation. Nanoemulsions, though more stable in biological applications, require energy-intensive preparation and lack inherent thermodynamic stability, potentially impacting their long-term storage and shelf life [16].

Micro- and nanoemulsions play a pivotal role in modern drug delivery systems due to their ability to improve the solubility, stability, and bioavailability of therapeutic compounds [17]. Their small droplet size provides an increased surface area that enhances the dissolution and absorption of poorly water-soluble drugs. This makes them particularly valuable for oral formulations of hydrophobic drugs, like curcumin and paclitaxel, which face bioavailability challenges. Additionally, these emulsions facilitate controlled and sustained drug release, especially in topical and transdermal applications, where they enable steady drug penetration for extended therapeutic effects [18]. In targeted drug delivery, nanoemulsions can be engineered with specific ligands that recognize and bind to target cells, such as cancer cells, enhancing efficacy and reducing systemic toxicity [19]. Parenteral applications also benefit from nanoemulsions, which allow for stable, injectable formulations of lipophilic drugs, offering rapid onset in critical care settings. By encapsulating sensitive drugs, such as peptides and proteins, micro- and nanoemulsions also extend drug stability and shelf life. Overall, these versatile systems address a broad spectrum of pharmaceutical challenges, making them indispensable in drug formulation and delivery innovations [20].

This chapter reviews various techniques for characterizing nanoemulsions and microemulsions, focusing on the critical aspects required to optimize these nanodroplet systems for pharmaceutical and therapeutic applications. The characterization techniques discussed include dynamic light scattering (DLS) and electron microscopy, which are essential for accurately determining droplet size distribution and morphology – key factors influencing bioavailability and stability. Additionally, solubilization capacity is examined, as it plays a crucial role in enhancing the bioavailability of hydrophobic drugs, ensuring efficient delivery. Colloidal stability is also explored, highlighting factors such as surfactant concentration, zeta potential, and pH that contribute to preventing coalescence and phase separation, which are essential for long-term formulation stability. The chapter further delves into the optical properties of nano- and microemulsions, which influence formulation transparency, crucial for aesthetic considerations in pharmaceutical and cosmetic applications. Rheological characteristics, such as viscosity and flow behavior, are also reviewed, as they significantly impact the ease of application, especially in topical and injectable formulations. By providing a comprehensive overview of these parameters, the chapter underscores the importance of precise characterization to achieve stable, effective, and tailored nano- and microemulsion formulations in drug delivery systems.

1.2 Physicochemical properties of micro/nanoemulsions

Micro- and nanoemulsions possess unique physicochemical properties that make them valuable in pharmaceutical and cosmetic applications. Understanding these properties is crucial for optimizing their formulation and enhancing their therapeutic effectiveness [1, 21].

1.2.1 Droplet size and size distribution

Microemulsions typically have droplet sizes ranging from 10 to 500 nm, while nanoemulsions generally range from 10 to 100 nm. Smaller droplet sizes contribute to increased surface area, which enhances the solubilization and bioavailability of APIs. Smaller droplet sizes are beneficial in applications like drug delivery, as they increase the surface area, which can enhance the bioavailability and absorption of encapsulated active ingredients [22]. In the food and cosmetic industries, smaller droplet sizes result in clear or translucent formulations [3]. A narrow size distribution indicates uniformity in droplet size, which is critical for stability. A broader size distribution may lead to coalescence and phase separation, affecting the emulsion's performance. Nanoemulsions are often polydisperse, that is, they have a range of droplet sizes as shown in Figure 1.1, whereas microemulsions, due to their thermodynamic stability, tend to be more monodisperse [23].

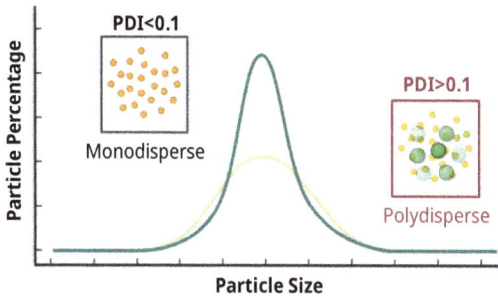

Figure 1.1: Particle size distribution.

Techniques such as DLS, transmission electron microscopy (TEM), and cryo-TEM are commonly used to measure droplet size and distribution in emulsions [24, 25]. Narrow size distribution contributes to uniform properties, such as stability and predictable behavior in biological systems.

1.2.2 Interfacial tension and stability

Interfacial tension is the force that exists at the interface between oil and water phases. In emulsions, surfactants (surface-active agents) reduce this interfacial tension, enabling the formation of stable droplets of one phase within the other. Microemulsions are stabilized by very low interfacial tension due to high surfactant concentrations. The reduction in interfacial tension is so significant that microemulsions can form spontaneously, a property that contributes to their thermodynamic stability. However, nanoemulsions require an energy input, such as ultrasonication or high-pressure homogenization, to form stable droplets because they are only kinetically stable. Surfactants reduce the tendency of droplets to coalesce, enhancing the stability of the emulsion over time.

Stability mechanisms in emulsions can be broadly categorized into steric stabilization and electrostatic stabilization. In systems utilizing nonionic surfactants, steric stabilization occurs when the long polymeric chains create a physical barrier that prevents droplets from approaching each other closely, thereby reducing coalescence and enhancing stability. On the other hand, electrostatic stabilization is observed in emulsions stabilized by ionic surfactants, where the electrostatic repulsion between droplets carrying like charges prevents them from aggregating. This charge can be quantified by zeta potential measurements, with higher zeta potentials (whether positive or negative) indicating greater stability and resistance to aggregation. These mechanisms are crucial in maintaining the integrity and longevity of emulsion-based formulations [26, 27].

Destabilizing mechanisms in emulsions include Ostwald ripening and coalescence leading to phase separation. Ostwald ripening occurs when small droplets dissolve and redeposit onto larger droplets due to differences in chemical potential, resulting in an increase in average droplet size over time. This phenomenon is particularly problematic in nanoemulsions without adequate surfactant coverage. Coalescence and subsequent phase separation happen when droplets merge, especially in cases where surfactant concentration is insufficient. While nanoemulsions are susceptible to these issues, microemulsions are generally more stable due to their thermodynamic stability [28, 29].

1.2.3 Viscosity and rheology

The viscosity of micro- and nanoemulsions is influenced by several key factors, including oil content, droplet size, and surfactant concentration [30]. These emulsions typically exhibit lower viscosities compared to traditional emulsions, making them particularly suitable for applications like pharmaceutical injections and spray cosmetics. When the dispersed phase volume fraction, such as the oil-to-water ratio, is increased, viscosity tends to rise, especially in nanoemulsions where droplet size and distribution play a significant role. This increase in viscosity can be critical in designing formulations for specific applications, ensuring optimal performance and stability [31].

Rheologically, micro- and nanoemulsions can display varying behaviors depending on their concentration [32]. At low concentrations, these emulsions often exhibit Newtonian behavior, where viscosity remains constant regardless of the shear rate applied. However, at higher concentrations, they tend to exhibit non-Newtonian, shear-thinning behavior, where viscosity decreases as shear rate increases. This characteristic is particularly advantageous for topical or injectable applications, allowing the emulsion to spread or flow more easily under stress [33]. Additionally, in systems with high droplet concentrations or strong inter-droplet attractive forces, more complex rheological properties, such as viscoelasticity, can be observed. This viscoelastic nature is beneficial in applications that require retention at the application site, such as controlled drug release in tissues [34].

Several factors significantly impact the viscosity of these emulsions, including temperature, surfactant type, droplet size, and phase volume ratio. Temperature variations can alter the fluidity of the oil phase and the behavior of surfactants, thereby affecting both the viscosity and stability of the emulsion. Similarly, the choice of surfactant and its concentration can influence the interfacial properties and stability of the emulsion. Droplet size and distribution also play crucial roles, as smaller droplets with a narrow size distribution can enhance the stability and reduce the viscosity of the emulsion. Understanding these factors is essential for optimizing the formulation and application of micro- and nanoemulsions in various biomedical and industrial contexts [35].

1.2.4 Phase behavior and structure

Micro- and nanoemulsions can adopt different structures based on their composition and the balance between oil, water, and surfactants [11]. Common structures include:

a. *Oil-in-water (O/W):* Droplets of oil dispersed in water. This is typical in food, cosmetic, and pharmaceutical applications [36].
b. *Water-in-oil (W/O):* Water droplets dispersed in oil, often used in applications requiring a more hydrophobic continuous phase, like some topical medications or cosmetics [37].
c. *Multiple phases:* Both oil and water form interconnected continuous phases, creating a more complex, sponge-like structure. These multiple microemulsions have unique properties, such as enhanced diffusion of both polar and nonpolar compounds [38].

(i) *Phase diagrams*:
Ternary or quaternary phase diagrams are commonly used to map out the conditions under which each type of emulsion forms. These diagrams plot the proportions of oil, water, and surfactant, along with temperature or cosurfactant presence, to show the regions of stability for each phase [39].

(ii) *Influence of surfactants and cosurfactants*:
Higher surfactant concentrations generally stabilize the oil–water interface and allow the formation of smaller droplets. However, excessive surfactant levels can cause unfavorable phase transitions or toxicity in certain applications, especially in pharmaceuticals and food products [26]. In microemulsions, cosurfactants like short-chain alcohols or glycols reduce interfacial tension further and enhance flexibility of the surfactant film, enabling the formation of thermodynamically stable microemulsions. In nanoemulsions, cosurfactants can be used to improve stability and reduce droplet size [40].

(iii) *Temperature and phase transitions*:
Temperature changes can induce phase transitions between O/W, W/O, and discontinuous phases structures due to changes in surfactant solubility and interfacial tension. For example, nonionic surfactants tend to become less soluble in water at higher temperatures, shifting the emulsion type from O/W to W/O [41].

These physicochemical properties of micro- and nanoemulsions – droplet size and distribution, interfacial tension and stability, viscosity and rheology, and phase behavior – are fundamental to their behavior in various applications. By adjusting formulation parameters like surfactant concentration, oil-to-water ratio, and temperature, researchers and developers can design emulsions tailored to specific needs, making them invaluable in fields such as pharmaceuticals, food science, cosmetics, and environmental science. Some important differences between micro- and nanoemulsions are mentioned in Table 1.1.

Table 1.1: Key differences in the micro- and nanoemulsions.

Feature	Microemulsions	Nanoemulsions	References
Droplet size	Typically, 10–500 nm	Typically, 10–100 nm	[3]
Appearance	Transparent or slightly translucent	Usually transparent, but may vary with droplet size	[2]
Stability	Thermodynamically stable; formed spontaneously	Kinetically stable; requires energy input to form	[26]
Preparation method	Spontaneous emulsification using surfactants and cosurfactants	Requires high-energy methods (e.g., ultrasonication and high-pressure homogenization)	[8]
Surfactant requirement	High surfactant and cosurfactant concentrations needed	Requires surfactant but generally in lower amounts than microemulsions	[42]
Phase behavior	Forms microdomains of oil and water in continuous phase	More uniform phase distribution without coexisting phases	[39]
Thermodynamic properties	Thermodynamically stable (will not separate over time)	Not thermodynamically stable, but can be kinetically stable for a long period	[43]
Applications	Widely used in drug delivery, food industry, and cosmetics	Ideal for drug delivery, particularly for poorly soluble drugs	[44]
Advantages	Self-assembled, stable at low energy; useful for slow-release and targeted delivery	Enhanced bioavailability, controlled drug release, and ability to cross biological barriers	[45]
Limitations	High surfactant concentrations may cause toxicity or irritation; limited to certain conditions	Requires energy-intensive preparation; not inherently thermodynamically stable	[45]

1.3 Techniques for characterizing micro/nanoemulsions

1.3.1 Dynamic light scattering (DLS)

A method for characterizing dispersed systems using optical measurement is DLS. The technique evaluates high-frequency fluctuations in scattered light, illuminating the dynamics of microstructural phenomena such as elastic vibrations in gels, sol–gel transition, and particle agglomeration. In order to analyze particle size, DLS is most frequently employed to gauge each particle's Brownian motion within a liquid. Spectral measurement data from representative samples are converted into numerical size distributions using DLS, a spectroscopic measurement technique. Compared to counting or fractionating procedures, it is therefore essentially restricted and has a lower resolution of minute aspects of a size distribution. In spite of this, this technique has become the most used way to measure particle size in the sub micrometer range (1 nm to 1 μm). This is due to a number of practical and financial factors, including speedy analysis and inexpensive expenses per measurement. DLS is also an effective add-in for hyphenated measurement configurations, which integrate size analysis and fractionation. Not to mention, continuous advancements in data analysis and technology will strengthen the method's ability to determine size distribution details and work with very polydisperse materials [46].

1.3.1.1 Principles of DLS

The foundation of DLS is the measurement of coherent light scattered over time by scattering objects, such as large molecules or tiny particles.

Variations of the measurement signals, which may come from several sources, are considered in the evaluation. DLS handles these kinds of fluctuations, which are caused by the thermal motion of the scattering particles and occur at extremely small-time scales (microseconds and even nanoseconds). By examining the vibrations in particle networks, DLS, for instance, may investigate phase transitions in colloidal suspensions and quantify the elastic properties of gels.

The most popular use of DLS is the size analysis of particles in the sub micrometer range. Utilizing the particles' Brownian motion, it modifies the interference between the distinct scattering signals and generates a fluctuating signal at the detection by shifting the area indefinitely. The kinetics of the rearrangement is controlled by the particle diffusion coefficient (D_p), which is inversely proportional to particle size:

$$D_p = \frac{kBT}{3\pi\eta xh, t}$$

The Stokes–Einstein equation is represented by the variables k_B and T, which stand for temperature and Boltzmann's constant, respectively, and η, which stands for the fluid's dynamic viscosity xh, and t for the hydrodynamic diameter of translational motion. The latter is termed an equivalent diameter, and it might vary significantly from other equivalent diameters. It describes a random particle orientation. These variations are most noticeable for aggregates that resemble fractals, when xh, t is significantly larger than the individual particles, the volume equivalent diameter, and the Stokes diameter, but much less than the surrounding sphere's diameter [46].

1.3.1.2 Applications in emulsion characterization

Finding the size distribution of dispersed droplets in emulsions using a typical approach is DLS. This chapter describes the several ways that DLS is used in emulsion characterization, including size measurement, stability analysis, and formulation process optimization:

a. Measurement of droplet size: DLS is often used to ascertain the distribution of droplet sizes in emulsions, such as microemulsions and nanoemulsions. It gives the polydispersity index, which indicates size homogeneity, and the hydrodynamic diameter [47].
b. Stability analysis: In order to assess stability against flocculation, coalescence, and Ostwald ripening, DLS tracks changes in droplet size over time. It assesses the stability of emulsions under additive impact and thermal stress.
c. Formulation optimization: To obtain stable, homogeneous droplet sizes for food, cosmetic, and pharmaceutical formulations, DLS helps optimize surfactant concentration and emulsification process parameters (e.g., mixing speed) [48].
d. Drug delivery systems: DLS is used in the pharmaceutical industry to optimize droplet size for controlled drug release and bioavailability in nanoemulsions, liposomes, and other emulsion-based drug carriers [6].
e. Food emulsions: DLS influences texture, taste, and shelf life by ensuring droplet size uniformity in dairy products, sauces, and flavor emulsions [49].
f. Cosmetics: DLS regulates droplet size in sunblock, lotion, and cream formulations to provide a consistent UV protection, spread ability, and acceptable texture [50].

1.3.1.3 Limitations and challenges: several limitations and challenges associated with DLS

– DLS measurements exhibit high sensitivity to variations in temperature and solvent viscosity. Consequently, a dependable DLS experiment requires accurate knowledge of solvent viscosity and the maintenance of a steady temperature.

- DLS is a low-resolution method that usually cannot tell apart molecules that are closely linked, such as dimers and monomers.
- Plotting the intensity of scattered light [$K^*c/R\theta$, where K^* is an optical parameter, c denotes concentration, and $R\theta$ is an angle-dependent light-scattering parameter] allows one to determine molecular weight using DLS. This method lacks reliability and reproducibility. In these instances, alternative methods such as analytical ultracentrifugation and static light scattering combined with multi-angle laser light scattering ought to be used.
- DLS can only be used to create clear samples.
- The sixth power of the size of the macromolecule determines the scattering intensity ($I \propto d^6$, where d represents the diameter). Consequently, even a minimal presence of large aggregates can significantly influence the measurements. Consequently, it is essential to thoroughly clean the sample-holding cuvette and filter the sample before conducting DLS calculations.
- The DLS signal depends on the macromolecules' size and concentration. Consequently, optimizing the concentration range may be necessary to achieve reliable measurements [51].

1.3.2 Small-angle X-ray scattering (SAXS) and small-angle neutron scattering (SANS)

By examining the scattering of X-rays or neutrons at small angles as they go through a sample, small-angle X-ray scattering (SAXS) and small-angle neutron scattering (SANS) are efficient techniques for examining the nanoscale structure of materials. While both techniques share a similar principle, they employ distinct types of radiation (X-rays versus neutrons), resulting in unique advantages for each approach.

1.3.2.1 SAXS and SANS principles

By measuring the scattering of X-rays at tiny angles, often less than 5°, the technique known as SAXS provides information on the size, shape, and distribution of particles or structural elements within the nanometer to micrometer range. The method analyses the interaction of X-rays with the electron clouds of atoms in a sample, correlating scattering intensity with variations in electron density. SAXS exhibits high sensitivity to electron-dense structures, rendering it appropriate for the investigation of various materials, including proteins, polymers, nanoparticles, and colloids. Analyzing scattering patterns allows researchers to deduce structural parameters, including particle size, shape, and internal organization. This technique investigates structures ranging from 1nm to 100 nm and is frequently employed to examine macromolecular aggregation, nanostructures, and the internal organization of porous materials and biological complexes [52].

SANS employs neutron beams scattered by atomic nuclei at small angles to elucidate nanoscale structures. SANS, in contrast to SAXS, depends on the interaction of neutrons with atomic nuclei, providing distinct contrast mechanisms. The scattering intensity in SANS is contingent upon the contrast in scattering length density among different regions of the sample. This technique is advantageous for distinguishing between isotopes, such as hydrogen and deuterium, making it suitable for the study of biological systems and materials with low electron contrast. SANS enables contrast variation via isotope substitution, thereby facilitating comprehensive analyses of complex systems. This technique examines structures ranging from 1 to 100 nm, demonstrating heightened sensitivity to hydrogen. It is thus appropriate for the study of soft matter, biological macromolecules, and the internal architecture of composite materials and porous systems [53].

1.3.2.2 Use in structural characterization

In the fields of structural biology and materials science, two crucial techniques that provide insight into the nanoscale structure of a range of materials are SAXS and SANS. Each approach is used in structural characterization as follows [54]:

– The structural investigation of biological macromolecules, polymers, nanoparticles, and porous materials may be effectively accomplished with the help of SAXS. It provides details on the size, shape, flexibility, conformation, and aggregation of proteins and nucleic acids without crystallization. It is employed in the study of these substances in solution. SAXS also examines macromolecular complexes to get a deeper understanding of quaternary structures and protein-protein interactions. In polymer characterization, it makes the form, size distribution, and phase behavior of copolymers and mixes clearer. SAXS also analyses the pore structure of porous materials, including pore size, shape, and connectivity, and computes the size and shape of nanoparticles and colloidal systems to aid in the design of materials for uses like filtration and catalysis.

– SANS serves as an important method for investigating soft matter and biological systems, especially proficient in examining lipid bilayers, micelles, and macromolecular complexes such as proteins and DNA. This offers a deeper understanding of their composition, behavior, and relationships. SANS is utilized in self-assembly investigations involving polymers and surfactants, taking advantage of deuterated materials for contrast variation, enabling thorough examination of particular components within intricate systems. Furthermore, SANS provides insights into the internal structure and morphology of composite materials and porous structures, deepening the comprehension of their functional properties.

Through the use of various isotopes, SANS facilitates focused investigations of particular areas within diverse samples.

1.3.2.3 Data interpretation and analysis

To extract structural information from scattering data, SAXS and SANS require a variety of critical data interpretation and analysis processes. Initially, raw intensity data must be collected. Normalization and background removal techniques are then used to analyze the data. As a function of the scattering vector q, the scattering intensity I (q) provides crucial information on the sample's size and shape.

To characterize the overall structure, important parameters are found, such as the radius of gyration (R_g) and the distance distribution function ($P(r)$). Scattering profiles may be fitted to a variety of models using SAXS and SANS, enabling the characterization of the morphology of macromolecules, polymers, and nanoparticles. Furthermore, SANS takes advantage of isotope substitution to improve contrast, facilitating a thorough examination of particular elements within intricate systems. Ultimately, understanding structural characteristics and interactions in materials and biological systems is improved by combining data from SAXS and SANS with other methods [52].

1.3.3 Transmission electron microscopy (TEM) and scanning electron microscopy (SEM)

The scanning electron microscopy (SEM) and TEM instruments are both able to generate highly focused electron beams that are directed at the sample in a vacuum chamber. Nevertheless, the main area in which the SEM units are used is in the examination of the surfaces of different materials (just like a reflection light microscope). However, the primary use of TEM in monorail systems is in examining the internal structure of specimens (like a transmission light microscope).

In an SEM, the electron beam travels over the object one by one after being focused on a certain spot, captured by detectors. A detector signal is timed with the known beam position on the specimen and the signal is then used to modify corresponding image pixels. The signals that are collected one after the other are added together to create an image that has its own dimensions depending on pixels distribution, which is decided according to the scan pattern chosen, that is, 30 keV is the most common electron energy.

In TEM, an electron beam is made to hit a specific area of a sample. The electrons passing the sample are focused by lenses and then collected by a parallel detector to become a picture. The energy of the electrons in a TEM is higher than SEM with the average of 80– 300 keV, which allows them to penetrate the material [55].

1.3.3.1 Imaging techniques for micro/nanoemulsions

Micro/nanoemulsions are complex systems requiring visualization to understand their structure, stability, and behavior. The measurement of techniques using different imaging for both micro- and nanoemulsions is critical in determining their size, structure, stability, and morphology. Various imaging techniques enable researchers to characterize these systems; here are some of the most commonly used methods for these purposes:

–   TEM: This technique uses electrons to pass through thin materials to produce high-resolution pictures of emulsion droplets at the nanoscale. It is best suited for internal structure visualization, but before this success is achieved, it takes much time for complex sample preparation.
–   SEM: SEM is a special way of viewing surfaces in great detail. It illuminates structures finely with an electron beam and then displays high-quality images that show the morphology of the surface in detail. Conductive coating of the samples is needed.
–   Cryo-electron microscopy (cryo-EM): High-resolution imaging without the need for drying is possible if materials are rapidly frozen and kept undamaged. This technique will be useful to study emulsions in their natural hydrated form instead of dry solids.
–   Atomic force microscopy (AFM): Scans the surface of the material and determines its mechanical features through direct contact with a hard needle tip. AFM produces, in addition to all the 3D imagery of the 1 nm to 10 cm scale, the pictures of the surface topography, which is favorable to the liquid samples but can sometimes be very challenging.
–   DLS: Measuring light scattering through Brownian vibration is a procedure that is utilized to determine particle size distribution.
–   SAXS: One of the most recent techniques that have been developed in the field of medical imaging is the so-called grazing incidence X-ray scattering at small angle technique, which provides former crystallographic information on the object.

1.3.3.2 Sample preparation

To ensure the production of precise and top-notch images of micro- and nanoemulsions, the right sample preparation steps should be utilized. Diverse imaging methods demand certain advanced preparation methods to maintain the emulsion structure and prevent defects. Here are the most typical techniques for getting samples ready:

a)   **For TEM:**
–   Staining: Typically, electron-transparent emulsions are stained using heavy metals as uranyl acetate or osmium tetroxide to boost contrast.

- Thin sectioning: It is mandatory that the samples are ultra-thin (usually less than 100 nm) to allow the electron to penetrate them.
- Cryo-TEM: The samples are immediately frozen using liquid nitrogen or ethane, resulting in their direct imaging in their natural condition without any staining or dehydration.

b) **For SEM:**
- Emulsions are generally dried off from liquid substances. Critical point drying and freeze-drying are normally opted for to prevent the collapse of the structure.
- Conductive coating: A thin coating of carbon, platinum, or gold is applied to nonconductive sample surfaces to make them conductive and prevent the accumulation of negative charge.
- Cryo-SEM: Like in cryo-TEM, samples are frozen to maintain their natural state and are then imaged without dehydration.

c) **For cryo-EM:**
- Vitrification: To prevent ice crystals from growing, the samples are rapidly frozen. Vitrified samples are stored at a very low temperature of liquid nitrogen for imaging.
- No staining or drying: This method preserves the emulsion's characteristics without requiring it to be stained or dried.

d) **For AFM:**
- Substrates are used during cathodic deposition: The substrate is immersed into the plating solution, and the plating is carried out using cathodic method.
- The plating process includes wet bench system with spin coat, spray, rinse, and spin/dry processes in different solutions wherein lithium ion can be used, and the amorphous oxide barrier layer is deposited.
- The turned-out value for the actual deposited material in artificial cornea membranes is 167.6 nm in thickness. As indicated by several features to be described, such a trend may be anticipated by a set of equilibrium equations.

e) **For DLS:**
- To prevent multiple scattering and obtain accurate size measurement, samples are generally diluted in the spectrophotometer before observation can proceed.
- DLS directly measures particles in a liquid; therefore it does not need any drying or explicit preparation steps.

1.3.3.3 Analysis of emulsion morphology

The morphology of emulsions or the properties of the droplets as they are defined by droplet size, shape, distribution, and internal structure are significant factors that determine their stability, efficiency, and performance in sectors from pharmaceuticals to food products. Emulsion morphology is studied through different analytical techniques with each one of them giving its peculiar insight into the structure of the emulsion as in Table 1.2.

Table 1.2: Analysis of emulsion morphology through different analytical techniques.

Technique	Resolve	Information provided	Advantages	Challenges
Transmission electron microscopy (TEM)	Nanometer scale	Internal structure, droplet size, shape, and distribution	High resolution; useful for nanoemulsions	Complex sample preparation; staining may alter structure
Scanning electron microscopy (SEM)	Submicrometer scale	Surface morphology, droplet shape, and distribution	Clear surface images; cryo-SEM preserves natural state	Requires conductive coating; limited internal detail
Confocal laser scanning microscopy (CLSM)	Micrometer scale	3D images, distribution of components, and phase interactions	3D imaging; useful with fluorescent labeling	Lower resolution compared to TEM and SEM
Dynamic light scattering (DLS)	Nanometer to micrometer	Droplet size distribution and polydispersity	Quick, noninvasive, and real-time analysis	Provides average size; no structural detail
Small-angle X-ray scattering (SAXS)	Nanometer scale	Internal structure, droplet shape, and size distribution	Nondestructive; effective for internal morphology	Requires specialized equipment, and complex data interpretation
Atomic force microscopy (AFM)	Nanometer scale	Surface topography; mechanical properties	High-resolution 3D images; mechanical property analysis	Limited to surface analysis; challenging for liquid samples
Fluorescence microscopy	Micrometer scale	Droplet distribution; phase interactions	Real-time monitoring; fluorescent labeling possible	Lower resolution; requires labeling
Nuclear magnetic resonance (NMR)	Bulk structure	Molecular interactions, droplet size, and internal dynamics	Noninvasive; suitable for molecular-level analysis	Low resolution; no direct visualization
X-ray tomography	Micrometer to submicrometer	3D internal structure and droplet size distribution	Nondestructive 3D imaging	Limited resolution; requires transparent materials

1.3.4 Cryo-electron microscopy (cryo-EM)

Electron microscopy, often known as cryo-EM, is one of the main methods to point out the structures of new proteins produced by the methods of point mutations and unnatural amino acid incorporations. Single particle analysis (SPA), its most popular subfield, is a technique utilized in an increasing number of labs throughout the globe to ascertain high-resolution protein structures. Although cryo-EM has been very successful, we should still try to find ways to make it better. Sample preparation, sample screening, image processing, data collecting, and the structure validation procedures are a few of the phases that might be enhanced and added to the total procedure. Among these concerns include sensitivity, flexibility, and the control of beam focus [56].

1.3.4.1 Principles of cryo-EM

cryo-EM is a microscopy method. It is customary to get extremely powerful images of biological macromolecules, nanostructures, and emulsions in their native state. The frosting is nothing like TEM where the sample is always stained and moisturized; here, in the case of cryo-EM, the sample has been naturally preserved. The principles of cryo-EM include:

a) Rapid freezing (vitrification): The samples are quickly frozen by using liquids like liquid nitrogen and −150 °C that are below extreme temperatures like liquid nitrogen or ethane. This process, called vitrification, does not form ice crystals that would damage the sample's structure. The water molecules are not frozen in the ice crystals, apparently; instead they settle themselves in a solid glass-like structure.

b) Electron transmissions: The sample is frozen at a very low dose of electron. This leads to minimal distortion and radiation damage, which preserves the internal structures within it as the image is being taken.

c) No staining required: There is no heavy metal staining required in this process as the process utilizes the self-electron density of the sample itself for contrast in the creation of the image. Hence, the sample's original structure is completely preserved.

d) Direct electron detectors: They are sensitive enough to pick up high-quality images that show minimal noise; therefore, they offer resolution improvement.

e) SPA: Numerous 2D images of the same single particle (e.g., a protein) are taken and used to construct the 3D structure at near-atomic resolution.

f) Cryo-tomography: A tomographic reconstruction technique in which images recorded at different angles can be taken; well-suited for larger structures, such as cells.

g) Sample preservation: The sample is frozen without losing its native morphology or molecular interactions while preserving it throughout the freezing process with no artifacts.

1.3.4.2 Application in nanoemulsion studies

Cryo-EM is an excellent research tool in terms of structure and stability studies about nanoemulsions with high-resolution performance. Major applications are as follows:
a) Visualization of droplet morphology: With the help of cryo-EM, the inside and outside morphology of nanoemulsion droplets can be visualized at a nanoscale resolution, which will further allow researchers to study droplet size, shape, and distribution without staining or dehydration techniques.
b) Characterization of stabilization and coalescence: Direct imaging of the native structure of frozen nanoemulsions allows cryo-EM to track, over time, droplet stability and coalescence, and thus provide insight into how emulsions respond to different conditions.
c) Layered structures and interfaces: The technique can visualize extremely complex internal nanoemulsion structures that often manifest in terms of layered systems or phenomena related to interfaces between different phases, such as the oil and water phases.
d) Nanoencapsulated systems: Cryo-EM may now yield the nanoimages of how nanoparticles or the active agent is encapsulated into the emulsion droplets, thus allowing perfect formulation for release at targeted sites.
e) Phase behavior: Since the phase transitions or separation in nanoemulsions can be observed in real time, cryo-EM yields information on how the formulations respond to changes in temperature and pH.

1.3.4.3 Principles of cryo-EM

The major parameter used to analyze the stability of colloidal systems, such as emulsions and nanoemulsions, has the possibility for zeta. It explains the electrical potential generated at the plane of slippage, which is the boundary between the stationary fluid layer attached to a particle and the surrounding fluid in motion as described in Figure 1.2.

1.3.5 Significance of zeta potential in stability

Another important parameter that could be referred to as analyzing and managing nanoemulsion stability has the possibility for zeta. Electrostatic repulsion between droplets increases with zeta potential; thus, aggregation is hindered and ensures a stable homogeneous dispersion over time. A formulator can significantly improve the shelf life and performances in given applications by monitoring and adjusting the zeta potential. Zeta potential can be critically assessed to be an indicator of stability in a colloidal system, such as nanoemulsions. The value of zeta potential indicates the

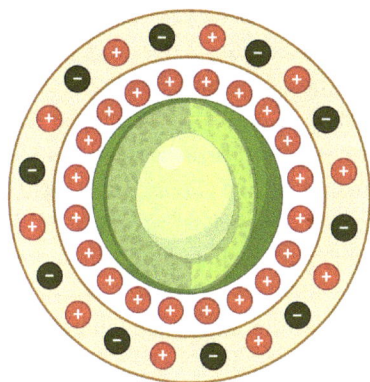

Figure 1.2: Nanoemulsion particle surrounded by zeta potential charge.

electrostatic repulsion of particles or droplets within the emulsion. This is against the following important points:

- High zeta potential (±30 mV and higher): Electrostatic repulsion that is strong enough keeps droplets from clinging to one another; hence it keeps the emulsion stable against settling, creaming, and phase separation.
- Low zeta potential (less than ±15 mV): Low chances of aggregation leads to weak repulsion. Due to these reasons, droplet agglomeration takes place, which causes instability, flocculation, coalescence, and sedimentation.
- Agglomeration and coalescence control: The high zeta potential prevents agglomeration of the droplets as the force of repulsion is significantly increased, which is critical for maintaining nanoemulsions intact.
- Electrostatic stabilization: Charge stabilization at the surface ensures that the zeta potential is high, and this will ensure that emulsions can be stable for a long period, especially in pharmaceutical and cosmetics applications.
- Surfactant selection: Zeta potential measurements guide the selection of surfactants whereby surface charge and droplet repulsion of the droplets increase with higher stability.
- Shelf life prediction: The shelf life will increase with increasing zeta potential levels. Low values predict faster degradation or phase separation.
- Environmental influence: pH, ionic strength and salts influence the zeta potential; thus the monitoring of changes in it can predict stability under varied conditions.

1.3.5.1 Techniques for measuring zeta potential

There exist two categories of measuring methods, characterizing distinctly different from each other. In nanodispersions, this includes zeta potential. The first is often based on electrophoresis. This makes use of optical methods to track the motion of particles. Ultrasound is used in place of light in the other class, which is based on electroacoustics.

Both of these types complement rather than compete with each other. Electrokinetic methods are useful only in dilute systems, while electroacoustic techniques focus solely on the work. About 1% of particles make up the border. There are signs of low conductivity dispersions or very translucent suspensions [57]. Methods based on electrophoresis can be used when the weight fraction is greater. However, electroacoustics may be used on nanodispersions of heavy enough particles, such gold, rutile, and alumina, down to 0.1 wt%. Verification tests, which will be covered in the material that follows, are made possible by the existence of this gray region, which is roughly 1 wt% where both approaches function. However, the general rule still stands:

In diluted systems, electrophoresis-based techniques are employed, whereas in concentrated systems, electroacoustics is employed.

Optical techniques and electrophoresis: Electric field application to the dispersion of nanoparticles that are electrically charged cause this dispersion to move in relation to the liquid. Electrophoresis is the term for the effect. The electric field intensity and the particle's motion velocity, V, are proportional. Electrophoretic mobility μ is the proportionate coefficient:

$$\mu = V/E$$

Electrophoretic mobility measurement: A cell equipped with electrodes holds a suspension of charged particles with specific surface charges. Between the electrodes, a potential is applied, and the net electric charge-bearing particles move towards the opposite electrode. Through an electrophoretic light scattering by laser Doppler arrangement, the particles' velocity is ascertained. The particle electrophoretic mobility distribution is determined by the frequency shift distribution, which in turn determines the zeta potential. ELS provides complicated particle sample electrophoretograms that are precise, automated, and repeatable.

Electroacoustics for concentrated nanodispersions: When the wavelength of the mechanical stress equals the sample size, electroacoustics is electro kinetics at high frequencies, usually in the megahertz frequency range [4, 10].

Depending on what plays a driving role in particles in motion with relation to the liquid, there are two modes of electroacoustics. Traditionally, colloid vibration current (CVI) [10] is the first type of AC electric current produced by charged colloidal particles moving in relation to a liquid while being affected by an ultrasonic pulse. An equation can be used to express this current's value:

$$CVI = \mu\phi \frac{\rho p - \rho s}{\rho s} \nabla P$$

where ϕ is the ultrasonic wave's pressure, ϕ is the particle volume fraction, and ρp and ρs are the particle and dispersion densities, respectively. Using this equation, measuring CVI enables the calculation of electrophoretic mobility, which may then be utilized for using the proper electroacoustic theory to calculate the zeta potential [58].

Electric sonic amplitude [https://doi.org/10.1006/jcis.1995.1341], another type of electro-acoustics, employs an AC electric field to move particles in relation to a liquid. An ultrasonic is produced by the motion wave. For such devices, the basic output data is the wave's amplitude and phase [59].

1.3.6 Rheological analysis

Exploring the flow characteristics of particle emulsions provides insights into how a blend stays intact and how droplets evolve over time, while also uncovering the effectiveness of drug release from them. Tackling the rheology of formulations holds the key in comprehending their flow behaviors and long term stability in micro/nanoemulsions where droplet dimensions and interactions play a role, in influencing viscosity and overall robustness [60].

1.3.6.1 Methods of rheological characterization

Rheological characterization of micro- and nanoemulsions is critical for understanding their stability, texture, and flow behavior, which are important for applications across industries such as pharmaceuticals, cosmetics, and food. The methods used to characterize the rheology of these systems can vary, but typically involve the following techniques as shown in Table 1.3.

Table 1.3: Methods of rheological characterization for micro/nanoemulsions.

Method	Description	Application to micro/ nanoemulsions	References
Steady-shear rheology	Determines viscosity under continuous stress or shear rates.	Aids in the understanding of shear-thickening or shear-thinning behavior, which is typical of non-Newtonian emulsions.	[61]
Oscillatory rheometric (dynamic rheology)	Measures viscoelastic qualities, namely storage (G') and loss (G'') moduli, using oscillating stress/strain.	Establishes the elasticity and stability of emulsions, which is essential for phase transition prediction.	[62]
Creep and recovery test	Evaluates the long-term deformation of a material under continuous tension and its recovery after removal of stress.	Aids in comprehending viscoelasticity and long-term deformation in emulsions.	[61]

Table 1.3 (continued)

Method	Description	Application to micro/ nanoemulsions	References
Stress relaxation test	Evaluates the dissipation of stress under continuous tension.	Beneficial for researching microemulsion relaxation processes, especially after deformation.	[62]
Dynamic light scattering (DLS)	Determines the dispersion of droplet sizes using the light scattering of the particle.	Essential for examining droplets the size of nanometers and connecting it to rheological characteristics like viscosity.	[62]
Micro rheology	Measures the local rheological characteristics by following the motion of the implanted particles.	Helpful in figuring out the heterogeneity and local viscosity of nanoemulsions.	[61]
Interfacial rheology	Evaluates the rheological characteristics of the oil/water contact, which is a dispersed phase.	Essential to comprehending the stability of emulsions, particularly in nanoemulsions where interfacial tension plays a crucial role.	[61]
Capillary rheometric	Measures viscosity by applying high shear rates as the substance travels via a small capillary.	Beneficial in industrial settings when high-shear processing methods, such as extrusion, are used on nanoemulsions.	[62]

1.3.6.2 Relationship between viscosity and stability

A new way that oil DV influences the stability and instability of emulsions was discovered, highlighting the viscosity of emulsions and droplet motion. As density becomes more crucial with greater oil DV, the droplet motion resistance, or the higher bulk and more viscous properties of the oil under any given volume are connected to friction. This promoted a more uniform inner network and increased the apparent/macroscopic viscosity of the emulsions by limiting the fluid inclination of the droplets and confining them to a small region. These two elements slowed the droplets' rate of travel while storing emulsions for a long time and increased the frequency/effect of droplet collision that favors reducing the growth of droplets and improving emulsion stability against several destabilizing processes (e.g., creaming, flocculation, and coalescence). To the best of our understanding, it is the first explanation of how oil DV influences MP emulsion stability using pure hydrophobic phases that share a high degree of structural similarity with model oils, so that all future research can be reproduced and made comparable.

1.3.7 Spectroscopic techniques

Spectroscopic techniques rely on how electromagnetic waves and matter interact; they are then applied for analysis and identification of substance structures and compositions by seeing how they behave when electromagnetic waves of various wavelengths are absorbed, scattered, or emitted. Energy dispersive X-ray spectroscopy (EDS), Raman spectroscopy (RS), Fourier-transform infrared spectroscopy (FTIR), and nuclear magnetic resonance (NMR) spectroscopy are a few of the primary methods employed. In order to identify electronic transitions, modifications in vibrational modes, and variations in the characteristic energy levels of the elements that investigate the properties of substances on the microscopic and macroscopic scales, these techniques make use of specific peculiarities of electromagnetic waves that range from infrared to X-ray regions [63].

1.3.7.1 NMR spectroscopy for emulsion structure

This is a very effective technique for emulsion structure research and dynamics, especially micro- and nanoemulsions. It furnishes detailed information at a molecular level regarding the nature of the dispersed phases and surfactants and about interactions within emulsions that makes clear statements both about their internal structures and stability. NMR encompasses many techniques that each yield different kinds of information and are rooted on a common principle; hence NMR spectroscopy is applied in the analysis of molecular structures and conformations [64].

Behavior of microemulsion droplets in intricate emulsion systems, but it was also employed to examine the microstructure and component distribution using various self-diffusion coefficients and chemical shifts that support the investigation of dynamic data on the surfactants inside, as diagrammatically represented in Figure 1.3 [65].

1.3.7.2 FTIR and Raman spectroscopy

FTIR can clarify intermolecular interactions, solvation effects, and component distributions in microemulsions by examining their molecular structures and the amount of dissolved water present. Microemulsions are often made with both water and oil, and their infrared spectroscopy often proves problematic because of the large water background signal. On the other hand, unsaturated C=C bonds in oil have a high Raman response, but water provides a very weak Raman signal that is often negligible at low wavenumbers; so, RS can be used to study the microemulsion [67]. In order to summarize the impact of refractive index-matching of liquid phases and enable research on W/O/W type droplets, the intensity of the Raman signal of the internal water phase was measured using a vari-

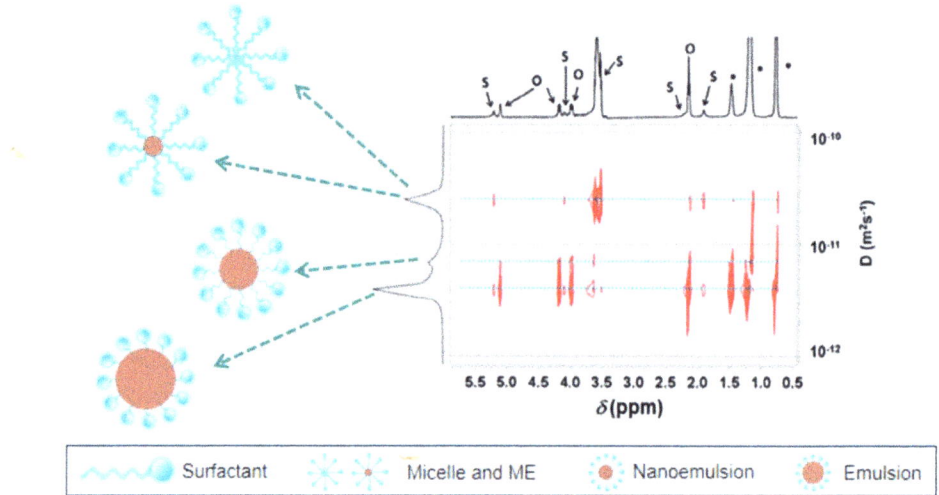

Figure 1.3: NMR for microemulsion structural characterization [66].

ety of permutations and combinations of refractive indices [68]. The use of Raman and FTIR spectroscopy in the investigation of microemulsions is illustrated in Figure 1.4.

1.3.8 Turbidity and optical microscopy

The nanoemulsions may be optically clear or even a little turbid depending on the droplet size. The turbidity τ that is the opacity expressed is determined through measurement of transmission. The droplet size affects the rheology and release behavior in addition to the optical property and stability. For this reason, nanoemulsions are more suited for a variety of applications than microemulsions. Compared to other forms, the nanoemulsions' tiny particle size has several significant implications. Nanoemulsions support stability and improve physicochemical characteristics [72].

1.3.8.1 Turbidity as an indicator of droplet size [73]

The most significant factor influencing the turbidity was particle size. It has been shown that the mean volume-surface diameter of 0.2 μm for the oil droplet size is where the most turbidity occurs. Additional considerations were the refractive index, the makeup of the excess emulsifier, and the aqueous phase [73].

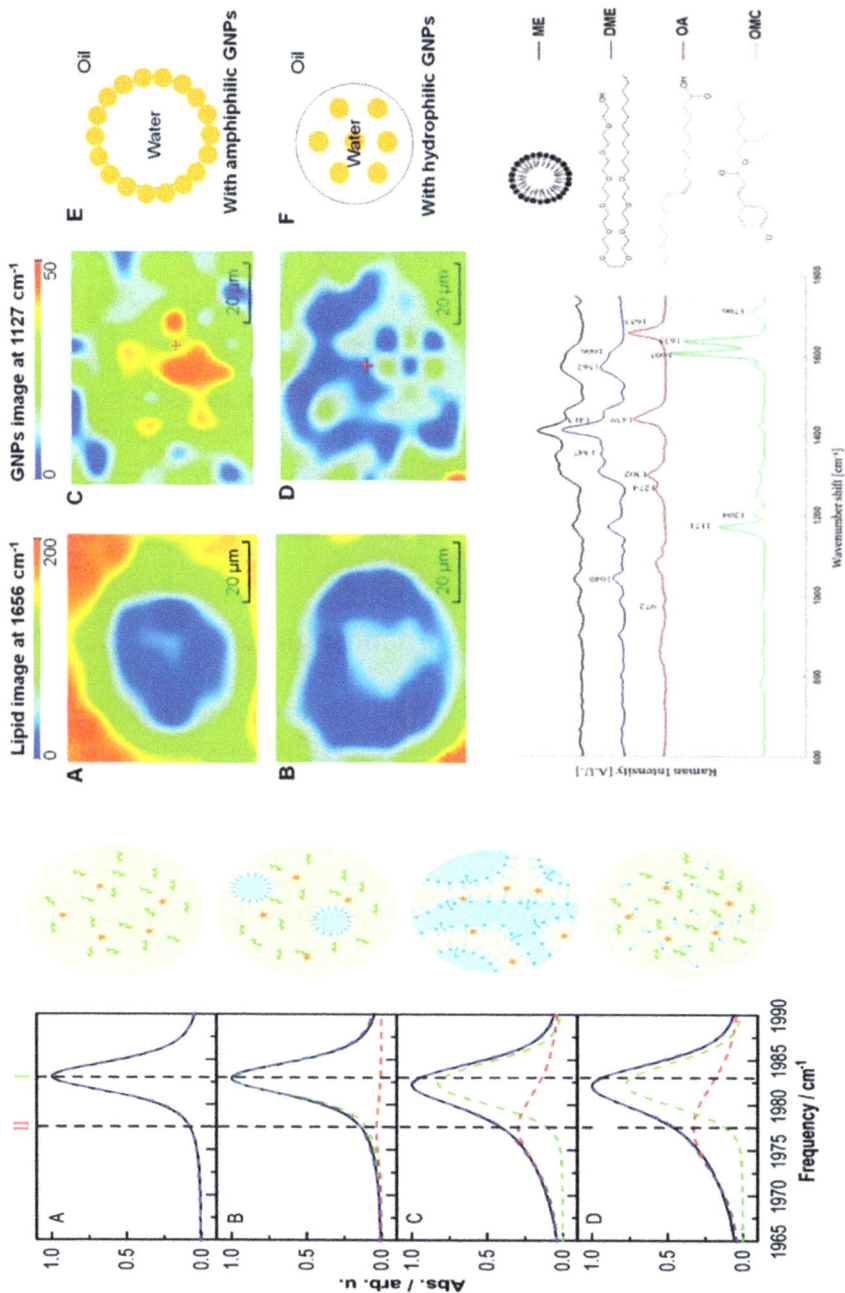

Figure 1.4: (a) Stretching mode FTIR spectra of C–O in various liquid phases [69]. (b) The production diagram of microemulsions (E, F) and Raman imaging of several chemicals (A, D). (c) Various Raman spectra microemulsion composition and forms [70, 71].

1.3.8.2 Optical microscopy for emulsion observation

A strong and flexible method for emulsion observation and characterization, optical microscopy provides information on droplet size, distribution, and shape. Emulsions are used extensively in a variety of sectors, including food technology and pharmaceuticals. They are composed of distributed liquid droplets in a continuous phase [74].

The fundamental idea behind optical microscopy is that light is transmitted through or around the material, and then the objective and eyepiece lenses are used to magnify the image. Because emulsions are optically heterogeneous, light is scattered or refracted, making droplet borders visible:

- Resolution: Because of the diffraction limit of light, the optical microscope's resolution limit is around 200 nm. Submicron droplet emulsions need sophisticated microscopy techniques or supplementary approaches such as electron microscopy.
- Contrast: Contrast-enhancing techniques like staining or phase-contrast microscopy are frequently employed to distinguish the droplets from the continuous phase.

1.3.9 Stability assessment of micro/nanoemulsions

On the other hand, besides its application, mainly in pharmaceutical formulations, consistency in drug delivery and shelf life, the question of stability is a very important issue in micro/nanoemulsions as presented in Figure 1.4 [75]. The evaluation of stability is necessary to ensure that the physicochemical properties are retained during the storage and use of such colloidal species [76]. Nanoemulsions do depend upon droplet size, surface charge, the viscosity of the dispersion medium, and even interaction between environmental factors such as temperature and pH [8]. Several methods can be employed to check for the stability of these systems – both thermodynamically and kinetically [77].

1.3.9.1 Methods to evaluate emulsion stability

Stability assessment is defined as the resistance of an emulsion against physical changes such as phase separation, creaming, flocculation, and coalescence [78]. The following techniques have been in common use in the assessment of the stability of nanoemulsions:

- **Centrifugation:** It is one of the most applied accelerated tests for the assessment of emulsion stability, which involves the treatment of emulsion by higher forces of gravitation. The methodology of the test imitates the action of long storage because it accelerates the movement of droplets, which allows phase separation or creaming in a relatively short period of time.

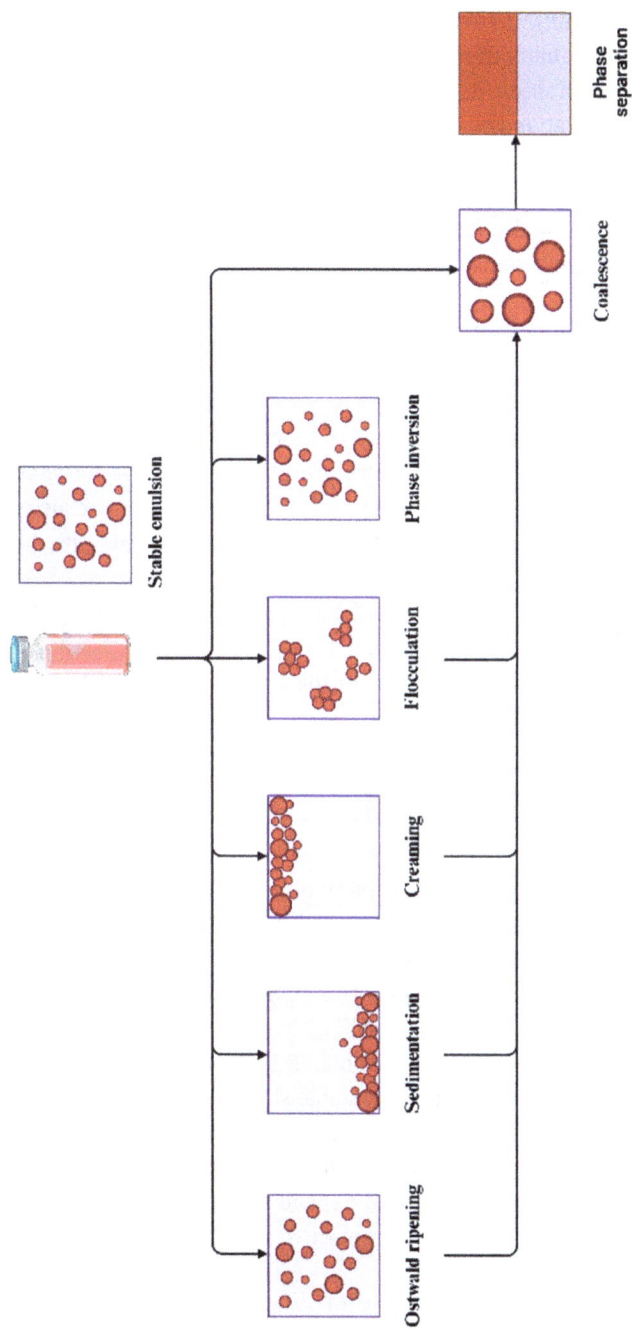

Figure 1.5: Instability of emulsion.

–   **Heating-cooling cycles:** Emulsions are subjected to cycles of extremes of variable temperature testing their stability in respect to thermal stress. The test will hardly benchmark the nanoemulsion on its capability to allow repeated heating and cooling without phase separation [79].

–   **Freeze-thaw stability:** This technique is the freezing of emulsion and its subsequent thawing in order to check whether it can resist a process that may lead to the formation of ice crystals and hence cause destabilization. The test is of great importance in the case of emulsions destined for use in temperature-fluctuating environments.

–   **Optical microscopy and turbidity measurements:** These methods therefore make it possible to detect changes in the distribution of droplet size as an indication of instability, while an increase in turbidity or the occurrence of larger droplets may indicate possible droplet aggregation or coalescence [80].

–   **Zeta potential analysis:** These methods therefore make it possible to detect changes in the distribution of droplet size as an indication of instability, while an increase in turbidity or the occurrence of larger droplets may indicate possible droplet aggregation or coalescence [81].

–   **Rheological measurements:** Therefore, rheological studies enable the evaluation of changes in the flow properties of emulsion. On one hand, a time increase in viscosity can be related to the instability due to the flocculation or gelation of droplets [82].

1.3.9.2 Thermodynamic and kinetic stability tests

Thermodynamic stability expresses the intrinsic ability of the emulsion system to retain its nonequilibrium status for a longer period, while kinetic stability refers to the time it takes when applicable instability processes start to develop, such as creaming, flocculation, or coalescence [83]. Both are very important in maintaining the functionality of nanoemulsions for a sufficiently long period [84]:

–   **Thermodynamic stability:** Therefore, thermodynamic stability tests provide evidence as to whether an emulsion is in its most stable state, with the least energy in maintaining its structure [85]. An emulsion is accepted as being thermodynamically stable if no spontaneous separation into their constituent phases occurs over a period of time [86]. The stability could depend on such variables as the composition of the oil and water phases, the concentration of surfactant, and the presence of cosurfactants [87].

–   **Kinetic stability:** Kinetic stability tests aim at the rate at which destabilization phenomena occur, such as coalescence, Ostwald ripening, or phase separation. Generally speaking, kinetic stability is better when droplet size is smaller and surfactant coverage of droplets is appropriate. This process of Ostwald ripening, where larger droplets grow at the expense of smaller ones, is an important cause of kinetic stability, which nanoemulsions in conditions of volatile or water-soluble oil phases can influence largely.

1.3.9.3 Factors influencing stability: temperature, pH, and ionic strength [88]

– **Temperature:** Temperature variations influence the stability of the nanoemulsion, further affecting oil phase solubility, aqueous phase viscosity, and interfacial tension between the phases. Droplet coalescence enhances with increased temperature due to decreased viscosity of the aqueous phase, while at low temperatures, phase separation may occur because of reduced molecular mobility.
– **pH:** The pH of the formulation can determine the ionization state of the surfactants or activity that could affect their ability to stabilize the emulsion; for example, an emulsion that may be stabilized by ionic surfactants could be destabilized at a pH-of-neutralization of such surfactants.
– **Ionic strength:** The presence of salts or electrolytes can influence stability to emulsions by the mechanism of electrostatic repulsion screening among charged droplets. High ionic strength would favor flocculation/coalescence through compressing the electrical double layer around droplets, decreasing zeta potential and thus reducing stability.

1.3.10 Characterization of drug release from micro/nanoemulsions

Characterization of the release mechanism of drugs from micro/nanoemulsions is important for their performance as drug delivery systems. The profile of the release will define the therapeutic efficacy, duration of action, and the bioavailability of the encapsulated drug [89]. Several in vitro techniques are employed for characterizing the way in which the drug is released from nanoemulsions and its interaction with biological membranes using diffusion models.

1.3.10.1 In vitro drug release techniques

In vitro release studies provide crucial data on the rate and extent of drug release from the emulsion. These techniques help in predicting the in vivo performance of nanoemulsions and guide formulation optimization:
– **Dialysis methods:** The techniques for studying it include mostly putting the nanoemulsion into a dialysis bag and allowing the drug to diffuse, by means of a semipermeable membrane, into an external medium. Following this, the concen-

tration of the drug is usually monitored in the release medium as a function of time to determine the release kinetics.

– **Franz diffusion cells:** In this setup, the nanoemulsion goes into a donor compartment, separated by a membrane from the receptor compartment. In this chamber is a release medium simulating physiological conditions. Drug concentration measured after regular time intervals in the release medium is done to study the release profile.

– **Membraneless methods:** These methods are based on the direct dispersion of nanoemulsion into the release media, in which time as a function examines the release of the drug. This method is more representative of in vivo conditions in which the emulsion interacts directly with biological fluids.

1.3.11 Diffusion and permeation studies

Diffusion studies help understand how the drug diffuses out of the nanoemulsion, while permeation studies assess the ability of the drug to penetrate biological barriers such as the skin or mucous membranes:

– **Fick's law of diffusion:** The drug release from nanoemulsion can often be explained by Fick's law, where the rate of diffusion depends on the gradient of concentration, area of surface, and coefficient of diffusion. These become important to be known with a view to optimizing drug delivery through nanoemulsion.

– **Permeation studies:** These studies simulate, in vitro, the interaction of the drug-loaded nanoemulsions with either the skin or the intestinal epithelium biological membrane for assessment of drug absorption and bioavailability. Excised animal tissues or artificial membranes, such as Strat-M membranes, are used to simulate in vivo conditions.

1.3.11.1 Interaction with biological membranes

The interaction of nanoemulsions with biological membranes is crucial for their effectiveness in drug delivery. Nanoemulsions must penetrate biological barriers to deliver their drug payload to the target site. The following mechanisms are typically involved:

– **Endocytosis:** Cells can take up nano-sized droplets by endocytosis, wherein the cell membrane engulfs the emulsion droplets, thus enabling intracellular drug delivery.

– **Membrane fusion:** Nanoemulsions can also fuse with the biological membrane's lipid bilayer, allowing the drug to be transferred directly into the cells. Whether fusion occurs would depend on the type of the emulsion and target cell membrane lipid composition.

- **Permeation enhancement:** The surfactants employed in nanoemulsions disrupt the breathing of the lipid bilayers in biological membranes, hence acting again to enhance permeation for the uptake of drugs.

1.4 Future perspective and challenges

Despite significant advancements in the characterization of nanoemulsions, several challenges remain in developing more efficient and scalable techniques for their study, especially as the field of nanomedicine grows.

1.4.1 Emerging techniques in micro/nanoemulsion studies

Advancements in analytical techniques have opened up new possibilities for the detailed characterization of nanoemulsions. These emerging techniques provide more precise, real-time data on emulsion properties:

- **AFM:** It produces nanoscale images of droplet morphology and surface characteristics at higher resolution than afforded by traditional electron microscopy. The technique is nondestructive, enabling analyses in liquid states with insight into emulsion behaviors in biological environments.
- **Nanoparticle tracking analysis (NTA):** Since NTA can track the Brownian motion of individual droplets, it can allow real-time monitoring of droplet size and concentration. In view of its high sensitivity, this technique is appropriate for carrying out studies in the size distribution and stability of nanoemulsions.
- **Cryo-TEM:** It has provided a lot of insight into the structure of nanoemulsion in its native form by providing high-resolution images of droplets in frozen conditions without fixation or staining.

1.4.2 Challenges in scaling up and industrial applications

Problematic conditions exist for scaling up nanoemulsions from lab to industrial settings. Nanoemulsions are sensitive systems, and slight variations in their preparation parameters may alter their properties; this may impede the capability of repeating the laboratorial results at a bigger scale. Quality, efficiency, and cost-effectiveness have to be ensured for both pharmaceutical and commercial applications. Below are some of the primary challenges in scaling up and industrializing nanoemulsion formulations.

1.4.2.1 Maintaining droplet size and distribution

While in the laboratory, fabrication with sophisticated equipment, stringent control of the parameters, and small batches usually produced nanoemulsion uniformity in both droplet size and distribution. During scale-up, uniform droplet size distribution becomes quite challenging because of changes in shear force, time of mixing, and energy input. Appropriately, the high-energy approaches, such as high-pressure homogenization and ultrasonication, have to be fitted for scale-up without altering the quality of the emulsion.

Poor droplet size distribution can lead to reduced stability, enhanced phase separation, and/or altered drug release profiles. The manufacturing process needs optimization of the processing conditions, surfactant concentration, and energy input in a balanced way with consideration of control over droplet size while limiting costs.

1.4.2.2 Process reproducibility and consistency

Another important challenge is the reproducibility at industrial production scale of nanoemulsion formulations. Variables such as temperature fluctuation, variation in pressure, and shear rate introduce inconsistencies in large-scale production. Nanoemulsions are sensitive to the slightest changes in such parameters; this leads to batch-to-batch variability of some key properties such as viscosity, surface charge, and stability.

These can be minimized by the preparation of strong SOP and employment of advanced process-monitoring technologies to ensure batch-to-batch reproducibility. Quality control at real time may be ensured through inline particle size analysis and rheological measurements, which are crucial in view of reproducibility in processes under scale-up studies.

1.4.2.3 Energy requirements and cost-effectiveness

The methods used for the preparation of nanoemulsions, such as ultrasonication and high-pressure homogenization, represent the class of high-energy techniques that are fairly energy-intensive. Scaling up these processes to industrial proportions often increases the cost of production, making it hard to work out an economically viable process. Industrial-scale continuous processing is often a necessity to increase efficiency, but again involves additional supporting infrastructures and equipment.

Among the proposed solutions for these challenges, low-energy methods would involve the development of either spontaneous emulsification or phase inversion techniques – both of which are far less energy-consuming. Nevertheless, these methods are still at the stage of optimization with regard to scaling up for higher-quantity productions.

1.4.2.4 Stability and shelf life concerns

Nanoemulsions are prone to time-induced destabilization due to processes such as Ostwald ripening, coalescence, and phase separation. However, while such emulsions may be described as stable under controlled laboratory conditions, the process of industrial formulation is far more challenging because it requires taking into account a wide range of environmental variables under different conditions of storage, transportation, and consumption.

The formulation at the time of manufacturing itself would have to be optimized to ensure a long-term stability in which droplet integrity is maintained through specifically chosen stabilizers and surfactants. Further, for commercial applications, the formulation of nanoemulsions has to take into consideration temperature, pH, and exposure to light as critical factors.

1.4.2.5 Regulatory challenges

In fact, from the regulatory standpoint, nanoemulsions, especially pharmaceutical ones, need to be thoroughly tested for their safety in order to satisfy different agencies like the FDA and EMA. Nanoemulsion-based drug delivery systems must be effective but also safe, stable, and biocompatible.

The production scale-up must be carried out under strict observation of good manufacturing practices to assure the product quality and compliance with the regulatory requirements, which may bring significant challenges in the case of innovative formulations with less-defined regulatory pathways.

1.5 Future perspectives and challenges in emulsion characterization

I. **Emerging characterization techniques:** Advanced techniques, such as AFM and NTA, are becoming more and more common, which enable higher resolution and far more detailed analysis of nanoemulsions. Accordingly, these techniques afford a much deeper understanding of the dynamics of nanoscale emulsions, including the ways to design more stable and effective formulations.

II. **Green and sustainable formulations:** Regarding this, the development of nanoemulsions with natural emulsifiers and biodegradable components is under increasing interest owing to the increasing demand for green pharmaceutical products and cosmetic formulations.

III. **Nanotechnology-enhanced drug delivery:** Future research will likely focus on further optimization of nanoemulsion systems for targeted drug delivery and personalized medicine applications. It is foreseen that the incorporation of function-alized nanoparticles, liposomes, or dendrimers will grant nanoemulsions with multifunctionality, as well as the capability for drugs to target a particular tissue or even a cell.

IV. **Improved stability and shelf life:** The ability to overcome nanoemulsion chal-lenges in stability will continue to remain at the forefront of research. Novel sta-bilizers, more efficient surfactants, and optimized formulation techniques will play a key role in the long-term stability of nanoemulsions in industrial applica-tions.

V. **Regulatory advancements:** This, in return, with an increase in participation of nanoemulsion-based drug delivery systems, calls for continuous changes in regu-latory frameworks that can accommodate such innovative formulations. In the near future, research will likely focus on generating the data required for the es-tablishment of clear regulatory pathways, which will assure the safety and effec-tiveness of nanoemulsions in pharmaceutical and cosmetic products.

Though the development and characterization of nanoemulsions have reached a high level, there are many issues that still need improvement, such as scale-up problems and regulatory issues. However, with continuous research and emerging new technol-ogy, nanoemulsions surely will prove to be game-changing drug delivery systems and find new areas of applications, from pharmaceuticals to cosmetics, and beyond.

1.6 Conclusion

Nanoemulsions have emerged as versatile drug delivery systems, enhancing the solu-bility, stability, and bioavailability of poorly soluble drugs. Over the last decade, major efforts have been undertaken to formulate and characterize these systems, pri-marily for pharmaceutical purposes. However, significant obstacles remain, mostly linked to scaling up production and ensuring long-term stability. Further advance-ments in such problems, as well as ongoing research into their potential usage in novel therapeutic applications, will ensure the future of nanoemulsions. Nanoemul-sions intended for medication delivery need thorough physicochemical characteriza-tion using diverse analytical techniques. DLS is used to evaluate droplet size and dis-persion, influencing medication release and stability. Zeta potential measurements reflect the surface charge of nanoemulsions, influencing colloidal stability. A high ab-solute zeta potential signifies improved stability and less aggregation. Modern imag-ing techniques such as cryo-EM and TEM enhance comprehension of the structural organization of nanoemulsions by generating high-resolution pictures of their shape. These techniques evaluate interactions among excipients, droplet uniformity, and bilayer configurations, enhancing formulation parameters and assuring consistent product performance.

The efficacy of nanoemulsions as drug-delivery devices may also be evaluated by in vitro drug release tests. Researchers use techniques such as dialysis bag diffusion and Franz diffusion cell approaches to simulate drug release rates and permeability via biological membranes. This discovery enables the prediction of the release profile of encapsulated medication inside the body, which is crucial for modifying the nano-emulsion formulation to meet specific therapeutic goals, including regulated and controlled drug delivery. The attainment of regulatory approval and the use of findings in clinical practice depend on the relationship between in vitro and in vivo release patterns. In addition to these characterization methods, evaluating stability is an essential component of nanoemulsion development. A series of stress tests is performed, encompassing thermal stability analyses with temperature variations, mechanical stability evaluations through centrifugation and agitation, and chemical stability assessments regarding pH fluctuations and oxidative stress, all aimed at forecasting the performance of nanoemulsions during storage and transport. These tests evaluate the formulation's resilience under various environmental conditions, therefore ensuring its durability and market feasibility.

These are essential to expecting the in vivo release pattern of a specific active principle and ensuring that the nanoemulsion formulation fits therapeutic needs. Aside from these methodologies, increased stability assessment methods such as thermal, mechanical, and chemical stability testing predict nanoemulsion performance during storage and transportation.

References

[1]   Preeti SS, Malik R, Bhatia S, Al Harrasi A, Rani C, et al. Nanoemulsion: An emerging novel technology for improving the bioavailability of drugs. Scientifica (Cairo), 2023;2023, 1–25.

[2]   Souto EB, Cano A, Martins-Gomes C, Coutinho TE, Zielińska A, Silva AM. Microemulsions and nanoemulsions in skin drug delivery. Bioengineering, 2022;9(4), 1–22.

[3]   Azmi NAN, Elgharbawy AAM, Motlagh SR, Samsudin N, Salleh HM. Nanoemulsions: Factory for food, pharmaceutical and cosmetics. Processes, 2019;7(9), 617.

[4]   Albert C, Beladjine M, Tsapis N, Fattal E, Agnely F, Huang N. Pickering emulsions: Preparation processes, key parameters governing their properties and potential for pharmaceutical applications. J Control Release, 2019;309, 302–332p.

[5]   Ait-Touchente Z, Zine N, Jaffrezic-Renault N, Errachid A, Lebaz N, Fessi H, et al. Exploring the versatility of microemulsions in cutaneous drug delivery: Opportunities and challenges. Nanomaterials, 2023;13(10), 1–19.

[6]   Wilson RJ, Li Y, Yang G, Zhao CX. Nanoemulsions for drug delivery. Particuology, 2022;64, 85–97.

[7]   Huang L, Huang XH, Yang X, Hu JQ, Zhu YZ, Yan PY, et al. Novel nano-drug delivery system for natural products and their application. Pharmacol Res, 2024;201(February), 107100.

[8]   Gupta A, Eral HB, Hatton TA, Doyle PS. Nanoemulsions: Formation, properties and applications. Soft Matter, 2016;12(11), 2826–2841.

[9]   Singh IR, Pulikkal AK Preparation, stability and biological activity of essential oil-based nano emulsions: A comprehensive review. OpenNano, 2022;8(August), 100066.

[10]   Marzuki NHC, Wahab RA, Hamid MA. An overview of nanoemulsion: Concepts of development and cosmeceutical applications. Biotechnol Equip, 2019;33(1), 779–797.

[11]   Pavoni L, Perinelli DR, Bonacucina G, Cespi M, Palmieri GF. An overview of micro-and nanoemulsions as vehicles for essential oils: Formulation, preparation and stability. Nanomaterials, 2020;10(1), 135.

[12]   Mushtaq A, Mohd Wani S, Malik AR, Gull A, Ramniwas S, Ahmad Nayik G, et al. Recent insights into nanoemulsions: Their preparation, properties and applications. Food Chem, 2023;18(March), 100684.

[13]   Julio LM, Copado CN, Diehl BWK, Tomás MC, Ixtaina VY. Development of chia oil-in-water nanoemulsions using different homogenization technologies and the layer-by-layer technique. Explor Foods Foodomics, 2024;2(2), 107–124.

[14]   Sutradhar KB, Amin L Nanoemulsions: Increasing possibilities in drug delivery. Eur J Nanomedicine, 2013;5(2), 97–110.

[15]   Sharma AK, Garg T, Goyal AK, Rath G. Role of microemuslsions in advanced drug delivery. Artif Cells Nanomed Biotechnol, 2016;44(4), 1177–1185.

[16]   Izquierdo P, Esquena J, Tadros TF, Dederen C, Garcia MJ, Azemar N, et al. Formation and stability of nano-emulsions prepared using the phase inversion temperature method. Langmuir, 2002;18(1), 26–30.

[17]   Al-Hussaniy HA, Almajidi YQ, Oraibi AI, Alkarawi AH. Nanoemulsions as medicinal components in insoluble medicines. Pharmacia, 2023;70(3), 537–547.

[18]   Bhalani DV, Nutan B, Kumar A, Singh Chandel AK. Bioavailability enhancement techniques for poorly aqueous soluble drugs and therapeutics. Biomedicines, 2022;10(9), 1–35.

[19]   Hiranphinyophat S, Otaka A, Asaumi Y, Fujii S, Iwasaki Y. Particle-stabilized oil-in-water emulsions as a platform for topical lipophilic drug delivery. Coll Surf B: Biointerfac, 2021;197(October 2020), 111423.

[20]   Jacob S, Kather FS, Boddu SHS, Shah J, Nair AB. Innovations in nanoemulsion technology: Enhancing drug delivery for oral, parenteral, and ophthalmic applications. Pharmaceutics, 2024;16(10), 1–41.

[21]   Ashaolu TJ. Nanoemulsions for health, food, and cosmetics: A review. Environ Chem Lett, 2021;19(4), 3381–3395.

[22]   Kupikowska-Stobba B, Domagała J, Kasprzak MM. Critical review of techniques for food emulsion characterization. Appl Sci, 2024;14(3), 1–50.

[23]   Fu Y, Xiao S, Liu S, Chang Y, Ma R, Zhang Z, et al. Atomistic insights into the droplet size evolution during self-microemulsification. Langmuir, 2022;38(10), 3129–3138.

[24]   Verma A, Gupta SK. History and Techniques of Bioimaging. In *Magnetic Quantum Dots for Bioimaging*, New York: CRC Press, 2023, pp. 111–132. doi: 10.1201/9781003319870-5

[25]   Yadav CK, Bhattarai A, Chaudhary Y. Transmission electron microscopy and dynamic light scattering-fundamental perspective. Cognition, 2024;6(1), 9–16.

[26]   Tian Y, Zhou J, He C, He L, Li X, Sui H. the formation, stabilization and separation of oil–water emulsions: A review. Processes, 2022;10(4), 738.

[27]   Costa C, Medronho B, Filipe A, Mira I, Lindman B, Edlund H, et al. Emulsion formation and stabilization by biomolecules: The leading role of cellulose. Polymers (Basel), 2019;11(10), 1570.

[28]   Li H, Zhang Q, Chong C, Yap R, Tay K, Hang T, et al. Accepted Muspt. Mater Today Proc, 2019;22(1), 16–20.

[29]   Whitby CP, Wanless EJ. Controlling pickering emulsion destabilisation: A route to fabricating new materials by phase inversion. Materials (Basel), 2016;9(8), 626.

[30]   Sarheed O, Dibi M, Ramesh KVRNS. Studies on the effect of oil and surfactant on the formation of alginate-based O/W lidocaine nanocarriers using nanoemulsion template. Pharmaceutics, 2020;12 (12), 1–21.

[31]	Bhatt P, Bhatt T, Jain V, Jain R, Bigoniya P. Nanoemulsion through cold emulsification: An advanced cold manufacturing process for a stable and advanced drug delivery system. J Appl Pharm Sci, 2024;14(5), 12–21.

[32]	Sanatkaran N, Zhou M, Foudazi R. Rheology of macro- and nano-emulsions in the presence of micellar depletion attraction. J Rheol (N Y N Y), 2021;65(3), 453–461.

[33]	Pal R. Non-Newtonian behaviour of suspensions and emulsions: Review of different mechanisms. Adv Colloid Interface Sci, 2024;333(July), 103299.

[34]	Budai L, Budai M, Fülöpné Pápay ZE, Vilimi Z, Antal I. Rheological considerations of pharmaceutical formulations: Focus on viscoelasticity. Gels, 2023;9(6).

[35]	Shafiei M, Kazemzadeh Y, Martyushev DA, Dai Z, Riazi M. Effect of chemicals on the phase and viscosity behavior of water in oil emulsions. Sci Rep, 2023;13(1), 1–14.

[36]	Khalid N, Kobayashi I, Neves MA, Uemura K, Nakajima M. Preparation and characterization of water-in-oil emulsions loaded with high concentration of l-ascorbic acid. Lwt, 2013;51(2), 448–454.

[37]	Dănilă E, Moldovan Z, Kaya MGA, Ghica MV. Formulation and characterization of some oil in water cosmetic emulsions based on collagen hydrolysate and vegetable oils mixtures. Pure Appl Chem, 2019;91(9), 1493–1507.

[38]	Salager JL, Marquez R, Rondón M, Bullón J, Graciaa A. Review on some confusion produced by the bicontinuous microemulsion terminology and its domains microcurvature: A simple spatiotemporal model at optimum formulation of surfactant-oil-water systems. ACS Omega, 2023;8(10), 9040–9057.

[39]	Han Y, Pan N, Li D, Liu S, Sun B, Chai J, et al. Formation mechanism of surfactant-free microemulsion and a judgment on whether it can be formed in one ternary system. Chem Eng J, 2022;437(P2), 135385.

[40]	Jha SK, Dey S, Karki R. Microemulsions-potential carrier for improved drug delivery. Asian J Biomed Pharm Sci, 2011;1(1), 5–9.

[41]	Prieto C, Calvo L. Performance of the biocompatible surfactant tween 80, for the formation of microemulsions suitable for new pharmaceutical processing. J Appl Chem, 2013;2013, 1–10.

[42]	Kale SN, Deore SL. Emulsion microemulsion and nanoemulsion. Syst Rev Pharm, 2017;8(1), 39–47.

[43]	Mohammed AN, Ishwarya SP, Nisha P. Nanoemulsion versus microemulsion systems for the encapsulation of beetroot extract: Comparison of physicochemical characteristics and betalain stability. Food Bioprocess Technol, 2021;14(1), 133–150.

[44]	Angelico R. Special issue Micro/nano emulsions: Smart colloids for multiple applications. Nanomaterials, 2022;12(21), 10–13.

[45]	McClements DJ. Nanoemulsions versus microemulsions: Terminology, differences, and similarities. Soft Matter, 2012;8(6), 1719–1729.

[46]	Frank B. Chapter 3.2.1 – Dynamic light scattering (DLS). In *Micro and Nano Technologies, Characterization of Nanoparticles*, V-D Hodoroaba, WES Unger, AG Shard Eds., Elsevier, 2020, pp. 137–172, ISBN 9780128141823. https://doi.org/10.1016/B978-0-12-814182-3.00010-9

[47]	Goddeeris C, Cuppo F, Reynaers H, Bouwman WG, Van Den Mooter G. Light scattering measurements on microemulsions: Estimation of droplet sizes. Int J Pharm, 2006;312(1–2), 187–195.

[48]	Montes de Oca-ávalos JM, Candal RJ, Herrera ML. Nanoemulsions: Stability and physical properties. Curr Opin Food Sci, 2017;16, 1–6.

[49]	Coupland JN, Julian Mcclements D. Droplet size determination in food emulsions: Comparison of ultrasonic and light scattering methods. J Food Eng, 2001;50(2), 117–120.

[50]	Chavda VP, Acharya D, Hala V, Daware S, Vora LK. Sunscreens: A comprehensive review with the application of nanotechnology. J Drug Deliv Sci Technol, 2023;86(June), 104720.

[51]	Stetefeld J, McKenna SA, Patel TR. Dynamic light scattering: A practical guide and applications in biomedical sciences. Biophys Rev, 2016;8(4), 409–427.

[52]	Gräwert TW, Svergun DI. Structural modeling using solution small-angle x-ray scattering (SAXS). J Mol Biol, 2020;432(9), 3078–3092.

[53] Papagiannopoulos A. Chapter 10 – Small-Angle Neutron Scattering (SANS). In *Micro and Nano Technologies, Microscopy Methods in Nanomaterials Characterization*, S Thomas, R Thomas, AK Zachariah, RK Mishra Eds., Elsevier, 2017, pp. 339–361, ISBN 9780323461412. https://doi.org/10.1016/B978-0-323-46141-2.00010-9

[54] Lombardo D, Calandra P, Kiselev MA. Structural characterization of biomaterials by means of small angle x-rays and neutron scattering (SAXS and SANS), and light scattering experiments. Molecules, 2020;25(23), 1–38.

[55] Inkson BJ. Scanning electron microscopy (SEM) and transmission electron microscopy (TEM) for materials characterization. Materials characterization using nondestructive evaluation (NDE) methods. Elsevier Ltd, 2016;17–43.

[56] Danev R, Yanagisawa H, Kikkawa M. Cryo-electron microscopy methodology: Current aspects and future directions. Trends Biochem Sci, 2019;44(10), 837–848.

[57] Xu R. Progress in nanoparticles characterization: Sizing and zeta potential measurement. Particuology, 2008;6(2), 112–115.

[58] 'Note : Page numbers followed by "f" indicate figures, "t" indicate tables.' :559–71.

[59] Dukhin AS, Xu R. Chapter 3.2.5 – Zeta-Potential Measurements. In *Micro and Nano Technologies, Characterization of Nanoparticles*, V-D Hodoroaba, WES Unger, AG Shard Eds., Elsevier, 2020, pp. 213–224, ISBN 9780128141823. https://doi.org/10.1016/B978-0-12-814182-3.00014-6

[60] Gonzalez Ortiz D, Pochat-Bohatier C, Cambedouzou J, Bechelany M, Miele P. Current trends in pickering emulsions: particle morphology and applications. Engineering, 2020;6(4), 468–482.

[61] Li L, Qu J, Liu W, Peng B, Cong S, Yu H, et al. Advancements in characterization techniques for microemulsions: From molecular insights to macroscopic phenomena. Molecules, 2024;29(12), 2901.

[62] Wang X, Anton H, Vandamme T, Anton N. Updated insight into the characterization of nano-emulsions. Expert Opin Drug Deliv, 2023 Jan;20(1), 93–114.

[63] Zhang X, Liao X, Gong Z, Li X, Jia C. Formation of fatty acid methyl ester based microemulsion and removal mechanism of PAHs from contaminated soils. J Hazard Mater, 2021;413(February), 125460.

[64] Conte P. Chapter Three – Applications of Fast Field Cycling NMR Relaxometry. In *Annual Reports on NMR Spectroscopy*, vol. 104, GA Webb Ed., Academic Press, 2021, pp. 141–188, ISSN 0066-4103, ISBN 9780128246207. https://doi.org/10.1016/bs.arnmr.2021.05.001

[65] Chatzidaki MD, Demisli S, Zingkou E, Liggri PGV, Papachristos DP, Balatsos G, et al. Essential oil-in-water microemulsions for topical application: Structural study, cytotoxic effect and insect repelling activity. Colloids Surfaces A Physicochem Eng Asp, 2022;654, 130159.

[66] Awad TS, Asker D, Romsted LS. Evidence of coexisting microemulsion droplets in oil-in-water emulsions revealed by 2D DOSY 1H NMR. J Colloid Interface Sci, 2018;514, 83–92.

[67] Huaizhou Jin Qifei Ma TDSJ, Jiang L. Raman spectroscopy of emulsions and emulsion chemistry. Crit Rev Anal Chem, 2023;0(0), 1–13.

[68] Hufnagel T, Rädle M, Karbstein HP. Influence of refractive index differences on the signal strength for Raman-spectroscopic measurements of double emulsion droplets. Appl Sci, 2022;12(18), 9056.

[69] Zang J, Feng M, Zhao J, Wang J. Micellar and bicontinuous microemulsion structures show different solute–solvent interactions: A case study using ultrafast nonlinear infrared spectroscopy. Phys Chem Chem Phys, 2018;20(30), 19938–19949.

[70] Li Y, Driver M, Winuprasith T, Zheng J, McClements DJ, He L. In situ SERS detection of emulsifiers at lipid interfaces using label-free amphiphilic gold nanoparticles. Analyst, 2014;139(20), 5075–5078.

[71] Do Nascimento Ds, Volpe V, Fernandez CJ, Oresti GM, Ashton L, Grünhut M. Confocal Raman spectroscopy assisted by chemometric tools: A green approach for classification and quantification of octyl p-methoxycinnamate in oil-in-water microemulsions. Microchem J, 2023;184, 108151.

[72] Nasr AM, Aboelenin SM, Alfaifi MY, Shati AA, Elbehairi SEI, Elshaarawy RFM, et al. Quaternized chitosan thiol hydrogel-thickened nanoemulsion: A multifunctional platform for upgrading the topical applications of virgin olive oil. Pharmaceutics, 2022;14(7), 1319.

[73] Linke C, Drusch S. Turbidity in oil-in-water-emulsions – Key factors and visual perception. Food Res Int, 2016;89, 202–210.

[74] Zhang J, Zhang H, Wu J, Zhang J. Fuel cell degradation and failure analysis. Pem Fuel Cell Testing and Diagnosis, 2013;283–335.

[75] Nikita SA. Nanoemulsions as advanced delivery systems of bioactive compounds for sustainable food preservation applications. Biocatal Agric Biotechnol, 2025;68, 103702, ISSN 1878–8181, https://doi.org/10.1016/j.bcab.2025.103702

[76] Jin W, Xu W, Liang H, Li Y, Liu S, Li B. 1 – Nanoemulsions for Food: Properties, Production, Characterization, and Applications. In *Nanotechnology in the Agri-Food Industry, Emulsions*, AM Grumezescu Ed., Academic Press, 2016, pp. 1–36, ISBN 9780128043066. https://doi.org/10.1016/B978-0-12-804306-6.00001-5

[77] Han HS, Koo SY, Choi KY. Emerging nanoformulation strategies for phytocompounds and applications from drug delivery to phototherapy to imaging. Bioact Mater, 2022;14, 182–205.

[78] Kumar D, Gupta SK, Verma A, Ajazuddin. Transferosome as nanocarrier for drug penetration enhancement across skin: A comparison with liposome and ethosome. Curr Nanomed, 2025;15, e24681873333341, doi: 10.2174/0124681873333341241030065955

[79] Banik BL, Fattahi P, Brown JL. Polymeric nanoparticles: The future of nanomedicine. Wiley Interdiscip Rev Nanomed Nanobiotechnol, 2016;8(2), 271–299.

[80] Aswathanarayan JB, Vittal RR. Nanoemulsions and Their Potential Applications in Food Industry. Front Sustain Food Syst, 2019;3. https://doi.org/10.3389/fsufs.2019.00095

[81] Maurya A, Singh VK, Das S, Prasad J, Kedia A, Upadhyay N, et al. Essential oil nanoemulsion as eco-friendly and safe preservative: Bioefficacy against microbial food deterioration and toxin secretion, mode of action, and future opportunities. Front Microbiol, 2021;12. https://doi.org/10.3389/fmicb.2021.751062

[82] Singh P, Khanna S, Chauhan ES Nanoemulsion: composition, preparation and its application in the food industry. Int J Pharm Sci Nanotechnol, 2024;17(3), 7398–7405.

[83] Porras M, Solans C, Gonzalez C, Martínez A, Guinart A, Gutiérrez J. Studies on W/O nano-emulsions. Colloids Surf A: Physicochem Eng Asp, 2004;249, 115–1118.

[84] Li J, Guo R, Hu H, Wu X, Ai L, Wu Y. Preparation optimization and storage stability of nanoemulsion-based lutein delivery systems. J Microencapsul, 2018;35, 570–583.

[85] Choi SJ, McClements DJ Nanoemulsions as delivery systems for lipophilic nutraceuticals: Strategies for improving their formulation, stability, functionality and bioavailability. Food Sci Biotechnol, 2020;29(2), 149–168.

[86] Nair AB, Singh B, Shah J, Jacob S, Aldhubiab B, Sreeharsha N, et al. Formulation and evaluation of self-nanoemulsifying drug delivery system derived tablet containing sertraline. Pharmaceutics, 2022;14(2).

[87] Mishra JS, Poonia SP, Kumar R, Dubey R, Kumar V, Mondal S, et al. An impact of agronomic practices of sustainable rice-wheat crop intensification on food security, economic adaptability, and environmental mitigation across eastern Indo-Gangetic Plains. F Crop Res, 2021;267, 108164.

[88] Helgeson ME. Colloidal behavior of nanoemulsions: Interactions, structure, and rheology. Curr Opin Colloid Interface Sci, 2016;25, 39–50.

[89] Djekic L, Jankovic J, Čalija B, Primorac M Development of semisolid self-microemulsifying drug delivery systems (SMEDDSs) filled in hard capsules for oral delivery of aciclovir. Int J Pharm, 2017;528(1), 372–380.

Ashish Kumar Pandey
Chapter 2
Micro/nanoemulsions in dermatology for enhanced therapeutic solutions

Abstract: Dermatological treatments often encounter obstacles such as poor drug penetration, instability of active ingredients, and undesirable side effects associated with
conventional topical formulations. Micro- and nanoemulsions have emerged as advanced delivery systems capable of overcoming these limitations. Owing to their
unique nanoscale architecture, these formulations enhance drug solubilization, stability, and controlled release, while minimizing irritation and systemic exposure. In dermatology, micro/nanoemulsions enable noninvasive, site-specific therapies for a
range of skin disorders, including chronic and inflammatory conditions. Innovations
such as temperature-, enzyme-, and light/magnetic-responsive emulsions further optimize drug delivery, tailoring treatment to individual patient needs. Ensuring formulation stability and extended shelf life requires careful selection of excipients, packaging materials, and manufacturing techniques. This work presents a comprehensive
overview of the structural design, penetration mechanisms, and clinical potential of
micro/nanoemulsions, underscoring their transformative role in dermatological therapeutics.

Keywords: Nanoemulsion, microemulsion, dermal delivery, topical formulation

2.1 Introduction

2.1.1 Rising challenges in dermatological care

Topical drug delivery is a critical aspect of dermatology. It involves delivering active
pharmaceutical ingredients (APIs) directly to the skin for local treatment or, in some
cases, systemic absorption. Despite its advantages, including convenience, reduced
systemic side effects, and direct targeting of affected skin areas, topical drug delivery
faces a number of challenges, particularly when managing chronic skin disorders.
These conditions, which can persist for long periods, not only affect patients' quality
of life but also present significant public health challenges, with a growing global
prevalence and considerable socioeconomic impact on individuals, healthcare systems, and society as a whole [1].

Ashish Kumar Pandey, School of Pharmacy, Shri Shankaracharaya Professional University, Bhilai,
490020, Durg, Chhattisgarh, India

https://doi.org/10.1515/9783111593654-003

Topical drug delivery, while beneficial for localized treatments, has several limitations. The skin's barrier properties, primarily the stratum corneum, restrict the penetration of many drugs, particularly larger or hydrophilic molecules [2]. Factors such as skin thickness, which varies across different body sites, further influence drug absorption. Additionally, the molecular size and lipophilicity of drugs play a crucial role in their ability to pass through the skin; smaller, lipophilic drugs tend to be more readily absorbed, whereas larger, hydrophilic drugs often face challenges in achieving effective topical delivery. These limitations become even more pronounced in the management of chronic skin disorders, where consistent and effective drug delivery is essential to improving patient outcomes and quality of life [3].

Skin conditions, such as eczema or psoriasis, can further complicate topical drug delivery by altering skin permeability, either enhancing or reducing drug absorption and making treatment outcomes less predictable. Additionally, patient compliance can be challenging, particularly when treatments require frequent application or have undesirable cosmetic properties like greasiness or staining. Moreover, the potential for irritation or allergic reactions adds another layer of complexity, especially for individuals with sensitive skin [4, 5]. These factors underscore the need for advanced delivery systems that can overcome these barriers while ensuring effective and patient-friendly treatment, particularly in managing chronic skin disorders.

2.1.2 Limitations of conventional topical formulations

Conventional topical formulations primarily act on the skin's outermost layers and often struggle to deliver therapeutic agents effectively to deeper layers of the skin. As a result, higher concentrations of bioactive compounds are required to achieve the desired therapeutic effect, which in turn increases the risk of side effects such as skin irritation, sensitization, and unwanted systemic absorption [6]. Upon application, the active ingredient tends to accumulate at the site, forming a localized depot that is susceptible to rapid absorption, evaporation, or removal due to external factors like washing, friction, or environmental exposure. These limitations highlight the need for strategies that prolong the retention of bioactive molecules within the epidermis and stratum corneum while minimizing unwanted transdermal diffusion [7–10].

2.1.3 Innovative approaches to skin treatment: a paradigm shift

Micro- and nanoemulsions represent a promising advancement in skin treatment, consisting of tiny droplets on the nanometer scale, typically smaller than 1 μm. Their pharmaceutical applications are governed by their chemical and physical characteristics. Microemulsions are considered thermodynamically stable dispersions, with mean droplet sizes ranging from 100 to 400 nm, while nanoemulsions are thermodynamically unstable dispersions, with mean droplet sizes of 1 to 100 nm, often requiring a co-surfactant for stabilization due to their higher Gibbs free energy [11, 12]. The average particle diameter and polydispersity index of these emulsions are influenced by both their quantitative and qualitative composition. Additionally, the interplay between the surfactant, which reduces the free interfacial energy of the system, and the selected lipid, which impacts interfacial tension, viscosity, and lipophilicity, plays a crucial role in determining the emulsion's stability and performance [13].

2.1.4 Micro/nanoemulsions: a game changer in dermatology

In recent years, dermatology has garnered significant attention from the pharmaceutical industry, with numerous clinical studies advancing through phase II, III, and preregistration stages. Among various dermatological conditions, plaque psoriasis remains the most widely targeted indication in this already saturated market. Amid these developments, microemulsions and nanoemulsions have emerged as highly effective delivery systems, offering distinct advantages over conventional topical formulations in terms of drug administration, bioavailability, and patient compliance [14]. These innovative delivery methods enhance the penetration of APIs into the skin, improve therapeutic outcomes, and provide a cosmetically elegant formulation that enhances patient adherence. As colloidal dispersions consisting of tiny droplets of one phase dispersed within another, microemulsions and nanoemulsions leverage their large surface area and small droplet size to offer unique benefits for topical drug delivery, making them promising tools for revolutionizing dermatological treatments [15].

2.2 Micro/nanoemulsions: structural design and functional mechanisms

Micro/nanoemulsions, with droplet sizes playing a pivotal role in determining optical clarity, stability, release patterns, and rheological properties, have emerged as crucial systems for drug delivery. Their turbidity (τ), governed by light transmission, directly reflects droplet size, making it a key parameter in formulation development [16]. Among various types, oil-in-water emulsions are particularly advantageous, enhancing drug absorption by minimizing intersubject variability and providing protection against oxidative and hydrolytic degradation. These systems not only improve therapeutic efficacy but also reduce required drug doses, thereby lowering the risk of side effects. Long-chain triacylglycerols are often preferred as the oil phase due to their abundance, cost-effectiveness, and additional nutritional benefits. The overall stability of micro/nanoemulsions is intricately linked to the physicochemical properties of the oil phase, including interfacial tension, refractive index, viscosity, polarity, chemical stability, density, water solubility, and phase behavior, all of which contribute to ensuring a robust and effective delivery system [17].

The skin, the body's largest organ, functions as a critical protective barrier, shielding internal systems from pathogens, chemicals, and physical harm [18]. This defense is primarily attributed to the **stratum corneum**, the outermost layer of the epidermis, which tightly regulates the penetration of both hydrophilic and lipophilic substances [19]. Drug delivery through the skin is inherently challenging due to its multilayered structure and selective permeability. The success of transdermal drug therapy depends on multiple factors, including the physicochemical properties of the drug, the formulation used, and the architecture of the skin itself [20].

Microemulsions and nanoemulsions have emerged as advanced delivery systems capable of bypassing the skin's natural defenses and enhancing drug penetration by utilizing multiple routes of entry. The three main pathways for dermal absorption include:

- **Intercellular route:** Drug molecules diffuse through the lipid-rich spaces between corneocytes, favoring small and lipophilic compounds.
- **Intracellular route:** Penetration occurs directly through keratin-filled cells, requiring a balance of hydrophilic and lipophilic properties.
- **Follicular route:** Drugs are transported through hair follicles and sebaceous glands, offering an efficient route for larger molecules or particles.

By exploiting these routes, micro- and nanoemulsions significantly enhance drug transport across the skin barrier. Their nanosized droplets and thermodynamically stable structures enable deeper and sustained delivery of therapeutic agents, ultimately improving efficacy, while minimizing systemic absorption [21]. Figure 2.1 provides a visual representation of the key skin penetration mechanisms utilized by topical drug delivery systems, highlighting how micro- and nanoemulsions enhance transport through intercellular, intracellular, and follicular pathways.

2.3 Therapeutic superiority of micro/nanoemulsions in dermatology

Microemulsions and nanoemulsions are particularly beneficial in the treatment of various skin conditions, offering superior drug solubilization, enhanced permeation,

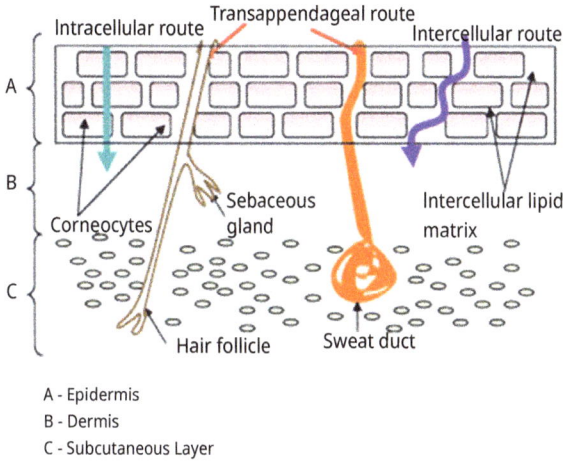

A - Epidermis
B - Dermis
C - Subcutaneous Layer

Figure 2.1: Representation of skin penetration mechanisms.

and reduced toxicity. Below are the key aspects linking their therapeutic superiority
to specific dermatological concerns [22].

2.3.1 Enhanced drug solubilization and stability

Enhanced drug solubilization and stability are key factors that contribute to the therapeutic superiority of micro/nanoemulsions in dermatology. These advanced delivery
systems excel at dissolving lipophilic (oil-soluble) medications within their oil phase,
creating an ideal environment for hydrophobic drugs that are typically challenging to
incorporate into aqueous-based formulations. Moreover, unlike conventional emulsions, micro- and nanoemulsions are thermodynamically stable and resistant to phase
separation over time. This remarkable stability is particularly beneficial for topical formulations in dermatology, which require a long shelf life and consistent therapeutic performance. As a result, enhanced solubilization and stability enable a steady, prolonged
drug release, making these systems especially effective in treating skin infections, psoriasis, acne, and even in cosmetic applications where sustained action is crucial [23].

2.3.2 Enhanced targeting and permeation

Skin penetration: Micro- and nanoemulsions significantly enhance drug penetration
through the skin barrier. Their extremely small droplet size, often in the nanometer
range, increases the surface area for interaction with the skin, thereby facilitating better diffusion through the stratum corneum – the skin's outermost, most resistant

layer. Additionally, these tiny droplets can temporarily disrupt the lipid matrix of the stratum corneum, allowing the encapsulated drug to traverse deeper into the viable epidermis and dermis more effectively. Compared to conventional formulations such as emulsions, creams, and ointment gels, numerous studies have shown that pharmaceuticals delivered as nanoemulsion systems have improved drug penetration [24].

Targeting: These delivery systems can be precisely engineered to target specific skin layers or cell types. For example, they can be modified to preferentially deposit drugs in sebaceous glands for acne treatment or in melanocytes, for addressing hyperpigmentation disorders. This enhanced targeting is primarily attributed to the small droplet size, which facilitates uniform distribution and deeper penetration at the intended site [25]. Furthermore, surface modifications – such as ligand conjugation or the incorporation of pH-responsive materials – can be employed to further refine the targeting capability, ensuring that the therapeutic agent is released precisely where it is needed for optimal efficacy.

Improved bioavailability: Improved bioavailability is achieved, as the permeability of the skin is enhanced by nano-emulsion, allowing drugs that are typically poorly absorbed to be delivered in higher concentrations at the target site [26]. Enhanced penetration and retention are facilitated by the small droplet size, resulting in increased local drug concentrations. As a consequence, therapeutic efficacy is improved in the treatment of chronic skin conditions, where elevated drug levels are required for optimal outcomes [26].

2.3.3 Reduction of drug toxicity and irritation

Conventional dermatological formulations, particularly those incorporating potent APIs, are often associated with skin irritation and adverse reactions. Micro- and nano-emulsions, however, present a sophisticated approach to drug delivery by enabling controlled, localized administration, thereby reducing systemic exposure and minimizing undesirable effects such as erythema and irritation. These emulsions encapsulate APIs within their internal phase, creating a protective barrier that prevents direct interaction with the skin surface, consequently mitigating the risk of dermal toxicity [27]. This characteristic is particularly advantageous in the delivery of corticosteroids or anti-inflammatory agents, where prolonged use in conventional formulations can result in significant side effects. Moreover, the enhanced permeation and retention offered by micro- and nanoemulsions facilitate the administration of higher drug concentrations, while reducing systemic absorption, making these systems especially suitable for treating sensitive skin conditions such as rosacea. The ability to provide sustained release, while minimizing irritation, highlights their potential in optimizing therapeutic outcomes and improving patient adherence [28].

2.3.4 Noninvasive therapeutics for chronic skin conditions

Micro- and nanoemulsions offer a noninvasive alternative to oral or injectable therapies for chronic skin conditions, mitigating the risk of systemic side effects, often associated with these conventional routes of administration. These emulsions have shown remarkable efficacy in managing chronic dermatological disorders such as vitiligo, eczema, and psoriasis, owing to their ability to provide sustained drug release directly at the site of the skin condition [29]. The prolonged retention and controlled release of therapeutic agents ensure long-lasting activity, reducing the need for frequent applications and enhancing treatment adherence. Additionally, the convenience and ease of application associated with these advanced delivery systems improve patient compliance, making them particularly suitable for long-term management of conditions like acne, which require consistent care over extended periods [30].

2.3.5 Smart micro/nanoemulsions for site-specific release

Smart microemulsions and nanoemulsions are particularly advantageous in applications where precise localization of therapeutic agents is essential, such as in dermatology, oncology, and targeted drug delivery [31].

2.3.5.1 Temperature-responsive emulsions

Temperature-responsive micro- and nanoemulsions represent a promising advancement in dermatological and transdermal drug delivery, leveraging structural or rheological changes in response to temperature variations to optimize drug deposition at the target site. Upon exposure to body temperature (37 °C), these emulsions can transition into a gel-like state, enhancing skin retention and providing sustained drug release. This transformation improves therapeutic outcomes by ensuring prolonged contact with the skin and reducing the need for frequent reapplication. Poloxamer-based emulsions, in particular, have been extensively investigated for their thermosensitive properties, making them highly suitable for delivering drugs in chronic skin conditions, where controlled and extended release is crucial [33].

2.3.5.2 Enzyme-responsive emulsions

Enzyme-responsive emulsions are engineered to exploit the enzymatic activity present in the skin environment. In lipase-responsive formulations, lipid-based carriers are degraded upon exposure to skin lipases, thereby ensuring controlled drug release and enhanced penetration [34]. Similarly, formulations incorporating peptide-based surfactants are designed to degrade in response to proteolytic enzymes, which facilitates targeted delivery in inflamed or diseased skin [34].

2.3.5.3 Magnetic and light-responsive emulsions

Stimuli-responsive micro- and nanoemulsions, activated by external magnetic fields or light exposure, offer a novel approach to enhancing drug penetration and therapeutic efficacy in dermatology and transdermal drug delivery. These emulsions undergo structural modifications upon exposure to specific stimuli, enabling precise control over drug release and improving treatment outcomes. In particular, light-responsive emulsions incorporate photosensitive compounds like spiropyrans or azobenzenes, which undergo reversible conformational changes when exposed to light, allowing for on-demand drug release [35, 36].

2.4 Formulation strategies for maximizing stability and shelf life

Formulating stable micro- and nanoemulsions is crucial for ensuring their efficacy and shelf life in dermatological applications. Key strategies focus on optimizing surfactant selection, oil phase composition, and droplet size to prevent phase separation and degradation over time. Proper pH adjustment, antioxidant incorporation, and temperature control further enhance stability by protecting sensitive active ingredients. These approaches ensure consistent drug delivery, making emulsions more reliable for long-term therapeutic use.

2.4.1 Material used for the preparation of nanoemulsion with their properties

The selection of materials for micro/nanoemulsion formulation plays a crucial role in determining their physicochemical properties and therapeutic efficiency. Micro/nanoemulsions are typically composed of oils, surfactants, co-surfactants, and aqueous phases. Each component contributes specific properties that influence the stability, droplet size, and drug release profile of the formulations listed in Table 2.1 [36].

Table 2.1: Formulation strategies for maximizing stability of micro- and nanoemulsions.

Materials used	Properties	Examples
Oil/lipids	The oil phase consists of lipophilic compounds, which can be essential oils, vegetable oils, or synthetic oils.	Essential oils (e.g., eucalyptus oil and peppermint oil) Vegetable oils (e.g., olive oil and coconut oil) Medium-chain triglycerides (MCTs) Mineral oils Synthetic oils (e.g., isopropyl myristate)
Aqueous phase	This is primarily water or an aqueous buffer that dissolves hydrophilic components.	Distilled water Phosphate-buffered saline (PBS) Glycerol or ethanol (sometimes used as co-solvents)
Polysaccharides	When dissolved in water, they form strong hydrogen bonds and impart favorable mechanical qualities to edible coatings and films.	Gum arabic, pectins, modified starch, modified alginate, modified celluloses, and some cellulose or galactomannans derivatives, gums, alginates, carrageenans, starches, and pectins
Surfactants	These help stabilize the nanoemulsion by reducing interfacial tension between oil and water.	Tween 20, Tween 80 (polysorbates) Span 80 (sorbitan monooleate) Lecithin (natural emulsifier) Sodium dodecyl sulfate (SDS) PEGylated surfactants
Active ingredients	Nanoemulsion is used for drug delivery, cosmetics, or food; bioactive compounds are included.	Pharmaceuticals (e.g., curcumin, ibuprofen, and cannabidiol) Vitamins (e.g., vitamin E and vitamin C) Antimicrobials (e.g., silver nanoparticles and essential oils)
Density modifiers	Substances that are added to oil droplets such that their densities correspond to the surrounding continuous phase.	Glycerol propylene glycol, sucrose, sodium chloride (NaCl), polyethylene glycol (PEG), mannitol, silicone oils, and fluorinated oils
Texture modifiers	Compounds that gel the aqueous phase or improve the viscosity. These can also be used to offer desired textural features, or prevent gravitational separation.	Gelatin, whey protein, soy protein, carbomers (e.g., Carbopol 940 and Carbopol 934), polyvinyl alcohol (PVA), polyacrylic acid, and polyethylene glycol (PEG)

Table 2.1 (continued)

Materials used	Properties	Examples
Stabilizers	These enhance the stability of the nanoemulsion by preventing phase separation.	Polymers (e.g., chitosan and xanthan gum) Proteins (e.g., whey protein and casein)
Plasticizers	Modifies the polymer's hardness and thermoplastic qualities, reduces brittleness and cohesiveness, and adds flexibility.	Monosaccharides, oligosaccharides (honey, glucose, sucrose, and fructose), polyols (sorbitol, polyethylene glycols, and glycerol), lipids and its derivatives (fatty acids and phospholipids), ethylene glycol, glycerol, sucrose, sorbitol, and xylitol

2.4.2 Formulation strategies

2.4.2.1 High-energy methods

High-energy methods utilize mechanical energy to break down larger droplets into nano-sized particles. Ultrasonication employs high-frequency sound waves to create intense cavitation, resulting in droplet disruption and size reduction [37, 38]. In high-pressure homogenization, a coarse emulsion is passed through a narrow orifice under high pressure, creating intense shear and turbulence that produce nanosized droplets [67]. These methods are particularly effective in producing small, uniform droplets, making them suitable for scalable production [68].

2.4.2.2 Low-energy methods

Low-energy methods rely on the spontaneous formation of nanoemulsions by exploiting the physicochemical properties of the system. Phase inversion involves altering the temperature or composition to achieve a phase transition that reduces interfacial tension, leading to the self-assembly of nanosized droplets [39]. Spontaneous emulsification, on the other hand, occurs when an organic phase containing oil, surfactants, and co-surfactants is introduced into an aqueous phase under specific conditions, resulting in the formation of nanoemulsions, without the need for external energy [40, 41]. These methods are cost-effective and suitable for thermosensitive drugs or compounds [42].

2.5 Clinical applications of micro/nanoemulsions in dermatological therapy

The clinical utility of micro- and nanoemulsion in dermatology continues to grow, with ongoing research exploring their applications in treating acne, psoriasis, and other chronic skin disorders. These advanced delivery systems not only improve drug absorption but also reduce side effects, offering a safer, more targeted approach to dermatological therapy [43].

Microemulsions and nanoemulsions are both types of emulsions that have unique properties useful in dermatological therapy. These emulsions have been increasingly utilized for the delivery of drugs, cosmetics, and other therapeutic agents due to their enhanced stability, skin penetration, and ability to deliver active ingredients more effectively than traditional formulations.

Numerous studies have demonstrated the potential of nanoemulsions in enhancing the solubility and bioavailability of poorly water-soluble drugs for topical administration. For instance, prednicarbate, a next-generation corticosteroid with low aqueous solubility but a favorable risk-benefit profile, has shown enhanced penetration and therapeutic efficacy when incorporated into nanoemulsion-based formulations. This advancement has proven particularly beneficial in managing conditions such as atopic dermatitis, where improved drug delivery to inflamed skin is crucial [44].

Moreover, the use of nanoemulsion extends to innovative applications such as topical caffeine delivery for skin cancer treatment. Recent studies have highlighted caffeine's anticancer properties, making it a potential candidate for topical intervention. Formulating caffeine in nanoemulsions has shown promise in enhancing its transdermal penetration, paving the way for more effective therapeutic outcomes [45].

The potential of oral retinoids, cyclosporines, and methotrexate (MTX) in treating psoriasis has been well established. However, systemic administration of MTX often leads to gastrointestinal side effects and liver toxicity, limiting its prolonged use. Nanoemulsion-based topical delivery of MTX offers a promising alternative, demonstrating reduced inflammation and epidermal cell proliferation, while minimizing systemic adverse effects. This enhanced efficacy is attributed to the nanoemulsion's ability to alter cellular organization and interact with membrane enzymes and transporters, facilitating improved drug penetration across psoriatic skin [45].

Nevertheless, psoriatic skin poses unique challenges to drug delivery, acting as a formidable barrier due to plaque formation, scaling, epidermal thickening, and elongation of epidermal ridges. These structural changes hinder nanoemulsion penetration, as confirmed by microdialysis and in vitro permeation studies. Compounds such as aceclofenac and capsaicin exhibited reduced penetration into the receptor compartment through inflammatory, psoriatic skin compared to healthy skin, underscoring the need for further optimization of nanoemulsion formulations to overcome these barriers and maximize therapeutic outcomes.

In addition to methotrexate, other poorly soluble drugs such as paclitaxel (PCL) have also been explored for nanoemulsion-based topical delivery in psoriasis treatment. Paclitaxel, classified as a Biopharmaceutics Classification System (BCS) Class IV drug, is characterized by poor solubility and limited permeability, posing challenges for effective delivery. To address these issues, TPGS-based (d-α-Tocopheryl polyethylene glycol succinate) nanoemulsions have been investigated for their ability to localize PCL in deeper skin layers for extended periods, thereby enhancing its therapeutic efficacy.

Due to its high molecular weight and lipophilic nature, PCL serves as an ideal candidate for skin localization using nanoemulsions. The goal is to increase the drug concentration directly at the site of action, while minimizing systemic exposure, a crucial aspect in managing psoriasis. TPGS-based o/w (oil-in-water) nanoemulsions have shown remarkable efficacy in achieving localized drug accumulation, significantly enhancing PCL's skin bioavailability. Pharmacokinetic studies revealed that upon dermal application, the nanoemulsion facilitated higher drug deposition across the stratum corneum, epidermis, and dermis, demonstrating a regional advantage over conventional delivery methods.

Notably, the maximum local concentration of PCL in the dermis reached 10 $\mu g/cm^2$ following topical nanoemulsion application, while systemic absorption remained minimal at approximately 100 ng/mL. In contrast, intravenous administration of the same formulation yielded a maximum local concentration of only 1 $\mu g/cm^2$, underscoring the superior localization achieved through nanoemulsion-mediated delivery. These findings highlight the potential of nanoemulsions not only in enhancing local drug concentrations but also in reducing systemic toxicity, making them a promising strategy for improving psoriasis management [46–48].

In the field of wound healing and dermatological therapy, micro- and nanoemulsions have been extensively studied as advanced delivery systems to enhance therapeutic outcomes. Coenzyme Q10 microemulsions and curcumin nanoemulsions have been identified as promising candidates for topical wound treatment due to their antioxidant and anti-inflammatory properties [63, 64]. Furthermore, nanoemulsion-based gels incorporating phytochemicals such as resveratrol and ferulic acid have been developed as innovative alternatives for UV protection, offering improved stability and controlled release [65, 66].

The formulation of clobetasol propionate (CP), a potent corticosteroid used in managing inflammatory skin conditions like atopic dermatitis and psoriasis, into a nanoemulsion (CP-NE) has been investigated to enhance its therapeutic efficacy. In a preclinical study involving a rat model of contact dermatitis, superior in vivo efficacy was demonstrated by CP-NE (0.05%) compared to conventional formulations. Enhanced NTPDase activity in rat lymphocytes was observed, indicating an improved anti-inflammatory response, likely attributed to enhanced drug penetration and controlled release.

The clinical utility of CP lies in its vasoconstrictive, anti-inflammatory, immuno-suppressive, and antiproliferative properties, making it valuable for the management of severe dermatological conditions. However, in traditional formulations, prolonged use has been associated with adverse effects such as skin atrophy and systemic absorption. Through nanoencapsulation, these risks were mitigated by reducing the frequency of administration, while optimizing dermato-kinetics. Sustained drug release and prolonged therapeutic action at the target site were ensured, resulting in an enhanced risk-benefit ratio. The successful incorporation of CP into nanoemulsions exemplifies the potential of micro- and nanoemulsion technologies to improve dermatological therapies, offering safer and more effective treatment options for chronic skin conditions [59].

The versatility of micro- and nanoemulsion-based delivery systems has paved the way for diverse applications in both medical and cosmetic fields. Table 2.2 highlights a broad spectrum of natural and synthetic drugs with significant pharmaceutical potential that can be incorporated into these advanced formulations. Such versatility not only facilitates the development of innovative therapeutic solutions but also offers new product opportunities in the highly competitive cosmetics market [56–58].

In dermatological therapy, these emulsions have been harnessed for the formulation of anti-inflammatory creams, such as diclofenac-containing Voltaren Emulgel (GSK, Brentford, UK), which provides localized relief from pain and inflammation. Similarly, emulsions have found use in the treatment of burns, exemplified by trolamine-based products like Biafine (Johnson & Johnson), renowned for their soothing and healing properties. Moreover, the cosmetic industry has embraced emulsion technology for the creation of antiaging products by well-known brands such as Mibelle, Avène, and Vichy, leveraging the enhanced penetration and sustained delivery benefits of these systems [52–55].

Several emulsion-based products are already commercially available for general consumer use, demonstrating the widespread acceptance and effectiveness of this technology. These formulations offer improved bioavailability, stability, and targeted delivery of active ingredients, ensuring prolonged therapeutic action and minimizing side effects. As research in this field advances, micro- and nanoemulsions continue to emerge as promising platforms for the next generation of dermatological and cosmetic products, combining efficacy, safety, and consumer appeal [62].

2.5.1 Fungal infections

Nanoemulsions have emerged as a promising technology in the treatment of fungal infections due to their enhanced antifungal activity, improved bioavailability, and potential to reduce drug resistance. Composed of nanosized oil and water droplets, stabilized by surfactants, these formulations have demonstrated superior drug delivery mechanisms, offering new therapeutic avenues against human-pathogenic fungi.

Table 2.2: Microemulsion and nanoemulsion technology applications to improve natural and synthetic drug delivery and bioactivity.

S. no.	Drug name	Bioactivity	Formulation	References
01	Ibuprofen	Anti-inflammatory analgesic activity; increased drug solubility	o/w Palmolein ester	50
02	Pioglitazone	Anti-inflammatory activity	Castor oil, Labrasol®, propylene glycol, and o/w Labrafac® lipophile	51
03	Retinyl palmitate	Increased permeability Anti-inflammatory	Labrasol® and Plurol® Oleiqueo/w	60
04	Capsaicin	Analgesic Increase solubility and bioavailability	Olive oil, Tween® 80, Span® 80, ethanol o/w	65
05	Betulin	Anti-inflammatory activity Anticarcinogenic activity	Flax-seed oil, egg, phosphatydilcholine, o/w	66
06	Eugenol	Anti-inflammatory activity	Tween® 80, Labrasol®, o/w Tween® 20, isopropyl alcohol, o/w	70
07	Curcumin	Increased permeability	Limonene, lecithin, ethanol, eucalyptol, lecithin, And ethanol	71
08	Clotrimazole	Antifungal activity	Labrafac® lipophile, Labrasol®, capryol	72
09	Lipoic acid	Antiaging activity	w/o and o/w Shea butter, squalane, stearol, Plurol® oleique CC947	74
10	Etoricoxib	Increased drug delivery	Triacetin, cremophor	75

Recent investigations have highlighted the ability of nanoemulsions to encapsulate and improve the effectiveness of antifungal agents. For instance, the incorporation of undecanoic acid into chitosan-based nanoemulsions significantly enhanced antifungal efficacy and biocompatibility, positioning them as a competitive alternative to conventional therapies [68]. Similarly, cationic nanoemulsions, loaded with miconazole, nitrate exhibited synergistic effects, increasing drug penetration and deposition in fungal-infected tissues [69]. Essential oil-based nanoemulsions have also shown remarkable antifungal properties. Studies on clove oil and thyme oil nanoemulsions revealed strong activity against *Candida* and *Aspergillus* species, outperforming traditional emulsions by enhancing biofilm disruption and improving antifungal agent penetration, crucial for addressing persistent infections [70].

By enhancing drug transport and targeting mechanisms, nanoemulsions present a viable strategy to overcome fungal resistance. Their capacity to improve drug solubility, stability, and controlled release further reinforces their suitability for antifungal applications [71]. Notably, ketoconazole-loaded cationic nanoemulsions demonstrated superior in vitro and in vivo efficacy in treating cutaneous fungal infections compared to traditional formulations [72].

Overall, nanoemulsions represent a highly effective approach to combating human–pathogenic fungi, offering enhanced antifungal activity, reduced drug resistance, and improved bioavailability.

2.5.2 Pigmentation disorders: understanding melasma

Pigmentation disorders, such as melasma, vitiligo, and post-inflammatory hyperpigmentation, present significant challenges in dermatology. Conventional treatments, including hydroquinone, corticosteroids, and laser therapy, often exhibit limitations such as instability, side effects, and inconsistent efficacy. Recent advances in nanotechnology have positioned microemulsions as an innovative delivery system for managing pigmentation disorders due to their enhanced skin penetration, stability, and controlled drug release [73].

Microemulsions have demonstrated significant potential in improving the efficacy of depigmenting agents. A study focused on hydroquinone-loaded microemulsions, highlighting their superior stability and dermal penetration compared to conventional formulations. The optimized microemulsion achieved prolonged drug release and enhanced permeability, making it a promising vehicle for hyperpigmentation treatment. Similarly, investigation done on the use of *Pouteria macrophylla* fruit extract in a microemulsion for cutaneous depigmentation revealed increased tyrosinase inhibition and improved melanin reduction compared to standard formulations [74].

Beyond hydroquinone, natural bioactives have been explored for their depigmenting potential. Curcumin-loaded microemulsions, formulated with geranium oil, exhibited significant antioxidant and anti-inflammatory properties, which are essential in reducing oxidative stress-related hyperpigmentation. These findings align with broader discussions on melanin biopolymers in pharmacology, where the importance of targeting melanogenesis pathways to achieve effective pigmentation control was emphasized [75].

In addition to therapeutic applications, microemulsions play a crucial role in cosmeceuticals. A review explored various pigmentation treatment strategies, noting the advantages of microemulsion-based delivery for active ingredients such as tyrosinase inhibitors and botanical extracts. These formulations offer a safer and more effective alternative to traditional skin-lightening treatments [76].

2.5.3 Wound healing

Wound healing is a complex biological process involving inflammation, tissue regeneration, and remodeling. Traditional wound treatments encounter challenges such as infection risks, slow healing rates, and limited drug penetration. Microemulsions, as advanced drug delivery systems, have demonstrated significant potential in enhancing wound healing by improving the stability, permeability, and bioavailability of therapeutic agents [77–79].

Recent studies have explored the role of microemulsions in wound healing, particularly in delivering bioactive compounds with antibacterial, anti-inflammatory, and regenerative properties. The thermodynamic stability and small droplet size of microemulsions facilitate deeper skin penetration and sustained drug release, offering a promising avenue for wound and burn healing [80].

Microemulsions incorporating plant-based extracts have shown remarkable effectiveness in promoting wound repair. For instance, *Sophora gibbosa* extract, delivered via a chitosan/gelatin-based microemulsion dressing, significantly accelerated full-thickness wound repair. This formulation provided enhanced moisture retention, antimicrobial activity, and histological improvements, highlighting its clinical applicability.

Furthermore, fusidic acid-loaded microemulsion gel demonstrated efficacy in burn wound healing by promoting re-epithelialization and reducing bacterial infections in in vivo studies with Wistar rats. These findings reinforce the importance of microemulsions in delivering antimicrobial agents to prevent infections and improve wound recovery [81].

Beyond antimicrobial applications, microemulsions have also been employed for anti-inflammatory purposes. Curcumin encapsulated in geranium oil microemulsions exhibited superior antioxidant, anti-inflammatory, and antibacterial properties. Given the role of oxidative stress in delayed wound healing, microemulsions, enhancing the bioavailability of curcumin and similar compounds, could offer substantial therapeutic benefits [82].

Overall, microemulsions represent a promising strategy for wound healing by providing controlled drug release, improving drug solubility, and enhancing skin penetration. Future research should focus on optimizing formulations, conducting clinical trials, and exploring their synergistic potential with other nanotechnology-based therapies.

2.5.4 Diabetic ulcers

In addition to moisture retention, nanoemulsions have shown remarkable potential in delivering therapeutic agents that target infection control and inflammation reduction in diabetic ulcers. A study involving a silver sulfadiazine-loaded nanoemulsion demonstrated significant antibacterial activity against multidrug-resistant bacteria commonly found in chronic wounds, thereby preventing infection and facilitating faster tissue regeneration. Moreover, nanoemulsions incorporating natural bioactives such as tea tree oil and eucalyptus oil exhibited synergistic antimicrobial effects, effectively combating biofilm formation, while minimizing cytotoxicity to healthy tissues [82]

The ability of nanoemulsions to modulate inflammatory responses further contributes to their efficacy in treating chronic wounds. For instance, curcumin-loaded nanoemulsions exhibited potent anti-inflammatory activity, reducing oxidative stress and enhancing fibroblast proliferation, key factors in promoting tissue repair in diabetic wounds. Furthermore, the controlled release of therapeutic agents ensured a sustained effect, preventing drug degradation and maintaining optimal drug concentrations at the wound site over prolonged periods [83].

Another promising approach involved the use of growth factors encapsulated in nanoemulsions, which accelerated cellular proliferation and angiogenesis, crucial for tissue remodeling in chronic wounds. This strategy not only enhanced the bioavailability of growth factors but also provided protection against enzymatic degradation in the wound environment, thereby ensuring prolonged therapeutic action [84].

Overall, the integration of micro- and nanoemulsions in diabetic ulcer treatment offers a multifaceted approach by addressing infection control, inflammation, hydration, and tissue regeneration. Future research should focus on clinical validation and exploring combinations of bioactive agents to optimize therapeutic outcomes, paving the way for the development of more effective and personalized treatments for chronic wounds [85].

2.5.5 Quality evaluation of micro/nanoemulsions in dermatology

Micro- and nanoemulsions are widely used in dermatology due to their enhanced skin penetration, stability, and ability to encapsulate active ingredients. Evaluating their quality is essential to ensure efficacy, safety, and stability.

2.5.5.1 In vitro studies

- **Skin models**: Human or animal skin models are used to evaluate the permeability and penetration of drugs delivered through nanoemulsions. These studies help determine whether the active ingredients in a formulation can effectively reach the target tissue (e.g., dermis and epidermis) [86].
- **Cell culture**: Cell-based assays are used to assess the cytotoxicity, bioactivity, and potential therapeutic effects of micro/nanoemulsions on skin cells, such as keratinocytes, fibroblasts, and melanocytes.

- **Release profile**: In vitro testing is also used to investigate the release profile of active ingredients encapsulated in nanoemulsions, ensuring that the drugs are released in a controlled and sustained manner [87].

2.5.5.2 In vivo studies

- **Animal models**: Animal models (e.g., mice, rats, and rabbits) are employed to assess the safety, irritation potential, and therapeutic efficacy of emulsion-based formulations. These models simulate various skin diseases, such as wound healing, psoriasis, acne, and fungal infections.
- **Toxicology**: Studies on skin irritation, allergic responses, and general toxicity are conducted to ensure the safety of formulations before human testing. For example, formulations like nanoemulsions are tested for dermal irritation and sensitization in preclinical animal models [88].
- **Efficacy testing**: Animal models are used to assess the efficacy of nanoemulsions in treating specific dermatological conditions. For instance, the reduction of acne lesions or the healing of diabetic ulcers may be tested by applying a nanoemulsion containing active ingredients (e.g., retinoids, antibiotics, and growth factors) [89].

2.5.6 Toxicity concerns in dermatological applications

Toxicity in dermatological applications refers to the potential adverse effects that substances may exert on biological systems, emphasizing the importance of ensuring that topical treatments do not cause long-term harm to the skin or underlying tissues [90]. Personal care products often comprise a diverse range of chemical compounds, some of which have raised significant concerns regarding their safety. Certain ingredients have been associated with skin irritation, endocrine disruption, and, in some cases, carcinogenic risks, highlighting the need for thorough evaluation and regulatory oversight [91].

2.5.6.1 Common toxic ingredients and their dermatological impacts

2.5.6.1.1 Heavy metalss
Heavy metals such as lead and mercury are often found in skin-lightening creams and cosmetics. Lead exposure is linked to neurotoxicity, while mercury can cause nephrotoxicity and dermatitis. A study using laser-induced breakdown spectroscopy detected high concentrations of these metals in nonbranded cosmetic products, exceeding safe limit [92].

2.5.6.1.2 Endocrine disruptors (parabens, phthalates, and bisphenol A)

These chemicals are widely used as preservatives and plasticizers in skincare products. Studies have linked them to hormonal imbalances, reproductive toxicity, and breast cancer risk. A high-performance liquid chromatography study identified endocrine-disrupting phenolic compounds in various cosmetic formulations, highlighting regulatory concerns [93].

2.5.6.1.3 Fragrance compounds and allergens

Synthetic fragrances and preservatives such as formaldehyde-releasing agents can cause allergic dermatitis and respiratory issues. A recent review examined the impact of fragrance chemicals on indoor air quality and their role in exacerbating conditions like asthma and eczema [94].

2.5.6.1.4 Skin-lightening agents (hydroquinone and steroids)

Hydroquinone, often used in skin-lightening creams, has been associated with ochronosis, a condition leading to skin discoloration and long-term damage. A comprehensive dermatological review outlined the health risks of hydroquinone and steroid-based depigmenting agents [95–98].

2.5.6.1.5 Nanoparticles in sunscreens and skincare

Titanium dioxide and zinc oxide nanoparticles are commonly used in sunscreens. While effective as UV filters, concerns exist regarding their ability to penetrate the skin and cause oxidative stress [99, 100].

2.5.6.2 Toxicity testing for nano- and microemulsions

2.5.6.2.1 In vitro cytotoxicity testing

Cytotoxicity assessments evaluate the impact of nanoemulsions on cell viability and function. Hemp (*Cannabis sativa* L.) essential oil nanoemulsions were tested on human cell lines, revealing no significant increase in proinflammatory cytokine levels, indicating a favorable safety profile [101].

2.5.6.2.2 Genotoxicity and ecotoxicity

Genotoxicity studies assess DNA damage potential, while ecotoxicity examines environmental impact. Mancozeb and eugenol nanoemulsions underwent cytotoxicity and genotoxicity evaluations, showing low toxicity in human cells but raising environmental concerns [102].

2.5.6.2.3 In vivo toxicity studies

Animal models provide insights into systemic toxicity and organ accumulation. Gluta-thione-loaded nanoemulsions demonstrated hepatoprotective effects, suggesting biocompatibility and reduced toxicity in liver cells [103].

2.5.6.2.4 Biocompatibility and long-term safety

The safety of nanoemulsions in dermatological and pharmaceutical applications requires extended stability testing. Palm tocotrienol-rich fraction nanoemulsions showed no signs of dermal or ocular irritation in reconstructed human epidermis models, supporting their potential use in topical formulations [104].

2.5.6.2.5 Toxicity mitigation strategies

Nanoemulsion formulation adjustments, such as using biocompatible surfactants, reducing particle size, and incorporating natural stabilizers, can minimize toxicity. A study on cinnamaldehyde nanoemulsions optimized formulation parameters to enhance antifungal activity while maintaining safety [105].

2.5.6.2.6 Skin sensitization

Sensitization studies are conducted to identify any potential allergic reactions that might result from exposure to the formulation [106].

2.5.6.2.7 Dermal penetration studies

Preclinical studies test the extent of drug penetration through the skin using methods like **tape stripping** or Franz diffusion cells to determine if harmful concentrations of active ingredients or excipients can reach deeper layers of the skin or systemic circulation [107].

2.6 Stability testing and packaging considerations for micro/nanoemulsions

Stability testing is crucial in determining the shelf life and ensuring the quality of **micro/nanoemulsions** over time, as shown in Figures 2.2 and 2.3. Since these formulations are prone to phase separation, degradation, or changes in particle size, stability testing evaluates their ability to maintain consistency under various conditions [45, 108].

Oxygen-impermeable and UV-resistant packaging can prevent oxidation and photo-degradation. Research on transdermal nanoemulsions emphasized the importance of light-protective packaging to maintain formulation stability. Smart packaging materials with moisture-absorbing and antimicrobial properties extend product shelf life [109, 110].

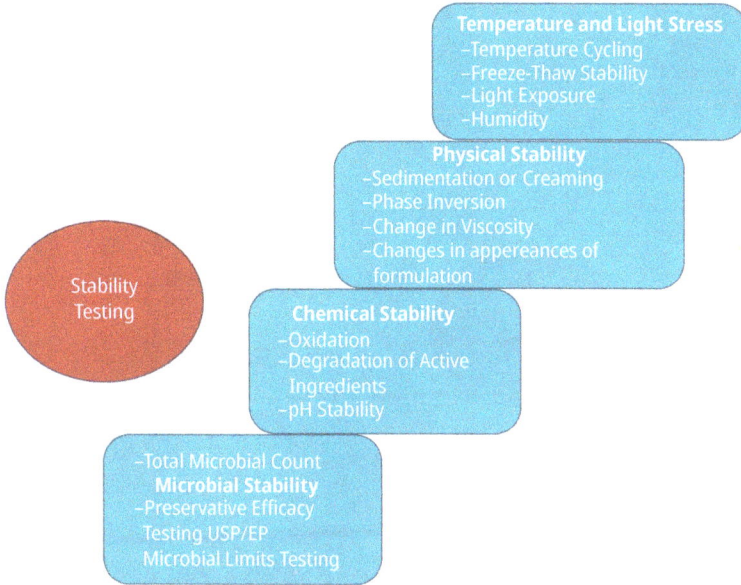

Figure 2.2: Stability considerations for micro/nanoemulsions.

2.7 Future trends and innovations in dermatological micro/nanoemulsions

Future trends and innovations in dermatological micro/nanoemulsions aim to revolutionize skin treatments by enhancing therapeutic efficacy, ensuring greater safety, and improving patient compliance [111].

2.7.1 Personalized medicine

With advancements in genomics and proteomics, there is a shift toward personalized dermatological treatments. Micro/nanoemulsions can be customized to deliver specific active ingredients tailored to an individual's skin type, condition, and genetic makeup, improving treatment outcomes and minimizing adverse reactions [112].

2.7.2 Enhanced skin targeting

Innovations in formulation technologies are focused on improving the **selectivity** and **targeting** of active ingredients within the skin layers. Nanoparticles, such as

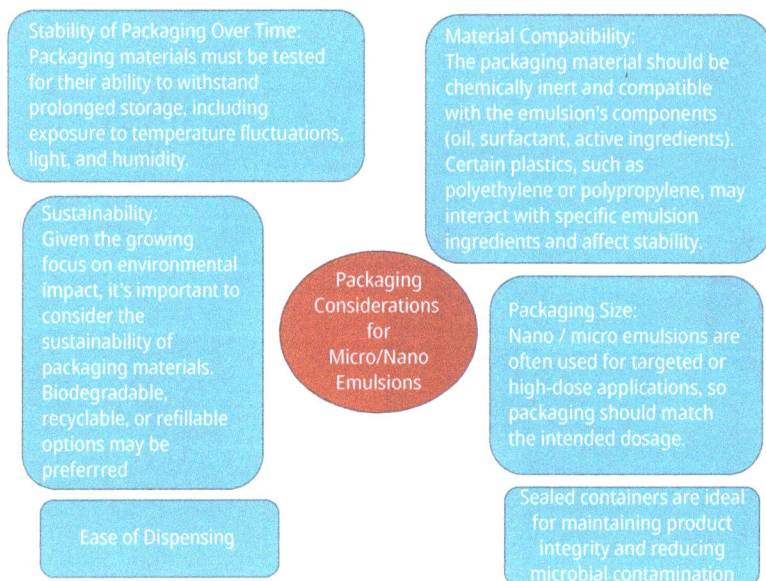

Figure 2.3: Packaging considerations for micro/nanoemulsions.

lipid based formulations, can be engineered to deliver drugs more precisely to the affected site, such as the dermis or subcutaneous layers, enhancing therapeutic efficacy, while minimizing systemic exposure [113].

2.7.3 Sustained and controlled release

Future formulations are likely to focus on **long-lasting and controlled drug release** mechanisms. Nanoemulsions can be designed to release active ingredients over extended periods, improving the convenience of use and reducing the frequency of application for chronic conditions like psoriasis and acne [114–115].

2.7.4 Biodegradable and eco-friendly emulsions

There is a growing trend toward **eco-friendly** and **biodegradable** emulsions that minimize environmental impact. Natural surfactants, plant-based oils, and biodegradable polymers are gaining prominence as sustainable alternatives in formulation development [116, 117].

2.7.5 Combination therapies

The integration of **multidrug systems** in micro/nanoemulsions is an emerging trend. These systems could simultaneously deliver multiple active ingredients targeting various pathways, enhancing the overall therapeutic effect for conditions such as **melasma**, **eczema**, and **acne vulgaris** [118, 119].

2.8 Conclusion

Microemulsions and nanoemulsions represent innovative delivery systems in dermatological therapy, enhancing the efficacy and bioavailability of various therapeutic agents, while minimizing toxicity. Their unique properties and production methods offer significant potential for developing advanced formulations in both medical and cosmetic applications. The future of dermatological treatments is poised for significant advancements through the use of micro/nanoemulsions, which promise enhanced drug delivery, personalized medicine, and eco-friendly formulations, ultimately leading to improved patient outcomes and treatment efficacy. Innovations such as smart delivery systems and the integration of artificial intelligence further underscore the potential for tailored and effective dermatological care.

References

[1]   Alexander SD, Ajazuddin TK, Giri S, Saraf S, Tripathi DK. Approaches for breaking the barriers of drug permeation through transdermal drug delivery. J Control Rel, 2012;164(1), 26–40.
[2]   Nastiti TP, Abd E, Grice JE, Benson HAE, Roberts MS. Topical nano and microemulsions for skin delivery. Pharmaceutics, 2017;9(4), 37.
[3]   Moser K, Kriwet K, Naik A, Kalia YN, Guy RH. Passive skin penetration enhancement and its quantification in vitro. Eur J Pharm Biopharm, 2001;52(2), 103–112.
[4]   Zhai Y, Zhai G. Advances in lipid-based colloid systems as drug carrier for topic delivery. J Control Release, 2014;193, 90–99.
[5]   Touitou E. Drug delivery across the skin. Expert Opin Biol Ther, 2002;2(7), 723–733.
[6]   Cevc G. Lipid vesicles and other colloids as drug carriers on the skin. Adv Drug Deliv Rev, 2004;56 (5), 675–711.
[7]   Lin CH, Aljuffali IA, Fang JY. Lasers as an approach for promoting drug delivery via skin. Expert Opin Drug Deliv, 2014;11(4), 599–614.
[8]   Bouwstra JA, Honeywell-Nguyen PL, Gooris GS, Ponec M. Structure of the skin barrier and its modulation by vesicular formulations. Prog Lipid Res, 2003;42(1), 1–36.
[9]   Shinoda K, Lindman B. Organized surfactant systems: Microemulsions. Langmuir, 1987;3(2), 135–149.
[10]   Neubert RHH, Sommer E, Schölzel M, Tuchscherer B, Mrestani Y, Wohlrab J. Dermal peptide delivery using enhancer molecules and colloidal carrier systems. Part II: Tetrapeptide PKEK. Eur J Pharm Biopharm, 2018;124, 28–33.

[11] Danielsson BL The definition of microemulsion. Colloids Surf, 1981;3(4), 391–392.

[12] Sunitha S, Jitendra W, Sujatha D, Kumar MS. Design and evaluation of hydrogel-thickened microemulsion for topical delivery of minoxidil, Iran. J Pharm Sci, 2013;9(4), 1–14.

[13] Bonabello MRG, Canaparo R, Isaia GC, Serpe L, Muntoni E, Zara GP. Dexibuprofen (S+-isomer ibuprofen) reduces gastric damage and improves analgesic and anti-inflammatory effects in rodents, Anesth Analg, 2003 table of contents;97(2), 402–408.

[14] Kaehler ST, Phleps W, Hesse E. Dexibuprofen: Pharmacology, therapeutic uses and safety. Inflammopharmacology, 2003;11(4), 371–383.

[15] Ali FR, Shoaib MH, Yousuf RI, Ali SA, Imtiaz MS, Bashir L, Naz S. Design, development, and optimization of dexibuprofen microemulsion based transdermal reservoir patches for controlled drug delivery. BioMed Res Int, 2017;2017(1), 4654958.

[16] Microemulsion based transdermal reservoir patches for controlled drug delivery, 20172017 4654958

[17] Cohen S, Flescher E. Methyl jasmonate: A plant stress hormone as an anti-cancer drug. Phytochemistry, 2009;70(13–14), 1600–1609.

[18] Farmer EE, Ryan CA. Interplant communication: Airborne methyl jasmonate induces synthesis of proteinase inhibitors in plant leaves. Proc Natl Acad Sci USA, 1990;87(19), 7713–7716.

[19] Cesari IM, Carvalho E. Methyl jasmonate: Putative mechanisms of action on cancer cells cycle, metabolism, and apoptosis. Int J Cell Biol, 2014 ;2014, 572097.

[20] Fingrut O, Flescher E. Plant stress hormones suppress the proliferation and induce apoptosis in human cancer cells. Leukemia, 2002;16(4), 608–616.

[21] Fingrut O, Reischer D, Rotem R, Goldin N, Altboum I, Zan-Bar I, Flescher E. Jasmonates induce nonapoptotic death in high-resistance mutant p53-expressing B-lymphoma cells. Br J Pharmacol, 2005;146(6), 800–808.

[22] Yehia R, Hathout RM, Attia DA, Elmazar MM, Mortada ND. Anti-tumor efficacy of an integrated methyl dihydrojasmonate transdermal microemulsion system targeting breast cancer cells: In vitro and in vivo studies. Colloids Surf B Biointerfaces, 2017;155, 512–521.

[23] Ryan E, Garland MJ, Singh TR, Bambury E, O'Dea J, Migalska K, Gorman SP, McCarthy HO, Gilmore BF, Donnelly RF. Microneedle-mediated transdermal bacteriophage delivery. Eur J Pharm Sci, 2012;47(2), 297–304.

[24] Rastogi V, Pragya NV, Mishra AK, Nath G, Gaur PK, Verma A. An overview on bacteriophages: A natural nanostructured antibacterial agent. Curr Drug Deliv, 2018;15(1), 3–20.

[25] Taj MK, Ling JX, Bing LL, Qi Z, Taj I, Hassani TM, Samreen Z, Yunlin W. Effect of dilution, temperature and pH on the lysis activity of T4 phage against e.cOli BL21. J Anim Plant Sci, 2014;24(4), 1252–1255.

[26] Hegde RR, Bhattacharya SS, Verma A, Ghosh A. Physicochemical and pharmacological investigation of water/oil microemulsion of non-selective beta blocker for treatment of glaucoma. Currt Eye Res, 2014;39(2), 155–163.

[27] Rastogi V, Yadav P, Verma A, Pandit JK. Ex vivo and in vivo evaluation of microemulsion based transdermal delivery of E. Coli specific T4 bacteriophage: A rationale approach to treat bacterial infection. Eur J Pharm Sci, 2017;107, 168–182.

[28] Matsumoto M, Inoue M, Ueda H. NSAID zaltoprofen possesses novel anti-nociceptive mechanism through blockage of B2-type bradykinin receptor in nerve endings. Neurosci Lett, 2006;397(3), 249–253.

[29] Kawai S. Cyclooxygenase selectivity and the risk of gastro-intestinal complications of various non-steroidal anti-inflammatory drugs: A clinical consideration. Inflamm Res, 1998;47(Suppl 2), S102–S106.

[30] Li L, Ma P, Cao Y, Tao L, Tao Y. Single-dose and multiple-dose pharmacokinetics of zaltoprofen after oral administration in healthy Chinese volunteers. J Biomed Res, 2011;25(1), 56–62.

[31] Mishra R, Prabhavalkar KS, Bhatt LK. Preparation, optimization, and evaluation of Zaltoprofen-loaded microemulsion and microemulsion-based gel for transdermal delivery. J Liposome Res, 2016;26(4), 297–306.

[32] Frampton JE, Keating GM. Celecoxib: A review of its use in the management of arthritis and acute pain. Drugs, 2007;67(16), 2433–2472.

[33] Kim DH, Lee SY, Park JH. Thermo-gelling emulsions in skin drug delivery. Adv Drug Deliv Rev, 2018;127, 36–49.

[34] Zhang R, Wang H, Zhao J. Lipase-sensitive emulsions in transdermal drug delivery. Expert Opin Drug Delivery, 2018;15(7), 771–782.

[35] Zhang X, Wang J, Zhao Q. Mechanisms and applications of light-responsive emulsions. Colloids Surf B Biointerfaces, 2020;190, 273–289.

[36] Wang D, Lee J, Kim T. UV-sensitive emulsions for topical treatment. Expert Opin Drug Delivery, 2018;15(9), 1133–1145.

[37] Sjöblom J, Lindberg R, Friberg SE. Microemulsions – Phase equilibria characterization, structures, applications and chemical reactions. Adv Colloi Interface Sci, 1996;65, 125–287.

[38] Lagourette JP, Boned C, Clausse M. Percolative conduction in microemulsion type systems. Nature, 1979;281(5726), 60.

[39] Moreno MA, Ballesteros MP, Frutos P. Lecithin-based oil-in-water microemulsions for parenteral use: Pseudoternary phase diagrams, characterization and toxicity studies. J Pharm Sci, 2003;92(7), 1428–1437.

[40] Chiappisi L, Noirez L, Gradzielski M. A journey through the phase diagram of a pharmaceutically relevant microemulsion system. J Colloid Interface Sci, 2016;473, 52–59.

[41] Shah DO. *Surface Phenomena in Enhanced Oil Recovery*. Springer; New York: plenum press, 1981, pp. 53–72.

[42] Thacharodi D, Rao KP. Transdermal absorption of nifedipine from microemulsions of lipophilic skin penetration enhancers. Int J Pharm, 1994;111(3), 235–240.

[43] Lee J, Lee Y, Kim J, Yoon M, Choi YW. Formulation of microemulsion systems for transdermal delivery of aceclofenac. Arch Pharm Res, 2005;28(9), 1097–1102.

[44] Guay DR. Repaglinide, a novel, short-acting hypoglycemic agent for type 2 diabetes mellitus. Pharmacotherapy, 1998;18(6), 1195–1204.

[45] Changez M, Anwar MF, Alrahbi H. Olive oil-based reverse microemulsion for stability and topical delivery of methotrexate: In vitro. ACS Omega, 2024 Feb 2;9(6), 7012–21.

[46] Patel MR, Patel RB, Parikh JR, Solanki AB, Patel BG. Effect of formulation components on the in vitro permeation of microemulsion drug delivery system of S. Peltola, P. Saarinen-Savolainen, J. Kiesvaara, T.M. Suhonen, A. Urtti, Microemulsions for topical delivery of estradiol. Int J Pharm, 2003;254(2), 99–107.

[47] Shinde UA, Modani SH, Singh KH. Design and development of repaglinide microemulsion gel for transdermal delivery. AAPS Pharm Sci Tech, 2018;19(1), 315–325.

[48] Mura S, Manconi M, Valenti D, Sinico C, Vila AO, Fadda AM. Transcutol-containing vesicles for topical delivery of minoxidil. J Drug Target, 2011;19(3), 189–196.

[49] Sinclair R, Wewerinke M, Jolley D. Treatment of female pattern hair loss with oral antiandrogens. Br J Dermatol, 2005;152(3), 466–473.

[50] Meisheri KD, Cipkus LA, Taylor CJ. Mechanism of action of minoxidil sulfate-induced vasodilation: A role for increased K+ permeability. J Pharmacol Exp Ther, 1988;245(3), 751–760.

[51] Abd E, Benson HA, Roberts MS, Grice JE. Minoxidil skin delivery from nanoemulsion formulations containing eucalyptol or oleic acid: Enhanced diffusivity and follicular targeting. Pharmaceutics, 2018;10(1), 19.

[52] Sunitha S, Jitendra W, Sujatha D, Kumar MS. Design and evaluation of hydrogel-thickened microemulsion for topical delivery of minoxidil, Iran. J Pharm Sci, 2013;9(4), 1–14.

[53] Esmaeili F, Baharifar H, Amani A. Improved Anti-inflammatory Activity and Minimum Systemic Absorption from Topical Gels of Ibuprofen Formulated by Micelle or Nanoemulsion. J Pharm Innov, 2022;17(4), 1314–1321.

[54] Espinoza LC, Silva-Abreu M, Calpena AC, Rodríguez-Lagunas MJ, M.-j. F, Garduño-Ramírez ML, Clares B. Nanoemulsion strategy of pioglitazone for the treatment of skin inflammatory diseases. Nanomed Nanotechnol Biol Med, 2019;19, 115–125.

[55] Rosso JQD. Management of facial erythema of rosacea: What is the role of topical α-adrenergic receptor agonist therapy?. J Am Acad Dermatol, 2013;69(6), S44–S56.

[56] Yu M, Ma H, Lei M, Li N, Tan F. In vitro/in vivo characterization of nanoemulsion formulation of metronidazole with improved skin targeting and anti-rosacea properties. Eur J Pharm Biopharm, 2014;88(1), 92–103.

[57] Yu M, Ma H, Lei M, Li N, Tan F. In vitro/in vivo characterization of nanoemulsion formulation of metronidazole with improved skin targeting and anti-rosacea properties. Eur J Pharm Biopharm, 2014;88(1), 92–103.

[58] Arida AI, Al-Tabakha MM, Hamoury HAJ. Improving the high variable bioavailability of griseofulvin by SEDDS. Chem Pharm Bull, 2007;55(12), 1713–1719.

[59] Alam MS, et al. Enhancement of anti-dermatitis potential of clobetasol propionate by DHA [docosahexaenoic acid] rich algal oil nanoemulsion gel. Iran J Pharm Res, 2016;15(1), 35.

[60] Aggarwal N, Goindi S, Khurana R. Formulation, characterization and evaluation of an optimized microemulsion formulation of griseofulvin for topical application. Colloids Surf B Biointerfaces, 2013;105, 158–166.

[61] Kui GS, Chen KJ, Qian ZH, Weng WL, Qian MY. Tetramethylpyrazine in the treatment of cardiovascular and cerebrovascular diseases. Pl Med, 1983;47, 89.

[62] Hamed R, Basil M, AlBaraghthi T, Sunoqrot S, Tarawneh O. Nanoemulsion-based gel formulation of diclofenac diethylamine: Design, optimization, rheological behavior and in vitro diffusion studies. Pharm Dev Technol, 2016 Dec;21(8), 980–989.

[63] Clares B, Calpena AC, Parra A, Abrego G, Alvarado H, Fangueiro JF, Souto EB. Nanoemulsions (NEs), liposomes (LPs) and solid lipid nanoparticles (SLNs) for retinyl palmitate: Effect on skin permeation. Int J Pharm, 2014;473, 591–598. 10.1016/j.ijpharm.2014.08.001

[64] Kao T-K, Ou Y-C, Kuo J-S, Chen W-Y, Liao S-L, Wu C-W, Chen C-J, Ling NN, Zhang Y-H, Peng W-H. Neuroprotection by tetramethylpyrazine against ischemic brain injury in rats. Neurochem Int, 2006;48(3), 166–176.

[65] Ma L, Fan Y, Wu H, Na S, Wang L, Lu C. Tissue distribution and targeting evaluation of TMP after oral administration of TMP-loaded microemulsion to mice. Drug Dev Ind Pharm, 2013;39(12), 1951–1958.

[66] Zhao J-H, Ji L, Wang H, Chen Z-Q, Zhang Y-T, Liu Y, Feng N-P. Microemulsion-based novel transdermal delivery system of tetramethylpyrazine: Preparation and evaluation in vitro and in vivo. Int J Nanomed, 2011;6, 1611.

[67] Sritularak B, De-Eknamkul W, Likhitwitayawuid K. Tyrosinase inhibitors from Artocarpus lakoocha. Thai J Pharm Sci, 1998;22, 149–155.

[68] Ghiasi Z, Esmaeli F, Aghajani M, Ghazi-Khansari M, Faramarzi MA, Amani A. Enhancing analgesic and anti-inflammatory effects of capsaicin when loaded into olive oil nanoemulsion: An in vivo study. Int J Pharm, 2019;559, 341–347. 10.1016/j.ijpharm.2019.01.043

[69] Dehelean CA, Feflea S, Gheorgheosu D, Ganta S, Cimpean AM, Muntean D, Amiji MM. Anti-angiogenic and anti-cancer evaluation of betulin nanoemulsion in chicken chorioallantoic membrane and skin carcinoma in Balb/c mice. J Biomed Nanotechnol, 2013;9, 577–589.

[70] Malakar J, Nayak AK, Basu A. Ondansetron HCl microemulsions for transdermal delivery: Formulation and in vitro skin permeation. ISRN Pharm, 2012 ;2012(1), 428396.

[71] Elewski BE. Once-weekly fluconazole in the treatment of onychomycosis: Introduction. J Am Acad Dermatol, 1998;38(6), S73–S76.

[72] Elewski BE, Hay RJ. Update on the management of onychomycosis: Highlights of the third annual international summit on cutaneous antifungal therapy. Clin Infect Dis, 1996;23(2), 305–313.

[73] Salimi A. Enhanced stability and dermal delivery of hydroquinone using microemulsion-based system. Asian J Pharm, 2017;11(04), S773–781.

[74] Brathwaite ACN, Alencar-Silva T, Carvalho LA, Branquinho MS, Ferreira-Nunes R, Cunha-Filho M, Gratieri T. Pouteria macrophylla fruit extract microemulsion for cutaneous depigmentation: Evaluation using a 3D pigmented skin model. Molecules, 2022;27(18), 5982.

[75] Hassan SF, Asghar S, Ullah Khan I, Munir R, Khalid SH. Curcumin encapsulation in geranium oil microemulsion elevates its antibacterial, antioxidant, anti-inflammatory, and anticancer activities. ACS Omega, 2024;9(5), 5624–5636.

[76] Karkoszka M, Rok J, Wrześniok D. Melanin biopolymers in pharmacology and medicine – Skin pigmentation disorders, implications for drug action, adverse effects and therapy. Pharmaceuticals, 2024;17(4), 521.

[77] Ahmad N, Ahmad R, Al-Qudaihi A, Alaseel SE, Fita IZ, Khalid MS, Pottoo FH. Preparation of a novel curcumin nanoemulsion by ultrasonication and its comparative effects in wound healing and the treatment of inflammation. RSC Adv, 2019;9, 20192–20206.

[78] Manimaran V, Sivakumar PM, Narayanan J, Parthasarathi S, Prabhakar PK. Nanoemulsions: A better approach for antidiabetic drug delivery. Current Diabetes Rev, 2021 May 1;17(4), 486–95.

[79] Okur ME, Ayla Ş, Yozgatlı V, Aksu NB, Yoltaş A, Orak D, . . . Okur NÜ. Evaluation of burn wound healing activity of novel fusidic acid loaded microemulsion based gel in male Wistar albino rats. Saudi Pharm J, 2020;28(3), 338–348.

[80] Dos Santos Ramos MA, de Toledo LG, Sposito L, Marena GD, de Lima LC, Fortunato GC, . . . Chorilli M. Nanotechnology-based lipid systems applied to resistant bacterial control: A review of their use in the past two decades. Int J Pharm, 2021;603, 120706.

[81] Hassan SF, Asghar S, Ullah Khan I, Munir R, Khalid SH. Curcumin encapsulation in geranium oil microemulsion elevates its antibacterial, antioxidant, anti-inflammatory, and anticancer activities. ACS Omega, 2024;9(5), 5624–5636.

[82] Soriano-Ruiz JL, Calpena-Capmany AC, Cañadas-Enrich C, Febrer NB-D, Suñer-Carbó J, Souto EB, Clares-Naveros B. Biopharmaceutical profile of a clotrimazole nanoemulsion: Evaluation on skin and mucosae as anticandidal agent. Int J pharm, 2019;554, 105–115.

[83] Hassan SF, Asghar S, Ullah Khan I, Munir R, Khalid SH. Curcumin encapsulation in geranium oil microemulsion elevates its antibacterial, antioxidant, anti-inflammatory, and anticancer activities. ACS Omega, 2024 Jan 23;9(5), 5624–36.

[84] Ansari, Mojtaba, and Ahmad Darvishi. "A review of the current state of natural biomaterials in wound healing applications." Frontiers in bioengineering and biotechnology 12 (2024): 1309541.

[85] Chu X, Xiong Y, Knoedler S, Lu L, Panayi AC, Alfertshofer M, Jiang D, Rinkevich Y, Lin Z, Zhao Z, Dai G. Immunomodulatory nanosystems: Advanced delivery tools for treating chronic wounds. Research, 2023 Jul 14;6, 0198.

[86] Zhou Z, Liu C, Wan X, Fang L. Development of a w/o emulsion using ionic liquid strategy for transdermal delivery of anti-aging component α-lipoic acid: Mechanism of different ionic liquids on skin retention and efficacy evaluation. Eur J Pharm Sci, 2020;141, 105042.

[87] Lala RR, Awari NG. Nanoemulsion-based gel formulations of COX-2 inhibitors for enhanced efficacy in inflammatory conditions. Appl Nanosci, 2014;4, 143–151.

[88] Kong M, Chen XG, Kweon DK. Investigationson skin permeation of hyaluronic acid based nanoemulsion as transdermal carrier. Carbohydr Polym, 2011;86, 837–843.

[89] Kale SN, Deore SL. Emulsion micro emulsion and nano emulsion: A review. Syst Rev Pharm, 2016;8, 39–47.

[90]   Talegaonkar S, Azeem A, Ahmad FJ. Microemulsions: A novel approach to enhanced drug delivery. Recent Pat Drug Deliv Formul, 2008;2, 238–257.

[91]   Hengge UR, Currie BJ, Jäger G, Lupi O, Schwartz RA. Scabies: A ubiquitous neglected skin disease. Lancet Infect Dis, 2006;6(12), 769–779.

[92]   Alberici F, Pagani L, Ratti G, Viale P. Ivermectin alone or in combination with benzyl benzoate in the treatment of human immunodeficiency virus-associated scabies. Br J Dermatol, 2000;142(5), 969–972.

[93]   Bachewar NP, Thawani VR, Mali SN, Gharpure KJ, Dakhale VPSGN. Comparison of safety, efficacy, and cost effectiveness of benzyl benzoate, permethrin, and ivermectin in patients of scabies. Indian J Pharmacol, 2009;41(1), 9.

[94]   Hathout RM, Mansour S, Mortada ND, Geneidi AS, Guy RH. Uptake of microemulsion components into the stratum corneum and their molecular effects on skin barrier function. Mol Pharm, 2010;7(4), 1266–1273.

[95]   Rigg PC, Barry BW. Shed snake skin and hairless mouse skin as model membranes for human skin during permeation studies. J Invest Dermatol, 1990;94(2), 235–240.

[96]   Sharma G, Dhankar G, Thakur K, Raza K, Katare OP. Benzyl benzoate-loaded microemulsion for topical applications: Enhanced dermatokinetic profile and better delivery promises. AAPS PharmSciTech, 2015;17(5), 1221–1231.

[97]   Sharma G, Dhankar G, Thakur K, Raza K, Katare OP. Benzyl benzoate-loaded microemulsion for topical applications: Enhanced dermatokinetic profile and better delivery promises. AAPS PharmSciTech, 2016;17(5), 1221–1231.

[98]   Mishra RV, Jain PK, Dixit VK, Agrawal RK. Synthesis, characterization and pharmacological evaluation of amide prodrugs of ketorolac. Eur J Med Chem, 2008;43(11), 2464–2472.

[99]   Chen H, Mou D, Du D, Chang X, Zhu D, Liu J, Xu H, Yang X. Hydrogel thickened microemulsion for topical administration of drug molecule at an extremely low concentration. Int J Pharm, 2007;341(1–2), 78–84.

[100] Naeem M, Rahman NU, Tavares GD, Barbosa SF, Chacra NB, Lobenberg R, Sarfraz MK. Physicochemical, in vitro and in vivo evaluation of flurbiprofen microemulsion. An Acad Bras Ciênc, 2015;87(3), 1823–1831.

[101] Chi L, Liu R, Guo T, Wang M, Liao Z, Wu L, Li H, Wu D, Zhang J. Dramatic improvement of the solubility of pseudolaric acid B by cyclodextrin complexation: Preparation, characterization and validation. Int J Pharm, 2015;479(2), 349–356.

[102] Sun Q, Li Y. Inhibitory effect of pseudolaric acid B on gastric cancer in vivo and multidrug resistance via Cox-2–Pkc-α–P-gp pathway. J Dig Dis, 2014;15, 45–46.

[103] Yu B, Li M-H, Wang W, Wang Y-Q, Jiang Y, Yang S-P, Yue J-M, Ding J, Miao Z-H. Pseudolaric acid B-driven phosphorylation of c-Jun impairs its role in stabilizing HIF-1alpha: A novel function-converter model. J Mol Med, 2012;90(8), 971–981.

[104] Wan T, Xu T, Pan J, Qin M, Pan W, Zhang G, Wu Z, Wu C, Xu Y. Microemulsion based gel for topical dermal delivery of pseudolaric acid B: In vitro and in vivo evaluation. Int J Pharm, 2015;493(1–2), 111–120.

[105] Sperry P, Cua D, Wetzel S, Adler-Moore J. Antimicrobial activity of AmBisome and non-liposomal amphotericin B following uptake of Candida glabrata by murine epidermal Langerhans cells. Med Mycol, 1998;36(3), 135–141.

[106] Richardson SE, Bannatyne RM, Summerbell RC, Milliken J, Gold R, Weitzman SS. Disseminated fusarial infection in the immunocompromised host. Clin Infect Dis, 1988;10(6), 1171–1181.

[107] Moreno MA, Frutos P, Ballesteros MP. Lyophilized lecithin based oil-water microemulsions as a new and low toxic delivery system for amphotericin B. Pharma Res, 2001;18(3), 344–351.

[108] Butani D, Yewale C, Misra A. Amphotericin B topical microemulsion: Formulation, characterization and evaluation, Colloids Surf. B: Biointerfaces, 2014;116, 351–358.

[109] Gordon ML. The role of clobetasol propionate emollient 0.05% in the treatment of patients with dry, scaly, corticosteroid-responsive dermatoses. Clin Ther, 1998;20(1), 26–39.
[110] Kumari J. Vitiligo treated with topical clobetasol propionate. Arch dermatol, 1984;120(5), 631–635.
[111] Wiedersberg S, Leopold CS, Guy RH. Bioavailability and bioequivalence of topical glucocorticoids. Eur J Pharm Biopharm, 2008;68(3), 453–466.
[112] Patel HK, Barot BS, Parejiya PB, Shelat PK, Shukla A. Topical delivery of clobetasol propionate loaded microemulsion based gel for effective treatment of vitiligo–part II: Rheological characterization and in vivo assessment through dermatopharmacokinetic and pilot clinical studies. Colloids Surf B Biointerfaces, 2014;119, 145–153.
[113] Hemmila MR, Mattar A, Taddonio MA, Arbabi S, Hamouda T, Ward PA, Baker Jr JR. Topical nanoemulsion therapy reduces bacterial wound infection and inflammation after burn injury. Surgery, 2010;148(3), 499–509.
[114] da Silva WJ, Seneviratne J, Parahitiyawa N, Rosa EA, Samaranayake LP, Del Bel Cury AA. Improvement of XTT assay performance for studies involving Candida albicans biofilms. Braz Dent J, 2008;19, 364–369.
[115] Alanio A, Vernel-Pauillac F, Sturny-Leclère A, Dromer F, Alanio A, Vernel-Pauillac F, Sturny-Leclère A, Dromer F. Cryptococcus neoformans host adaptation: Toward biological evidence of dormancy. mBio, 2015;6, e02580–14.
[116] Rasoanirina BNV, Lassoued MA, Kamoun A, Bahloul B, Miladi K, Sfar S. Voriconazole-loaded self-nanoemulsifying drug delivery system (SNEDDS) to improve transcorneal permeability. Pharm Dev Technol, 2020;25, 694–703.
[117] Weir A, Westerhoff P, Fabricius L. Titaniumdioxide nanoparticles in food and personal careproducts. Environ Sci Technol, 2012;46, 2242–2250.
[118] Boonme P, Junyaprasert VB, Suksawad N. Microemulsions and nanoemulsions: Novel vehiclesfor whitening cosmeceuticals. J Biomed Nanotechnol, 2009;5, 373–383.
[119] Talegaonkar S, Azeem A, Ahmad FJ. Microemulsions: A novel approach to enhanced drugdelivery. Recent Pat Drug Deliv Formul, 2008;2, 238–257.

Shweta Ramkar, Binayak Mishra, and Preeti K. Suresh*

Chapter 3
Harnessing micro/nanoemulsion for transnasal drug delivery with enhanced efficacy and precision

Abstract: Transnasal drug delivery has been a subject of significant interest as a useful and efficient pathway for the delivery of therapeutic drugs. This chapter discusses the possibilities of transnasal delivery systems, highlighting the use of micro- and nano-emulsions in improved drug absorption and bioavailability. The nasal pathway has several important benefits such as quick onset of action, circumvention of first-pass metabolism, simplicity of administration, and possibility of both systemic and local action. Micro-/nano-emulsions, due to their minute particle size, thermodynamic stability, and capability for encapsulation of both hydrophilic and lipophilic drugs, are being used as great carriers for any number of different pharmaceutical substances. Their preparation enhances solubilization, stabilization, and muco-permeability, with room for controlled as well as target-specific delivery. Anatomy and physiology of the nasal cavity, absorption mechanisms of drug, and challenges to effective delivery are elaborated in the chapter. Strategies to transcend such obstacles, for example, utilization of mucoadhesive agents, permeation enhancers, and inhibitors of enzymes are discussed as well. While the chapter points toward the promise of direct nose-to-brain delivery in neurological diseases, its core discussion goes on to cover wide-ranging therapeutic fields such as pain management, hormone therapy, infections of the nasal cavity, and vaccine delivery. Trends in formulation science, regulatory insights, and directions for the future, for example, personalized medicine and smart delivery systems, are also covered. In general, the chapter highlights the multifaceted uses and future potential of transnasal micro/nanoemulsion systems in contemporary pharmaceutical delivery.

Keywords: Transnasal drug delivery, microemulsion, nanoemulsion, nasal absorption, noninvasive administration, mucosal drug delivery

*Corresponding author: Preeti K. Suresh, University Institute of Pharmacy, Pt. Ravishankar Shukla University, Raipur, Chhattisgarh, India, e-mail: suresh.preeti@gmail.com
Shweta Ramkar, Kamla Institute of Pharmaceutical Sciences, Shri Shankaracharya Professional University, Bhilai, Durg, Chhattisgarh, India, e-mail: shwetaramkar@gmail.com
Binayak Mishra, School of Pharmacy and Life Sciences, Centurion University of Technology and Management, Odisha, India

https://doi.org/10.1515/9783111593654-004

3.1 Introduction

3.1.1 Overview of transnasal drug delivery

Transnasal drug delivery involves administering pharmaceutical compounds through the nasal cavity to achieve localized or systemic effects. The nasal mucosa's rich vascularization enables rapid drug absorption into the bloodstream, bypassing first-pass metabolism associated with oral administration [1–3]. Additionally, the olfactory region offers a direct route to the central nervous system (CNS), making it particularly suitable for neuroactive compounds [4, 5]. This delivery method enhances therapeutic efficacy and patient compliance by offering faster onset of action and improved bioavailability compared to traditional routes, such as oral, intravenous, subcutaneous, or transdermal administration. Unlike oral administration, which can undergo significant first-pass metabolism that diminishes the drug's efficacy, nasally delivered medications provide rapid absorption directly into the bloodstream, often resulting in faster onset of action with better therapeutic effects (Table 3.1).

3.1.2 Advantages of transnasal drug delivery

– **Rapid onset of action:** Transnasal delivery can lead to the rapid absorption of drugs into the systemic circulation, providing quick therapeutic effects, which is particularly beneficial in acute situations (e.g., pain relief or emergency medication) [6–8].
– **Enhanced bioavailability:** By delivering drugs through the nasal route, the first-pass metabolism in the liver is avoided, potentially increasing the bioavailability of active compounds. This can allow for lower doses while achieving effective plasma concentrations [9–11].
– **Convenience and noninvasiveness:** Nasal delivery is generally more user-friendly and less invasive than injections, which can improve patient adherence and compliance. It is also suitable for self-administration [12].
– **Targeted delivery to the CNS:** Transnasal drug delivery can transport therapeutics directly to the CNS, bypassing the blood-brain barrier (BBB). This method is used for treating neurological conditions such as Alzheimer's disease, Parkinson's disease, and acute migraines [13–16].
– **Local treatment of nasal condition:** Transnasal delivery allows for targeted treatment of nasal and sinus conditions, such as allergies, congestion, and infections, reducing systemic side effects [17, 18].
– **Versatility in formulation**: The technique can accommodate various drugs, including small molecules, peptides, and vaccines, broadening the therapeutic landscape.

– **Minimized side effects:** Localized delivery minimizes systemic exposure to certain drugs, which can help reduce side effects compared to oral or intravenous route.

Table 3.1: Comparison of different routes of drug administration.

Drug delivery route	Advantages	Disadvantages
Transnasal	– Fast onset of action – Bypasses presystemic metabolism – Direct delivery to the central nervous system – Noninvasive and easy to self-administer	– Variable bioavailability – Potential irritation and discomfort; limited to smaller molecules
Oral	– Convenient and widely accepted – Suitable for a wide range of formulations – Generally low-cost and easy to manufacture	– Subject to first-pass metabolism – Slow onset of action – Absorption affected by gastrointestinal factors – Unsuitable for patients with swallowing difficulties
Intravenous (IV)	– Immediate and complete bioavailability – Precise control over drug levels in circulation – Suitable for larger volumes or irritating substances	– Invasive and needs skilled personnel – Risk of infection and injection-site complications – Higher equipment costs
Subcutaneous	– Relatively easy to self-administer – Slow, sustained release of medication – Less invasive than IV	– Slower and variable absorption – Limited to smaller volumes – Potential for local irritation or tissue reactions
Transdermal	– Noninvasive and can provide sustained release – Bypasses first-pass metabolism – Suitable for chronic conditions	– Limited to small, lipophilic drugs – Skin irritation or allergic reactions may occur – Variable absorption based on skin conditions

3.2 Role of micro/nanoemulsions

Microemulsions, identified by Hoar and Schulman in 1943, are unique dispersions that can be transparent or translucent. They discovered these emulsions during titration experiments with long-chain fatty acids and medium- and short-chain alcohols [19, 20]. Microemulsions can be defined as "a system of water, oil, and amphiphile which is a single optically isotropic and thermodynamically stable liquid solution." Generally, microemulsions are described as pseudo-homogeneous mixtures that consist of water, water-insoluble organic compounds, and a combination of surfactants and cosurfactants (Figure 3.1 and Table 3.2) [21–23]. Microemulsions can be prepared by varying the mixing ratios of water, oil, and surfactant/cosurfactant, depending on whether the desired type is oil-in-water or water-in-oil [24, 25]. The role of amphiphiles (having both hydrophilic and lipophilic properties), which include the surfactant and cosurfactant mixture, is crucial as they reduce the interfacial tension between oil and water through interfacial adsorption. This process minimizes the positive free energy change associated with the dispersion's surface formation, thereby contributing to the stability of the microemulsion [26–29].

Table 3.2: Classification of emulsifiers.

Type of emulsifier	Structure of system	Droplet size
Surfactants	Stabilized by molecules with hydrophilic heads and hydrophobic tails; forms both O/W and W/O emulsions.	Macroemulsions: >1 μm Microemulsions: 10–100 nm
Hydrophilic colloids (e.g., gelatine and gum arabic)	Forms a network within the continuous phase, stabilizing droplets by thickening the dispersion.	Macroemulsions: >1 μm
Solid particles (pickering emulsions)	Stabilized by solid particles that adsorb at the oil/water interface, creating a barrier against coalescence.	Varies: Microemulsions and nanoemulsions
Polymeric emulsifiers	Provides more stability through thickening and steric hindrance; can stabilize both O/W and W/O emulsions.	Macroemulsions: >1 μm Nanoemulsions: 100–1,000 nm
Proteins (e.g., whey protein)	Proteins adsorb at the oil/water interface, forming a film around droplets; used in both food and pharmaceutical applications.	Macroemulsions: >1 μm Microemulsions: 10–100 nm

3.2.1 Characteristics of microemulsion

Microemulsions possess several distinct properties that make them unique and valuable in various applications. Here are some key properties [30–32]:

- **Thermodynamic stability**: Microemulsions are thermodynamically stable and do not separate over time, unlike conventional emulsions, which can coalesce and break.

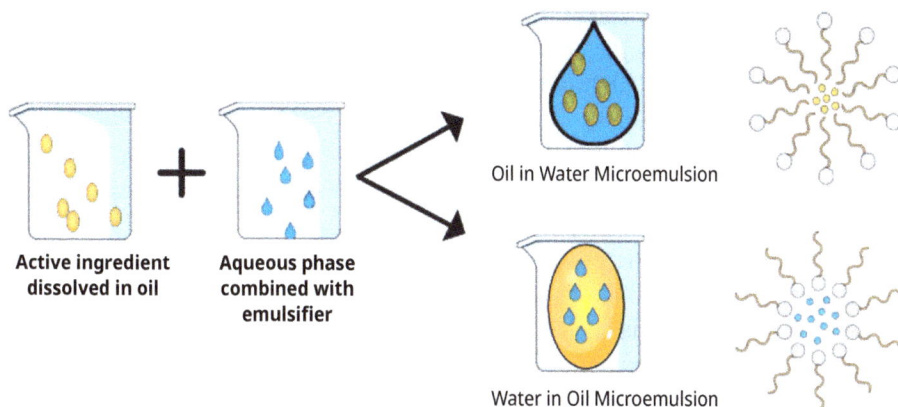

Figure 3.1: Schematic representation of the formation of microemulsions from oil and aqueous phases with emulsifiers.

-   **Transparency**: They are typically clear or translucent due to their small droplet size (usually less than 100 nm), which reduces light scattering.
-   **Low viscosity**: Microemulsions generally exhibit low viscosity, making them easy to handle and apply in various formulations.
-   **High surface area**: The small droplet size provides a large interfacial area that can enhance the solubilization of hydrophobic compounds.
-   **Solubilization properties**: They can solubilize both polar and nonpolar compo-nents, making them useful for delivering a wide range of active ingredients in pharmaceuticals, cosmetics, and food products.

3.2.2 Importance in drug formulation

Microemulsions are increasingly recognized as powerful vehicles in drug formulation, particularly for poorly water-soluble drugs. These thermodynamically stable systems consist of a mixture of oil, water, and surfactants, which facilitate the solubilization and delivery of active pharmaceutical ingredients. The importance of microemulsions in drug formulation can be attributed to several key factors:

-   **Enhanced bioavailability**: Microemulsions significantly improve the solubility and bioavailability of hydrophobic drugs, allowing for better absorption in the gas-trointestinal tract. This enhancement is crucial for drugs that traditionally face challenges in reaching therapeutic concentrations due to poor water solubility.
-   **Thermodynamic stability**: Unlike conventional emulsions, microemulsions are thermodynamically stable, which means they do not separate over time. This stabil-ity ensures consistent performance and reliability of drug formulations throughout their shelf life.

- **Ease of preparation and scalability**: The preparation of microemulsions is relatively simple and can be achieved through low-energy techniques. This ease of formulation and scalability makes them suitable for large-scale production in pharmaceutical industries.
- **Improved drug delivery**: Microemulsions enhance drug delivery for oral, topical, and parenteral use. Their small droplets increase absorption surface area, improving drug release and uptake.
- **Protective effect**: Microemulsions serve as protective carriers for drugs, safeguarding them from degradation due to oxidative and enzymatic processes. This protective capability is vital for maintaining the stability and efficacy of sensitive pharmaceutical compounds.
- **Controlled release**: By modulating the composition of microemulsions, formulators can achieve controlled release profiles, optimizing therapeutic effects, while minimizing side effects and dosing frequency.
- **Versatility**: Microemulsions are capable of encapsulating a diverse range of compounds, including small molecules, peptides, and biological macromolecules, making them a versatile tool for various therapeutic applications.

3.3 Anatomy and physiology of the nasal cavity

The nose serves as the primary entry point to the respiratory tract, enabling air intake for breathing. The nasal cavity, measuring 120–140 mm in depth and extending from the nasal vestibule to the nasopharynx, is divided by the nasal septum. With a surface area of approximately 160 cm^2 and a volume of 16–19 mL, it functions to warm and humidify inhaled air [33, 34]. Beyond its respiratory role, the nose filters airborne particles and provides immunological defense via a mucous-coated membrane [35]. The nasal cavity comprises three main regions: the vestibular, turbinate, and olfactory areas [34]. The anterior vestibular region, the narrowest part, contains vibrissae that filter large particles and features a transition from skin to stratified squamous epithelium. The vascularized turbinate region – divided into superior, middle, and inferior sections – is lined with pseudostratified columnar epithelium composed of mucus-secreting, ciliated, non-ciliated, and basal cells. Ciliated cells facilitate mucociliary clearance, while nonmotile microvilli increase the surface area for drug absorption. However, rapid mucociliary clearance can limit the residence time of drugs, reducing absorption efficiency.

3.3.1 Absorption

The initial phase of drug absorption in the nasal cavity involves passage through the mucus layer. Fine particles can penetrate this layer with ease, while larger particles may encounter resistance [36–38]. Mucus, which contains mucin, a solute-binding protein, can influence drug diffusion and its structure may be altered by environmental or physiological factors [39]. Once past the mucus, drugs are absorbed through the nasal epithelium via transcellular diffusion, paracellular transport, and transcytosis, with the former two being the primary mechanisms [40–42]. Paracellular transport is a passive and relatively slow process, with permeability inversely related to the molecular weight of water-soluble compounds. Molecules exceeding 1,000 Da typically exhibit poor bioavailability. In contrast, transcellular transport facilitates the movement of lipophilic drugs, with efficiency largely dependent on lipophilicity. Additionally, active transport via carrier proteins or through modulated tight junctions may aid drug passage across cell membranes.

3.4 Nose-to-brain delivery

The process of drug delivery through the nasal cavity involves several key steps that facilitate rapid access to the brain and systemic circulation. Initially, drugs are administered into the nasal cavity, where they can interact with the olfactory epithelium, a specialized tissue responsible for the sense of smell. This interaction allows some drugs to bypass the BBB and enter the brain directly, providing quick effects on the CNS [43–45]. Additionally, drugs may also engage with the respiratory epithelium, which lines the nasal passages and plays a role in mucociliary clearance. This mechanism helps remove mucus and foreign particles, potentially influencing drug absorption. Some drugs can enter systemic circulation through the respiratory epithelium, allowing them to be distributed throughout the body. Overall, this unique route of administration highlights the efficiency of nasal delivery in achieving both central and systemic effects (Figure 3.2).

3.4.1 Barriers to drug absorption

The nasal mucosal cavity presents a significant barrier to drug absorption due to its structural and protective mechanisms. The epithelial surface is lined with tightly joined cells that restrict the passage of large and hydrophilic molecules, limiting drug permeability into systemic circulation. Additionally, the mucus layer coating the nasal mucosa, while protective, acts as a viscous barrier that drugs must penetrate to reach the underlying epithelium. Ciliated epithelial cells contribute to mucociliary clearance,

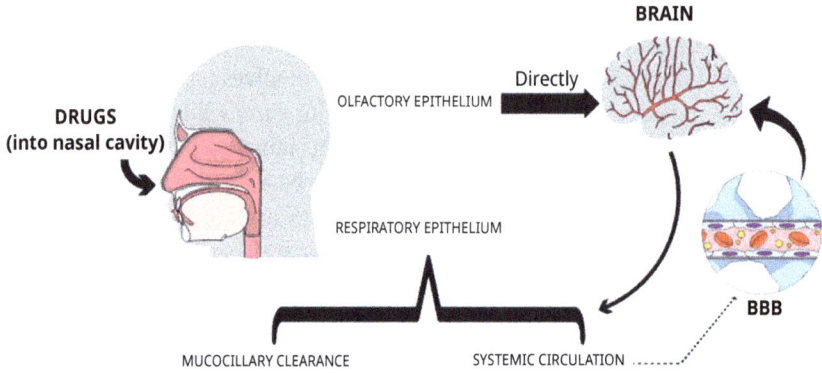

Figure 3.2: Schematic representation of drug delivery through the nasal cavity.

which rapidly removes mucus and entrapped substances, including nasally administered drugs, before sufficient absorption can occur. Enzymatic activity within the nasal
mucosa, including proteases and esterases, can degrade sensitive drugs such as peptides
and proteins, reducing their bioavailability. Moreover, the local pH and temperature
influence drug solubility and stability; some compounds require specific conditions for
optimal absorption, and any deviations can impair efficacy. Although the nasal cavity
offers a substantial surface area for absorption, it remains relatively limited compared
to other routes like the gastrointestinal tract, constraining the amount of drug that can
be effectively delivered, particularly for drugs requiring higher doses. Hydrophobic
drugs face additional challenges due to low permeability across the aqueous mucus
layer. Furthermore, drug transport may rely on active or facilitated mechanisms, which
are not universally available for all compounds. Without suitable carriers or transport
pathways, absorption can be significantly compromised.

In summary, while the nasal cavity offers an attractive route for systemic drug
delivery, it is impeded by multiple anatomical and biochemical barriers. Overcoming
these challenges requires strategic formulation approaches that enhance drug permeability, stability, and bioavailability within the nasal environment. Here are several
strategies to achieve this:

a) **Formulation modifications**: Utilizing drug formulations that improve solubility
and stability can enhance absorption. This can include the use of prodrugs, which
are modified versions of drugs that become active upon entering the body, and
nano-emulsions or solid lipid nanoparticles that enhance drug delivery.

b) **Permeation enhancers**: Incorporating permeation enhancers such as surfactants, fatty acids, or surfactant-like compounds can disrupt tight junctions and increase the permeability of the epithelial barrier. These agents can facilitate the
passage of certain drugs through the mucosal layer by altering the barrier properties.

c) **Mucoadhesive agents**: Using mucoadhesive polymers can enhance the retention time of drug formulations in the nasal cavity, allowing for prolonged contact with the mucosal surface and increasing the chances of absorption. These agents promote adhesion to the mucus layer and can also protect the drug from mucociliary clearance.

d) **Enzyme inhibitors**: Co-administering enzyme inhibitors can help prevent the degradation of sensitive drugs by the enzymes present in the nasal mucosa. This can enhance the overall bioavailability of drugs that are otherwise susceptible to enzymatic breakdown.

e) **Nanocarriers and microparticles**: Employing nanocarriers, such as liposomes, niosomes, or micelles, can improve drug encapsulation, protection against degradation, and controlled release. These carriers can also enhance cellular uptake and facilitate transport across mucosal barriers.

f) **Targeted delivery systems**: Designing targeted delivery systems that can specifically interact with the receptors or transport mechanisms in the nasal mucosa can enhance absorption. This can include ligands that promote endocytosis or transcytosis, allowing drugs to effectively cross the epithelial barrier.

g) **Ionization control**: Modifying the pH of the formulation to optimize the ionization state of the drug can improve solubility and absorption. Ensuring that drugs are in their optimal form for permeability can significantly influence the absorption rate.

h) **Thermal and chemical enhancers**: Thermal methods, such as electroporation or ultrasound, can increase permeability by creating transient pores in the membrane. Chemical enhancers or formulations that can temporarily disrupt the mucosal barrier may also be employed.

i) **Optimized dosing and administration techniques**: Adjusting the dosing regimen and administration techniques, such as using a metered-dose inhaler or nasal spray with specific spray patterns, can improve the distribution and absorption of the drug in the nasal cavity.

j) **Nanoparticle drug delivery**: Utilizing nanoparticles designed to bypass mucosal barriers can enhance drug delivery. These particles can exploit endocytic pathways or enhance diffusion across cellular membranes to improve absorption.

3.4.2 Role of micro/nanoemulsions in enhancing absorption

Microemulsions enhance drug absorption in the nasal cavity through multiple mechanisms that improve solubility, stability, and permeability. Their biphasic nature allows for the solubilization of both hydrophilic and hydrophobic drugs – hydrophobic compounds dissolve in the oil phase, while hydrophilic ones are accommodated in the aqueous phase. This versatility broadens the range of drugs suitable for nasal delivery and increases bioavailability by allowing higher concentrations of dissolved drug for effective mucosal uptake.

The small droplet size of microemulsions (typically 10–100 nm) facilitates diffusion across the mucosal barrier and supports efficient transport into systemic circulation. Additionally, surfactants and co-surfactants in the formulation can transiently open tight junctions between epithelial cells, further enhancing permeability. Many microemulsions possess mucoadhesive properties, improving their residence time on the nasal mucosa by resisting mucociliary clearance and allowing prolonged drug absorption.

Microemulsions can also leverage specific transport pathways, such as transcytosis, through interactions with receptors or transport proteins in the nasal epithelium. By encapsulating drugs, they provide protection from enzymatic degradation – particularly important for sensitive molecules like peptides and proteins. Moreover, microemulsions can be engineered for controlled or sustained release, minimizing fluctuations in drug levels and improving therapeutic outcomes.

Overall, these properties make microemulsions a highly promising strategy for nasal drug delivery, significantly enhancing absorption, bioavailability, and clinical efficacy.

3.5 Efficacy of transnasal drug delivery with emulsions

The choice of formulation for nasal drug delivery can greatly influence the drug's effectiveness. The nasal cavity is considered an appealing site for drug administration due to its extensive surface area and rich vascular network, which can transport active compounds through neural pathways crucial for rapid onset and efficacy. Nasally administered drugs can evade the liver's first-pass metabolism, resulting in enhanced bioavailability. Additionally, nasal formulations are typically well-received and user-friendly for patients. This route of administration also presents challenges that could affect the efficacy of nasal formulations, such as rapid removal of substances from the nasal mucous membrane, enzymatic breakdown of the active ingredient, potential irritation of the nasal mucosa, and inadequate permeability [46–49]. Consequently, the design of the delivery system is vital for ensuring safe and effective therapy.

Recent research has explored intranasal drug delivery systems, including polymer and lipid nanoparticles, micelles, nanoemulsions, liposomes, and other formulations. Mucoadhesive formulations and in situ gelling systems are also studied [50–53]. Mucoadhesive excipients can extend the residence time at the administration site, critical for the nasal membrane due to its high mucociliary clearance, which can reduce absorption and efficacy [54–56]. In situ forming gels contain stimuli-responsive excipients that react to nasal conditions, increasing viscosity. This allows the formulation to be administered as a liquid for optimal distribution, before transforming into a gel to prolong contact with the nasal membrane. These strategies often incorporate nanoparticulate carriers for encapsulating active ingredients.

Nose-to-brain drug delivery is a noninvasive alternative to traditional therapies. Benefits include easy administration, quick action, targeted brain delivery, and bypassing peripheral side effects. However, challenges like limited nasal volume, mucociliary clearance, and poor permeability can hinder absorption.

Microemulsions are considered carriers for improving drug delivery to the brain because they can solubilize various active ingredients and enhance permeation across biological membranes. Studies indicate that microemulsion-based systems can significantly increase the amount of drug delivered to the brain in animal models, confirming the effectiveness of the nose-to-brain pathway (Tables 3.3–3.5).

Despite these promising results, most studies have been conducted on animal models, and there is a lack of human clinical trials involving microemulsion-based carriers. Some human studies suggest inconsistent uptake of drugs in the CNS, indicating a need for further investigation into the efficacy and safety of these systems for human applications.

Table 3.3: Microemulsion-based systems investigated in nose-to-brain delivery for neurodegenerative diseases.

Active element	Microemulsion ingredients	Assessment techniques	Interpretations	References
Rivastigmine hydrogen tartrate	Capmul® MCM EP, Labrasol®, Transcutol® P, water, chitosan, cetyltrimethylammonium bromide	In vitro: Franz cells, cellulose acetate membrane (m.w. cut-off 12,000–14,000) Ex vivo: Franz cells, goat nasal mucosa	The chitosan-based microemulsion demonstrated enhanced ex vivo permeation.	[57]
Rivastigmine hydrogen tartrate	Capmul® MCM EP, Labrasol®	Ex vivo: Goat nasal mucosa	Increased brain concentration after intranasal administration versus solution.	[58]
Galantamine hydrochloride	Capmul® MCM EP, Tween® 80, Transcutol® P, water	In vivo pharmacokinetic studies	Enhanced brain delivery with fish oil and butter oil	[59]
Donepezil hydrochloride	Castor oil, Labrasol®, Transcutol® P, propylene glycol	Ex vivo: porcine nasal mucosa	Sustained release noted; possible reservoir effect in nasal mucosa.	[60]

Table 3.3 (continued)

Active element	Microemulsion ingredients	Assessment techniques	Interpretations	References
Huperzine A	1,2-Propanediol, castor oil, Cremophor RH40	In vivo pharmacokinetic study	Extended release, compared to solution.	[61]
Morin hydrate	Capmul® MCM, Cremophor EL, PEG-400, water	In vivo studies with Wistar rats	Higher concentrations in brain and blood than in drug solution.	[62]
Vinpocetine and piracetam	Tween® 20, oleic acid, ethanol, water	Ex vivo: sheep nasal mucosa	Increased piracetam release; notable cognitive function enhancement.	[63]
Ibuprofen	Capmul® MCM, Accenon® CC, Transcutol®	In vivo studies with mice	Neuroprotective effects noted; suitable for nasal delivery.	[64]
Clobazam	Capmul® MCM, Acconan® C6, Tween® 20, water, Carbopol® 940P	In vivo pharmacokinetic studies, behavioral tests	Improved cerebral delivery; enhanced performance of mucoadhesive formulations.	[65]
Carbamazepine	Oleic acid, Tween® 80, propylene glycol or Transcutol®, water	Ex vivo: sheep nasal mucosa; in vivo: induced convulsions in mice	Decreased seizure duration; increased drug levels in the brain.	[66]
Phenytoin	Capmul® MCM, Labrasol®, and Transcutol®	In vivo brain uptake study	Intranasal microemulsion achieves better brain uptake than an intraperitoneal solution.	[67]

Table 3.4: Microemulsion-based systems investigated in nose-to-brain delivery for epilepsy.

Drug	Microemulsion ingredients	Assessment techniques	Interpretations	References
Clobazam	Capmul® MCM, Acconan® C6, Tween® 20, water, and Carbopol® 940P	Ex vivo animal mucosa, in vivo gamma-scintigraphy, and pharmacodynamic tests	The mucoadhesive formulation exhibited superior efficacy; the intranasal delivery system demonstrated a prolonged effect in comparison to the intravenous method.	[68]
Lorazepam	Capmul® MCM, Nikkol PBC-34, Transcutol® 80, propylene glycol, water, and chitosan	In vivo pharmacokinetic studies and behavioral tests	Improved brain delivery via microemulsion systems; enhanced efficacy of mucoadhesive products.	[69]
Carbamazepine	Oleic acid, Tween® 80, propylene glycol or Transcutol®, and water	Ex vivo sheep nasal mucosa, in vivo pharmacokinetic studies, and induced convulsions in mice	Transcutol®-based microemulsion reduces seizure time, like intraperitoneal drug solution, and increases drug concentration in the brain.	[70]
Carbamazepine	Transcutol®, water, polycarbophil, Labrafil® M1944, and Cremophor® RH40	Ex vivo sheep nasal mucosa, pharmacokinetic studies, and gamma scintigraphy	There were no significant differences observed between microemulsion-based systems and drug solutions in the ex vivo study; however, higher concentrations in the brain were obtained for the microemulsion-based systems.	[71]
Phenytoin	Transcutol®, Capmul® MCM, and Labrasol®	Gamma scintigraphy imaging and in vivo brain uptake study	Improved outcomes were observed following intranasal microemulsion administration compared to intraperitoneal solution, with faster recovery noted after seizures.	[72]

Table 3.5: Summarizing the microemulsion-based systems investigated in nose-to-brain delivery for schizophrenia.

Drug	Microemulsion ingredients	Assessment techniques	Interpretations	References
Olanzapine	Transcutol®, water, polycarbophil, Oleic acid, and Kolliphor® RH40	Gamma scintigraphy; in vivo pharmacokinetic studies; pharmacodynamic tests	The brain shows higher concentration compared to intravenous microemulsion and intranasal solution, with no peripheral distribution.	[73]
Olanzapine	RH40, Labrafil® M1944CS, Cremophor® ethanol, water, poloxamer 407, and HPMC K4M	In vivo studies; ex vivo sheep nasal mucosa; and gamma scintigraphy	The permeation rate is higher compared to nanostructured lipid carrier (NLC). Drug concentrations in the brain are lower, relative to NLC. The selectivity of drug delivery is less than that of NLC. Nasal mucosa irritation has been observed.	[74]
Quetiapine	Capmul® MCM EP, Tween® 80, Transcutol® P, water, and chitosan	In vivo pharmacokinetic studies; ex vivo nasal and intestinal mucosa	Chitosan-loaded microemulsion showed the highest ex vivo permeation rate and the highest brain drug level in vivo.	[75]
Quetiapine	Capmul® MCM EP, Tween® 80, Transcutol® P, water, and butter oil	In vivo pharmacokinetic studies; ex vivo goat nasal mucosa	Butter oil-enriched microemulsion showed the highest ex vivo permeation rate and plasma drug levels.	[76]
Paliperidone	Water, polycarbophil; oleic acid, Cremophor® RH40, and Transcutol®	Gamma scintigraphy; behavioral studies; and pharmacokinetic in vivo studies	The mucoadhesive microemulsion demonstrated superior performance in behavioral studies and showed greater selectivity compared to the intravenous formulation.	[77]

3.6 Precision and targeted delivery

The micro- and nanoemulsion formulations are the new and highly promising drug delivery systems for the transnasal route. These formulations improve drug solubility, absorption, and bioavailability, thus aiding in targeted and effective drug delivery. The following sections discuss the mechanisms, applications, advances, and future aspects in this field of formulation [57–59].

3.6.1 Mechanisms of precision delivery

Micro- and nanoemulsions provide precision drug delivery through several mechanisms. First, small droplet sizes are associated with huge surface areas to absorb more for better drug penetration across the nasal mucosa. The decreased size of the droplets allows deep penetration within the nasal cavity where both local and systemic drug absorption is enhanced. This is supported further by the inclusion of permeation enhancers in the above preparations for enhanced passage of the drugs along the epithelium in the nasal mucosa. These enhancers allow loosening of epithelial cell tight junctions, hence facilitating drug movement. In addition, modifications of the emulsion surface such as ligand conjugation and surface charge modification allow for specific interactions with target proteins in the nasal mucosa, thereby ensuring localized drug delivery to specific cells and tissues [60–62].

3.6.2 Applications and advantages of targeted delivery

Transnasal drug delivery through micro and nanoemulsions offers several advantages in various therapeutic areas. These formulations allow the drugs to penetrate the BBB in CNS disorders, making them highly effective in treating conditions like Alzheimer's and Parkinson's disease. Local treatments of conditions like nasal inflammation and pain management are also benefited through the rapid absorption and localized action of these emulsions with minimized systemic side effects. Vaccine delivery through nasal emulsions enhances immune response due to rapid mucosal absorption because it is an easy, noninvasive mode of immunization [63–66]. Hormonal drugs, such as insulin and oxytocin, can be administered via the nasal route instead of injections; hence it is an alternative with improved patient compliance [67–70]. Anticancer drug delivery via the transnasal route minimizes systemic toxicity, especially for targeting brain tumors or metastatic sites. This route also benefits gene therapy applications, through which genetic material can be delivered non-topically directly into the systemic circulation. These micro- and nano-emulsions can improve drug stability and shelf life, thus being useful for sensitive drug molecules. Use of emulsions in adjuvant applications also improves the efficiency of the vaccine by initiating the immune response [71–74].

3.7 Advances and future work

Transnasal drug delivery using micro- and nano-emulsions is advancing rapidly, driven by continuous innovations in pharmaceutical technology. One of the most promising developments in this area is the emergence of smart drug delivery systems, which enable site-specific drug release in response to physiological stimuli such as pH or temperature fluctuations [75, 76]. These advancements align closely with the concept of personalized medicine, allowing treatments to be tailored to an individual's unique genetic profile.

The integration of nanotechnology into formulation techniques has significantly enhanced drug stability and provided improved control over release kinetics. Advanced methodologies, such as high-pressure homogenization and ultrasonic emulsification, facilitate uniform particle size distribution and contribute to increased bioavailability of therapeutic agents [77–79]. Furthermore, combination therapies that incorporate multiple active agents, including pharmaceuticals and adjuvants, within a single emulsion have demonstrated synergistic effects, offering greater therapeutic efficacy.

Looking ahead, transnasal drug delivery systems hold promise for managing chronic conditions such as diabetes, cancer, and neurodegenerative disorders [80–83]. Another area gaining momentum is the development of nasal vaccine formulations aimed at boosting both systemic and mucosal immunity. Regulatory frameworks must evolve to accommodate these novel approaches, ensuring efficient pathways for market approval.

Additionally, combining transnasal delivery with other routes, such as sublingual or transdermal administration, may enhance therapeutic outcomes. The development of user-friendly nasal delivery devices is essential to improve patient compliance and ease of use. Emphasis should also be placed on targeting rare diseases, especially those where conventional delivery systems fall short. Finally, the design of long-acting formulations is crucial, as reduced dosing frequency contributes to improved quality of life for patients.

Transnasal drug delivery via micro- and nano-emulsion systems holds immense potential for revolutionizing therapeutic strategies, particularly for complex and chronic diseases. Emerging technologies in this domain include the development of innovative layered nanoparticles capable of co-delivering multiple drugs in a single dose, thereby enhancing therapeutic efficacy and treatment outcomes. Additionally, the integration of 3D printing in the design of patient-specific nasal drug delivery devices offers opportunities for precision dosing and improved patient adherence.

Smart polymers that respond to environmental stimuli such as pH, temperature, or enzymatic activity enable controlled drug release, reducing the frequency of dosing. Furthermore, the application of personalized medicine – tailoring formulations based on genetic data – significantly improves therapeutic precision and efficacy. Advances in nanotechnology have contributed to the enhanced stability of formulations and more efficient drug delivery, expanding the applicability of transnasal emulsions, especially in the treatment of chronic and rare diseases [99].

3.8 Strategies toward targeting

To improve the specificity of transnasal drug delivery, a range of advanced strategies has been developed. One such approach involves ligand modification, wherein emulsions are conjugated with receptor-specific ligands – such as antibodies or peptides – to facilitate targeted drug binding within the nasal cavity [84–86]. Aptamers, which are short nucleic acid sequences with high affinity for specific targets, further augment drug-target interactions and enhance selectivity. Surface functionalization of nanoparticles with biologically active compounds is another effective strategy, significantly improving cellular uptake and targeting efficiency [87, 88]. The incorporation of stealth properties, achieved through polymer coatings on nanoparticles, extends circulation time and enhances drug transport across biological barriers.

Stimuli-responsive emulsions, designed to release their therapeutic payload in response to specific physiological triggers such as pH or temperature changes, represent a promising frontier in targeted delivery. Additionally, smart drug delivery systems that incorporate biosensors and biological recognition elements enable site-specific release, ensuring that the drug is activated only in the presence of its intended target [89–92].

The inclusion of permeation enhancers in emulsion formulations further improves drug absorption across the nasal mucosa, thereby reinforcing the efficiency of targeted delivery. Collectively, these strategies contribute to the advancement of precision medicine and the development of more effective and reliable transnasal therapeutic platforms.

3.8.1 Precision assessment

The evaluation of transnasal micro- and nano-emulsions requires a comprehensive framework encompassing both in vitro and in vivo studies to accurately assess their efficacy and safety. In vitro assessments commonly employ nasal epithelial cell lines to examine cytotoxicity, drug release profiles, and cellular uptake of the therapeutic agents [93, 94]. Flow cytometry is utilized to quantify targeting efficiency, particularly in formulations labeled with fluorescent markers. Additionally, advanced imaging modalities such as magnetic resonance imaging (MRI), positron emission tomography (PET), and fluorescence imaging are employed to visualize drug distribution and monitor in vivo biodistribution dynamics [95, 96].

Quantitative analysis of drug concentration within tissues, following nasal administration, is typically conducted using high-performance liquid chromatography (HPLC) and mass spectrometry. Functional outcome assessments in animal models serve to evaluate the therapeutic efficacy of the delivered drugs, providing crucial insights into their biological activity and effectiveness [97–99]. Pharmacokinetic studies involve the collection of blood and tissue samples over time to determine the absorption, distribution, and systemic availability of the drug. Furthermore, histological examinations – through tissue sectioning and imaging – are conducted to analyze the localization of the emulsion within the nasal mucosa and to evaluate potential interactions at the cellular level. Collectively, these precision assessment strategies ensure a thorough understanding of formulation behavior and therapeutic performance.

3.9 Challenges and strategies to overcome limitations

Despite promising advancements, transnasal drug delivery systems using micro- and nanoemulsions continue to face several significant challenges. One of the primary concerns is formulation stability, which can be compromised by environmental and storage conditions, as well as the interaction with nasal fluids [101]. These interactions may negatively impact structural integrity and reduce therapeutic performance. To address these issues, the development of robust formulation strategies is essential. The use of potent stabilizing agents – such as polysorbate 80, lecithin, and other nonionic surfactants – has proven effective in preventing droplet coalescence and phase separation. For instance, a nano-emulsion combining lecithin and polysorbate 80 for nasal delivery of levocetirizine can reduce droplet size and improve stability, emphasizing the importance of targeted stabilization approaches.

High-pressure homogenization stands out as one of the most effective emulsification techniques [102–104]. It facilitates the production of submicron emulsions with uniform droplet sizes, thereby significantly enhancing formulation stability and therapeutic efficacy. Additionally, the application of microfluidic technologies – using miniature channels for high-precision emulsion production – is gaining traction for industrial scalability [105–108]. These approaches have demonstrated high reproducibility and are being optimized for pharmaceutical-grade applications. For example, large-scale production of ibuprofen-loaded microemulsions using high-pressure homogenization underscores the technique's potential for commercial viability [109].

Scalability and reproducibility remain critical hurdles in the transition from laboratory-scale research to industrial-scale manufacturing. Addressing these concerns necessitates the adoption of cost-effective, reliable technologies and rigorous quality assurance measures. Standardization of formulation protocols, emulsification procedures, and storage conditions, alongside routine quality control assessments – including droplet size analysis, viscosity monitoring, and environmental stability testing – are key to ensuring consistent product performance.

Technical barriers such as limited mucosal penetration and potential systemic side effects also need to be addressed through advanced formulation refinement. Moreover, enhancing mucosal immunity – especially in nasal vaccine delivery – designing long-acting emulsions, and targeting therapeutic agents with poor oral bioavailability are critical areas requiring further research. Finally, regulatory pathways must evolve in tandem with technological progress. Streamlined approval processes are essential to accelerate the clinical translation of these novel drug delivery systems. With continued advancements in formulation science, nanotechnology, and regulatory frameworks, transnasal micro- and nanoemulsion-based delivery systems are poised to become a transformative modality in modern pharmaceutical care.

3.10 Research directions

Building on efforts to address existing challenges, future research should increasingly focus on leveraging micro- and nanoemulsion technologies for the transnasal delivery of poorly water-soluble drugs. This approach offers substantial promise for enhancing bioavailability and therapeutic efficacy, especially by bypassing hepatic first-pass metabolism and enabling rapid systemic absorption. The nasal route, with its highly vascularized mucosa and ease of access, stands out as a compelling alternative to conventional oral or parenteral drug delivery.

A prime example of this innovation is the chemotherapeutic agent paclitaxel, which is traditionally limited by its poor oral bioavailability. Reformulating paclitaxel into a nasal nano-emulsion can demonstrate significantly enhanced therapeutic performance, marked by a faster onset of action and improved clinical outcomes. This underscores the transformative potential of nanoemulsion-based transnasal delivery systems for drugs with challenging pharmacokinetic profiles.

Another promising avenue for future research is the development of combination therapies within a single emulsion system. Co-delivery of multiple active pharmaceutical ingredients allows simultaneous modulation of different biological pathways, potentially enhancing therapeutic outcomes, while reducing the risk of adverse effects. Such multifunctional formulations could play a key role in addressing complex and multifactorial diseases.

To enable clinical adoption and regulatory acceptance of these advanced delivery platforms, robust clinical validation is essential. Future investigations should prioritize comprehensive clinical trials that examine both short-term pharmacodynamic responses and long-term outcomes such as safety, efficacy, and patient compliance. Multicenter trials with diverse patient populations and extended follow-up periods will be critical in establishing the generalizability and durability of therapeutic benefits.

Real-world evidence, gathered through post-marketing surveillance and longitudinal observational studies, will further support these efforts. For example, the successful long-term use of a nasal spray formulation of sumatriptan for migraine management highlights both the practicality and sustained efficacy in real clinical settings.

Continued exploration across multiple fronts – including formulation optimization, enhancement of mucosal penetration, pharmacokinetic profiling, and regulatory alignment – will be vital to transitioning transnasal micro-/nano-emulsion drug delivery systems from experimental innovation to established therapeutic practice. The integration of emerging tools such as artificial intelligence for predictive modeling and patient-specific formulation design may further accelerate progress in this rapidly evolving field [110, 111].

3.11 Conclusion

In summary, the future of transnasal drug delivery using micro- and nano-emulsions appears highly promising, driven by ongoing advances in formulation science and drug delivery technologies. These systems offer significant advantages, including enhanced bioavailability, rapid onset of action, and the potential for noninvasive, patient-friendly administration.

However, key challenges – such as formulation stability, industrial scalability, and reproducibility – must be systematically addressed to facilitate widespread clinical adoption. The development of robust stabilization strategies and scalable manufacturing techniques will be pivotal in translating laboratory successes into market-ready products.

Moreover, the exploration of novel drug candidates, including combination therapies, and the accumulation of long-term clinical data will provide a strong foundation for the next generation of nasal drug delivery systems. As scientific evidence continues to support their efficacy and safety, transnasal micro/nanoemulsions are poised to play a transformative role in improving therapeutic outcomes and expanding treatment possibilities across a wide range of diseases.

References

[1] Kisku A, Nishad A, Agrawal S, Paliwal R, Datusalia AK, Gupta G, Singh SK, Dua K, Sulakhiya K. Recent developments in intranasal drug delivery of nanomedicines for the treatment of neuropsychiatric disorders. Front Med (Lausanne), 2024;11. https://doi.org/10.3389/fmed.2024.1463976

[2] Keller L-A, Merkel O, Popp A. Intranasal drug delivery: Opportunities and toxicologic challenges during drug development. Drug Deliv Transl Res, 2022;12:735–757. https://doi.org/10.1007/s13346-020-00891-5

[3] Kumar NN, Gautam M, Lochhead JJ, Wolak DJ, Ithapu V, Singh V, Thorne RG. Relative vascular permeability and vascularity across different regions of the rat nasal mucosa: Implications for nasal physiology and drug delivery. Sci Rep, 2016;6:31732. https://doi.org/10.1038/srep31732

[4] Patharapankal EJ, Ajiboye AL, Mattern C, Trivedi V. Nose-to-Brain (N2B) delivery: An alternative route for the delivery of biologics in the management and treatment of central nervous system disorders. Pharmaceutics, 2023;16:66. https://doi.org/10.3390/pharmaceutics16010066

[5] Akita T, Oda Y, Kimura R, Nagai M, Tezuka A, Shimamura M, Washizu K, Oka J-I, Yamashita C. Involvement of trigeminal axons in nose-to-brain delivery of glucagon-like peptide-2 derivative. J Contr Release, 2022;351:573–580. https://doi.org/10.1016/j.jconrel.2022.09.047

[6] Lofts A, Abu-Hijleh F, Rigg N, Mishra RK, Hoare T. Using the intranasal route to administer drugs to treat neurological and psychiatric illnesses: Rationale, successes, and future needs. CNS Drugs, 2022;36:739–770. https://doi.org/10.1007/s40263-022-00930-4

[7] Trenkel M, Scherließ R. Optimising nasal powder drug delivery – Characterisation of the effect of excipients on drug absorption. Int J Pharm, 2023;633:122630. https://doi.org/10.1016/j.ijpharm.2023.122630

[8] Kim J, De Jesus O. Medication Routes of Administration, 2025.

[9] Jeong S-H, Jang J-H, Lee Y-B. Drug delivery to the brain via the nasal route of administration: Exploration of key targets and major consideration factors. J Pharm Investig, 2023;53:119–152. https://doi.org/10.1007/s40005-022-00589-5

[10] Wang JT, Rodrigo AC, Patterson AK, Hawkins K, Aly MMS, Sun J, Al Jamal KT, Smith DK. Enhanced delivery of neuroactive drugs via nasal delivery with a self-healing supramolecular gel. Adv Sci, 2021;8. https://doi.org/10.1002/advs.202101058

[11] Formica ML, Real DA, Picchio ML, Catlin E, Donnelly RF, Paredes AJ. On a highway to the brain: A review on nose-to-brain drug delivery using nanoparticles. Appl Mater Today, 2022;29:101631. https://doi.org/10.1016/j.apmt.2022.101631

[12] Gandhi S, Shastri DH, Shah J, Nair AB, Jacob S. Nasal delivery to the brain: Harnessing nanoparticles for effective drug transport. Pharmaceutics, 2024;16:481. https://doi.org/10.3390/pharmaceutics16040481

[13] Gao H, Pang Z, Jiang X. Targeted delivery of nano-therapeutics for major disorders of the central nervous system. Pharm Res, 2013;30:2485–2498. https://doi.org/10.1007/s11095-013-1122-4

[14] Zhang X, Wang M, Liu Z, Wang Y, Chen L, Guo J, Zhang W, Zhang Y, Yu C, Bie T, Yu Y, Guan B. Transnasal-brain delivery of nanomedicines for neurodegenerative diseases. Front Drug Deliv, 2023;3. https://doi.org/10.3389/fddev.2023.1247162

[15] Du L, Chen L, Liu F, Wang W, Huang H. *Nose-to-brain Drug Delivery for the Treatment of CNS Disease: New Development and Strategies*, 2023, pp. 255–297. https://doi.org/10.1016/bs.irn.2023.05.014

[16] Raghav M, Gupta V, Awasthi R, Singh A, Kulkarni GT. Nose-to-brain drug delivery: Challenges and progress towards brain targeting in the treatment of neurological disorders. J Drug Deliv Sci Technol, 2023;86:104756. https://doi.org/10.1016/j.jddst.2023.104756

[17] Cingi C, Bayar Muluk N, Mitsias DI, Papadopoulos NG, Klimek L, Laulajainen-Hongisto A, Hytönen M, Toppila-Salmi SK, Scadding GK. The nose as a route for therapy: Part 1. Pharmacotherapy. Front Allergy, 2021;2. https://doi.org/10.3389/falgy.2021.638136

[18] Patel GB, Kern RC, Bernstein JA, Hae-Sim P, Peters AT. Current and future treatments of Rhinitis and Sinusitis. J Allergy Clin Immunol Pract, 2020;8:1522–1531. https://doi.org/10.1016/j.jaip.2020.01.031

[19] Suhail N, Alzahrani AK, Basha WJ, Kizilbash N, Zaidi A, Ambreen J, Khachfe HM. Microemulsions: Unique properties, pharmacological applications, and targeted drug delivery. Front Nanotechnol, 2021;3. https://doi.org/10.3389/fnano.2021.754889

[20] Rakshit AK, Naskar B, Moulik SP. Commemorating 75 years of microemulsion. Curr Sci, 2019;116:898. https://doi.org/10.18520/cs/v116/i6/898-912

[21] Suhail N, Alzahrani AK, Basha WJ, Kizilbash N, Zaidi A, Ambreen J, Khachfe HM. Microemulsions: Unique properties, pharmacological applications, and targeted drug delivery. Front Nanotechnol, 2021;3. https://doi.org/10.3389/fnano.2021.754889

[22] Lu M, Lindman B, Holmberg K. Effect of polymer addition on the phase behavior of oil–water–surfactant systems of Winsor III type. Phys Chem Chem Phys, 2024;26:3699–3710. https://doi.org/10.1039/D3CP04730J

[23] Lawrence MJ, Rees GD. Microemulsion-based media as novel drug delivery systems. Adv Drug Deliv Rev, 2012;64:175–193. https://doi.org/10.1016/j.addr.2012.09.018

[24] Eastoe J, Hatzopoulos MH, Tabor R. Microemulsions. In *Encyclopedia of Colloid and Interface Science*, Berlin Heidelberg, Berlin, Heidelberg: Springer, 2013, pp. 688–729. https://doi.org/10.1007/978-3-642-20665-8_25

[25] Suhail N, Alzahrani AK, Basha WJ, Kizilbash N, Zaidi A, Ambreen J, Khachfe HM. Microemulsions: Unique properties, pharmacological applications, and targeted drug delivery. Front Nanotechnol, 2021;3. https://doi.org/10.3389/fnano.2021.754889

[26] Ande SN, Sonone KB, Bakal RL, Ajmire PV, Sawarkar HS. Role of surfactant and co-surfactant in microemulsion: A review. Res J Pharm Technol, 2022;4829–4834. https://doi.org/10.52711/0974-360X.2022.00811

[27] Moulik SP, Chakraborty I, Rakshit AK. Role of surface-active materials (amphiphiles and surfactants) in the formation of nanocolloidal dispersions, and their applications. J Surfactants Deterg, 2022;25:703–727. https://doi.org/10.1002/jsde.12612

[28] Lombardo D, Kiselev MA, Magazù S, Calandra P. Amphiphiles self-assembly: Basic concepts and future perspectives of supramolecular approaches. Adv Condens Matter Phys, 2015;2015:1–22. https://doi.org/10.1155/2015/151683

[29] Marques EF, Silva BFB. Surfactant Self-Assembly. In *Encyclopedia of Colloid and Interface Science*, Berlin Heidelberg, Berlin, Heidelberg: Springer, 2013, pp. 1202–1241. https://doi.org/10.1007/978-3-642-20665-8_169

[30] Kanwar R, Rathee J, Tanaji Patil M, Kumar Mehta S. Microemulsions as Nanotemplates: A Soft and Versatile Approach. In *Microemulsion – A Chemical Nanoreactor [Working Title]*, IntechOpen, 2018. https://doi.org/10.5772/intechopen.80758

[31] Tartaro G, Mateos H, Schirone D, Angelico R, Palazzo G. Microemulsion microstructure(s): A tutorial review. Nanomaterials, 2020;10:1657. https://doi.org/10.3390/nano10091657

[32] Malik MA, Wani MY, Hashim MA. Microemulsion method: A novel route to synthesize organic and inorganic nanomaterials. Arab J Chem, 2012;5:397–417. https://doi.org/10.1016/j.arabjc.2010.09.027

[33] Freeman SC, Karp DA, Kahwaji CI. Physiology, Nasal, 2025.

[34] Sobiesk JL, Munakomi S. Anatomy, Head and Neck, Nasal Cavity, 2025.

[35] Fokkens WJ, Scheeren RA. Upper airway defence mechanisms. Paediatr Respir Rev, 2000;1:336–341. https://doi.org/10.1053/prrv.2000.0073

[36] Mali AH, Shaikh AZ. A short review on nasal drug delivery system. AJPTech, 2021;289–292. https://doi.org/10.52711/2231-5713.2021.00048

[37] Chavda VP, Jogi G, Shah N, Athalye MN, Bamaniya N, Vora LK, Cláudia Paiva-Santos A. Advanced particulate carrier-mediated technologies for nasal drug delivery. J Drug Deliv Sci Technol, 2022;74:103569. https://doi.org/10.1016/j.jddst.2022.103569

[38]   Ghadiri M, Young PM, Traini D. Strategies to enhance drug absorption via nasal and pulmonary routes. Pharmaceutics, 2019;11:113. https://doi.org/10.3390/pharmaceutics11030113

[39]   Lock JY, Carlson TL, Carrier RL. Mucus models to evaluate the diffusion of drugs and particles. Adv Drug Deliv Rev, 2018;124:34–49. https://doi.org/10.1016/j.addr.2017.11.001

[40]   Khalil A, Barras A, Boukherroub R, Tseng C-L, Devos D, Burnouf T, Neuhaus W, Szunerits S. Enhancing paracellular and transcellular permeability using nanotechnological approaches for the treatment of brain and retinal diseases. Nanoscale Horiz, 2024;9:14–43. https://doi.org/10.1039/D3NH00306J

[41]   Kaur P, Garg T, Rath G, Goyal AK. In situ nasal gel drug delivery: A novel approach for brain targeting through the mucosal membrane. Artif Cells Nanomed Biotechnol, 2015;1–10. https://doi.org/10.3109/21691401.2015.1012260

[42]   Boegh M, Nielsen HM. Mucus as a barrier to drug delivery – Understanding and mimicking the barrier properties. Basic Clin Pharmacol Toxicol, 2015;116:179–186. https://doi.org/10.1111/bcpt.12342

[43]   Aderibigbe B. In situ-based gels for nose to brain delivery for the treatment of neurological diseases. Pharmaceutics, 2018;10:40. https://doi.org/10.3390/pharmaceutics10020040

[44]   Djupesland PG, Messina JC, Mahmoud RA. The nasal approach to delivering treatment for brain diseases: An anatomic, physiologic, and delivery technology overview. Ther Deliv, 2014;5:709–733. https://doi.org/10.4155/tde.14.41

[45]   Jeong S-H, Jang J-H, Lee Y-B. Drug delivery to the brain via the nasal route of administration: Exploration of key targets and major consideration factors. J Pharm Investig, 2023;53:119–152. https://doi.org/10.1007/s40005-022-00589-5

[46]   Godge GR, Shaikh AB, Bharat SC, Randhawan BB, Raskar MA, Hiremath SN. Nasal drug delivery system and its applicability in therapeutics: A review. RGUHS – J Pharm Sci, 2023;13. https://doi.org/10.26463/rjps.13_4_7

[47]   Djupesland PG. Nasal drug delivery devices: Characteristics and performance in a clinical perspective – A review. Drug Deliv Transl Res, 2013;3:42–62. https://doi.org/10.1007/s13346-012-0108-9

[48]   Clementino A, Climani G, Bianchera A, Buttini F, Sonvico F. Polysaccharides: New frontiers for nasal administration of medicines. Polysaccharides, 2025;6:6. https://doi.org/10.3390/polysaccharides6010006

[49]   Pires PC, Rodrigues M, Alves G, Santos AO. Strategies to improve drug strength in nasal preparations for brain delivery of low aqueous solubility drugs. Pharmaceutics, 2022;14:588. https://doi.org/10.3390/pharmaceutics14030588

[50]   Mehta M, Bui TA, Yang X, Aksoy Y, Goldys EM, Deng W. Lipid-based nanoparticles for drug/gene delivery: An overview of the production techniques and difficulties encountered in their industrial development. ACS Mater Au, 2023;3:600–619. https://doi.org/10.1021/acsmaterialsau.3c00032

[51]   Onugwu AL, Nwagwu CS, Onugwu OS, Echezona AC, Agbo CP, Ihim SA, Emeh P, Nnamani PO, Attama AA, Khutoryanskiy VV. Nanotechnology based drug delivery systems for the treatment of anterior segment eye diseases. J Contr Release, 2023;354:465–488. https://doi.org/10.1016/j.jconrel.2023.01.018

[52]   Duong V-A, Nguyen T-T-L, Maeng H-J. Recent advances in intranasal liposomes for drug, gene, and vaccine delivery. Pharmaceutics, 2023;15:207. https://doi.org/10.3390/pharmaceutics15010207

[53]   Islam SU, Shehzad A, Ahmed MB, Lee YS. Intranasal delivery of nanoformulations: A potential way of treatment for neurological disorders. Molecules, 2020;25:1929. https://doi.org/10.3390/molecules25081929

[54]   Shaikh R, Raj Singh T, Garland M, Woolfson A, Donnelly R. Mucoadhesive drug delivery systems. J Pharm Bioallied Sci, 2011;3:89. https://doi.org/10.4103/0975-7406.76478

[55] Kaur G, Goyal J, Behera PK, Devi S, Singh SK, Garg V, Mittal N. Unraveling the role of chitosan for nasal drug delivery systems: A review. Carbohydr Polym Technol Appl, 2023;5:100316. https://doi.org/10.1016/j.carpta.2023.100316

[56] Chaturvedi M, Kumar M, Pathak K. A review on mucoadhesive polymer used in nasal drug delivery system. J Adv Pharm Technol Res, 2011;2:215. https://doi.org/10.4103/2231-4040.90876

[57] Preeti SS, Malik R, Bhatia S, Al Harrasi A, Rani C, Saharan R, Kumar S, Geeta RS. Nanoemulsion: An emerging novel technology for improving the bioavailability of drugs. Scientifica (Cairo), 2023;2023:1–25. https://doi.org/10.1155/2023/6640103

[58] Liu Y, Liang Y, Yuhong J, Xin P, Han JL, Du Y, Yu X, Zhu R, Zhang M, Chen W, Ma Y. Advances in nanotechnology for enhancing the solubility and bioavailability of poorly soluble drugs. Drug Des Devel Ther, 2024;18:1469–1495. https://doi.org/10.2147/DDDT.S447496

[59] Misra SK, Pathak K. Nose-to-Brain targeting via nanoemulsion: Significance and evidence. Colloids Interfaces, 2023;7:23. https://doi.org/10.3390/colloids7010023

[60] Patra JK, Das G, Fraceto LF, Campos EVR, Rodriguez-Torres MDP, Acosta-Torres LS, Diaz-Torres LA, Grillo R, Swamy MK, Sharma S, Habtemariam S, Shin H-S. Nano based drug delivery systems: Recent developments and future prospects. J Nanobiotechnol, 2018;16:71. https://doi.org/10.1186/s12951-018-0392-8

[61] Singh Y, Meher JG, Raval K, Khan FA, Chaurasia M, Jain NK, Chourasia MK. Nanoemulsion: Concepts, development and applications in drug delivery. J Contr Release, 2017;252:28–49. https://doi.org/10.1016/j.jconrel.2017.03.008

[62] Kumar A, Singh AK, Chaudhary RP, Sharma A, Yadav JP, Pathak P, Grishina M, Pathak K, Kumar P. Unraveling the multifaceted role of nanoemulsions as drug delivery system for the management of cancer. J Drug Deliv Sci Technol, 2024;100:106056. https://doi.org/10.1016/j.jddst.2024.106056

[63] Kehagia E, Papakyriakopoulou P, Valsami G. Advances in intranasal vaccine delivery: A promising non-invasive route of immunization. Vaccine, 2023;41:3589–3603. https://doi.org/10.1016/j.vaccine.2023.05.011

[64] Ramvikas M, Arumugam M, Chakrabarti SR, Jaganathan KS. Nasal Vaccine Delivery. In *Micro and Nanotechnology in Vaccine Development*, Elsevier, 2017, pp. 279–301. https://doi.org/10.1016/B978-0-323-39981-4.00015-4

[65] Mangla B, Javed S, Sultan MH, Ahsan W, Aggarwal G, Kohli K. Nanocarriers-assisted needle-free vaccine delivery through oral and intranasal transmucosal routes: A novel therapeutic conduit. Front Pharmacol, 2022;12. https://doi.org/10.3389/fphar.2021.757761

[66] Cho C-S, Hwang S-K, Gu M-J, Kim C-G, Kim S-K, Ju D-B, Yun C-H, Kim H-J. Mucosal vaccine delivery using mucoadhesive polymer particulate systems. Tissue Eng Regen Med, 2021;18:693–712. https://doi.org/10.1007/s13770-021-00373-w

[67] Wong CYJ, Baldelli A, Hoyos CM, Tietz O, Ong HX, Traini D. Insulin delivery to the brain via the nasal route: Unraveling the potential for Alzheimer's disease therapy. Drug Deliv Transl Res, 2024;14:1776–1793. https://doi.org/10.1007/s13346-024-01558-1

[68] Lofts A, Abu-Hijleh F, Rigg N, Mishra RK, Hoare T. Using the intranasal route to administer drugs to treat neurological and psychiatric illnesses: Rationale, successes, and future needs. CNS Drugs, 2022;36:739–770. https://doi.org/10.1007/s40263-022-00930-4

[69] Maeng J, Lee K. Systemic and brain delivery of antidiabetic peptides through nasal administration using cell-penetrating peptides. Front Pharmacol, 2022;13. https://doi.org/10.3389/fphar.2022.1068495

[70] Luo D, Ni X, Yang H, Feng L, Chen Z, Bai L. A comprehensive review of advanced nasal delivery: Specially insulin and calcitonin. Eur J Pharm Sci, 2024;192:106630. https://doi.org/10.1016/j.ejps.2023.106630

[71]   Kashyap K, Shukla R. Drug delivery and targeting to the brain through nasal route: Mechanisms, applications and challenges. Curr Drug Deliv, 2019;16:887–901. https://doi.org/10.2174/1567201816666191029122740

[72]   Prabahar K, Alanazi Z, Qushawy M. Targeted drug delivery system: Advantages, carriers and strategies. Indian J Pharm Educ Res, 2021;55:346–353. https://doi.org/10.5530/ijper.55.2.72

[73]   Won S, An J, Song H, Im S, You G, Lee S, Koo K, Hwang CH. Transnasal targeted delivery of therapeutics in central nervous system diseases: A narrative review. Front Neurosci, 2023;17. https://doi.org/10.3389/fnins.2023.1137096

[74]   Won S, An J, Song H, Im S, You G, Lee S, Koo K, Hwang CH. Transnasal targeted delivery of therapeutics in central nervous system diseases: A narrative review. Front Neurosci, 2023;17. https://doi.org/10.3389/fnins.2023.1137096

[75]   Katare P, Pawar Medhe T, Nadkarni A, Deshpande M, Tekade RK, Benival D, Jain A. Nasal drug delivery system and devices: An overview on health effects. ACS Chem Health Saf, 2024;31:127–143. https://doi.org/10.1021/acs.chas.3c00069

[76]   Koo J, Lim C, Oh KT. Recent advances in intranasal administration for brain-targeting delivery: A comprehensive review of lipid-based nanoparticles and stimuli-responsive gel formulations. Int J Nanomed, 2024;19:1767–1807. https://doi.org/10.2147/IJN.S439181

[77]   de Barros C, Portugal I, Batain F, Portella D, Severino P, Cardoso J, Arcuri P, Chaud M, Alves T. Formulation, design and strategies for efficient nanotechnology-based nasal delivery systems. RPS Pharm Pharmacol Rep, 2022;1. https://doi.org/10.1093/rpsppr/rqac003

[78]   Jaiswal M, Dudhe R, Sharma PK. Nanoemulsion: An advanced mode of drug delivery system. 3 Biotech, 2015;5:123–127. https://doi.org/10.1007/s13205-014-0214-0

[79]   Kaur G, Panigrahi C, Agarwal S, Khuntia A, Sahoo M. Recent trends and advancements in nanoemulsions: Production methods, functional properties, applications in food sector, safety and toxicological effects. Food Phys, 2024;1:100024. https://doi.org/10.1016/j.foodp.2024.100024

[80]   Yuan S, Ma T, Zhang Y-N, Wang N, Baloch Z, Ma K. Novel drug delivery strategies for antidepressant active ingredients from natural medicinal plants: The state of the art. J Nanobiotechnol, 2023;21:391. https://doi.org/10.1186/s12951-023-02159-9

[81]   Toader C, Dumitru AV, Eva L, Serban M, Covache-Busuioc R-A, Ciurea AV. Nanoparticle strategies for treating CNS disorders: A comprehensive review of drug delivery and theranostic applications. Int J Mol Sci, 2024;25:13302. https://doi.org/10.3390/ijms252413302

[82]   Meredith ME, Salameh TS, Banks WA. Intranasal delivery of proteins and peptides in the treatment of neurodegenerative diseases. AAPS J, 2015;17:780–787. https://doi.org/10.1208/s12248-015-9719-7

[83]   Wu H, Zhou Y, Wang Y, Tong L, Wang F, Song S, Xu L, Liu B, Yan H, Sun Z. Current state and future directions of intranasal delivery route for central nervous system disorders: A scientometric and visualization analysis. Front Pharmacol, 2021;12. https://doi.org/10.3389/fphar.2021.717192

[84]   Bourganis V, Kammona O, Alexopoulos A, Kiparissides C. Recent advances in carrier mediated nose-to-brain delivery of pharmaceutics. Eur J Pharm Biopharm, 2018;128:337–362. https://doi.org/10.1016/j.ejpb.2018.05.009

[85]   Ayub A, Wettig S. An overview of nanotechnologies for drug delivery to the brain. Pharmaceutics, 2022;14:224. https://doi.org/10.3390/pharmaceutics14020224

[86]   Sonvico F, Clementino A, Buttini F, Colombo G, Pescina S, Stanisçuaski Guterres S, Raffin Pohlmann A, Nicoli S. Surface-modified nanocarriers for Nose-to-Brain delivery: From bioadhesion to targeting. Pharmaceutics, 2018;10:34. https://doi.org/10.3390/pharmaceutics10010034

[87]   Seidu TA, Kutoka PT, Asante DO, Farooq MA, Alolga RN, Bo W. Functionalization of nanoparticulate drug delivery systems and its influence in cancer therapy. Pharmaceutics, 2022;14:1113. https://doi.org/10.3390/pharmaceutics14051113

[88]   Upadhyay K, Tamrakar RK, Thomas S, Kumar M. Surface functionalized nanoparticles: A boon to biomedical science. Chem Biol Interact, 2023;380:110537. https://doi.org/10.1016/j.cbi.2023.110537

[89] El-Husseiny HM, Mady EA, Hamabe L, Abugomaa A, Shimada K, Yoshida T, Tanaka T, Yokoi A, Elbadawy M, Tanaka R. Smart/stimuli-responsive hydrogels: Cutting-edge platforms for tissue engineering and other biomedical applications. Mater Today Bio, 2022;13:100186. https://doi.org/10.1016/j.mtbio.2021.100186

[90] Hajebi S, Rabiee N, Bagherzadeh M, Ahmadi S, Rabiee M, Roghani-Mamaqani H, Tahriri M, Tayebi L, Hamblin MR. Stimulus-responsive polymeric nanogels as smart drug delivery systems. Acta Biomater, 2019;92:1–18. https://doi.org/10.1016/j.actbio.2019.05.018

[91] Ahmadi S, Rabiee N, Bagherzadeh M, Elmi F, Fatahi Y, Farjadian F, Baheiraei N, Nasseri B, Rabiee M, Dastjerd NT, Valibeik A, Karimi M, Hamblin MR. Stimulus-responsive sequential release systems for drug and gene delivery. Nano Today, 2020;34:100914. https://doi.org/10.1016/j.nantod.2020.100914

[92] Wu J, Xue W, Yun Z, Liu Q, Sun X. Biomedical applications of stimuli-responsive "smart" interpenetrating polymer network hydrogels. Mater Today Bio, 2024;25:100998. https://doi.org/10.1016/j.mtbio.2024.100998

[93] Clementino AR, Pellegrini G, Banella S, Colombo G, Cantù L, Sonvico F, Del Favero E. Structure and fate of nanoparticles designed for the nasal delivery of poorly soluble drugs. Mol Pharm, 2021;18:3132–3146. https://doi.org/10.1021/acs.molpharmaceut.1c00366

[94] Ladel S, Schlossbauer P, Flamm J, Luksch H, Mizaikoff B, Schindowski K. Improved in vitro model for intranasal mucosal drug delivery: Primary olfactory and respiratory epithelial cells compared with the permanent nasal cell line RPMI 2650. Pharmaceutics, 2019;11:367. https://doi.org/10.3390/pharmaceutics11080367

[95] McKinnon KM. Flow cytometry: An overview. Curr Protoc Immunol, 2018;120. https://doi.org/10.1002/cpim.40

[96] Sklar L, Carter M, Edwards B. Flow cytometry for drug discovery, receptor pharmacology and high-throughput screening. Curr Opin Pharmacol, 2007;7:527–534. https://doi.org/10.1016/j.coph.2007.06.006

[97] Dighe S, Jog S, Momin M, Sawarkar S, Omri A. Intranasal drug delivery by nanotechnology: Advances in and challenges for Alzheimer's disease management. Pharmaceutics, 2023;16:58. https://doi.org/10.3390/pharmaceutics16010058

[98] Serralheiro A, Alves G, Falcão A. Bioanalysis of small-molecule drugs in nasal and paranasal tissues and secretions: Current status and perspectives. Open Chem, 2012;10:686–702. https://doi.org/10.2478/s11532-012-0021-6

[99] Van Woensel M, Wauthoz N, Rosière R, Amighi K, Mathieu V, Lefranc F, Van Gool S, De Vleeschouwer S. Formulations for intranasal delivery of pharmacological agents to combat brain disease: A new opportunity to Tackle GBM? Cancers (Basel), 2013;5:1020–1048. https://doi.org/10.3390/cancers5031020

[100] Uttreja P, Karnik I, Adel Ali Youssef A, Narala N, Elkanayati RM, Baisa S, Alshammari ND, Banda S, Vemula SK, Repka MA. Self-Emulsifying Drug Delivery Systems (SEDDS): Transition from Liquid to Solid – A comprehensive review of formulation, characterization, applications, and future trends. Pharmaceutics, 2025;17:63. https://doi.org/10.3390/pharmaceutics17010063

[101] Alharbi GG, Abdulhamid MA. Optimization of water/oil emulsion preparation: Impact of time, speed, and homogenizer type on droplet size and dehydration efficiency. Chemosphere, 2023;335:139136. https://doi.org/10.1016/j.chemosphere.2023.139136

[102] Salum P, Ulubaş Ç, Güven O, Cam M, Aydemir LY, Erbay Z. The impact of homogenization techniques and conditions on water-in-oil emulsions for casein hydrolysate–loaded double emulsions: A comparative study. Food Sci Nutr, 2024;12:9585–9599. https://doi.org/10.1002/fsn3.4525

[103] Juttulapa M, Piriyaprasarth S, Takeuchi H, Sriamornsak P. Effect of high-pressure homogenization on stability of emulsions containing zein and pectin. Asian J Pharm Sci, 2017;12:21–27. https://doi.org/10.1016/j.ajps.2016.09.004

[104] de O. Bianchi JR, de la Torre LG, Costa ALR. Droplet-based microfluidics as a platform to design food-grade delivery systems based on the entrapped compound type. Foods, 2023;12:3385. https://doi.org/10.3390/foods12183385

[105] Schultz S, Wagner G, Urban K, Ulrich J. High-pressure homogenization as a process for emulsion formation. Chem Eng Technol, 2004;27:361–368. https://doi.org/10.1002/ceat.200406111

[106] Juttulapa M, Piriyaprasarth S, Takeuchi H, Sriamornsak P. Effect of high-pressure homogenization on stability of emulsions containing zein and pectin. Asian J Pharm Sci, 2017;12:21–27. https://doi.org/10.1016/j.ajps.2016.09.004

[107] Yu LJ, Koh KS, Tarawneh MA, Tan MC, Guo Y, Wang J, Ren Y. Microfluidic systems and ultrasonics for emulsion-based biopolymers: A comprehensive review of techniques, challenges, and future directions. Ultrason Sonochem, 2025;114:107217. https://doi.org/10.1016/j.ultsonch.2024.107217

[108] Chen H, Chang X, Du D, Li J, Xu H, Yang X. Microemulsion-based hydrogel formulation of ibuprofen for topical delivery. Int J Pharm, 2006;315:52–58. https://doi.org/10.1016/j.ijpharm.2006.02.015

[109] Cimino C, Bonaccorso A, Tomasello B, Alberghina GA, Musumeci T, Puglia C, Pignatello R, Marrazzo A, Carbone C. W/O/W microemulsions for nasal delivery of hydrophilic compounds: A preliminary study. J Pharm Sci, 2024;113:1636–1644. https://doi.org/10.1016/j.xphs.2024.01.013

[110] Crowe TP, Hsu WH. Evaluation of recent intranasal drug delivery systems to the central nervous system. Pharmaceutics, 2022;14:629. https://doi.org/10.3390/pharmaceutics14030629

Mitali Singh*, Mukesh Kumar Singh*, and Arun Kumar Mishra

Chapter 4
Micro/nanoemulsions redefining cosmeceutical formulation for optimal skin health

Abstract: The evolving field of cosmeceuticals integrates skincare with pharmaceutical precision, addressing the growing demand for effective and safe solutions to enhance skin health. This chapter explores the transformative role of micro/nanoemulsions in modern cosmeceutical formulations. These advanced delivery systems exhibit unique properties, including enhanced stability, superior skin penetration, and controlled release of active ingredients, offering distinct advantages over conventional formulations. The formulation process involves a strategic combination of oil and aqueous phases, emulsifiers, and bioactive components, with plant-based extracts playing a pivotal role. Extracts like green tea, aloe vera, curcumin, and grape seed exhibit antioxidant, anti-inflammatory, and antiaging properties, while others such as licorice root and chamomile provide skin brightening and calming effects. The mechanisms of action emphasize the interaction of emulsions with the skin barrier, leveraging nanoscale droplet sizes for deeper penetration and synergistic effects with bioactive. Applications of micro/nanoemulsions span antiaging, hydration, acne treatment, skin brightening, and sunscreen formulations, showcasing their versatility. Despite their potential, challenges such as stability issues with natural extracts, phase separation, and droplet size growth require innovative stabilization techniques. Regulatory guidelines and safety evaluations ensure the development of products which is safe and effective, while recent patents highlight advancements in green and sustainable cosmeceuticals. This chapter underscores the transformative impact of micro/nanoemulsions in addressing consumer needs, advancing cosmeceutical science, and shaping the future of skincare.

Keywords: Cosmeceuticals, microemulsions, nanoemulsions, plant-based extracts, skin health, antiaging, stability

*Corresponding authors: Mitali Singh,** Faculty of Pharmacy, IFTM University, Moradabad, Uttar Pradesh, India
*Corresponding authors: Mukesh Kumar Singh,** Faculty of Pharmacy, IFTM University, Moradabad, Uttar Pradesh, India
Arun Kumar Mishra, Faculty of Pharmacy, IFTM University, Moradabad, Uttar Pradesh, India

https://doi.org/10.1515/9783111593654-005

4.1 Introduction

Cosmeceuticals denote a fusion of cosmetics and pharmaceuticals, designed to boost both the appearance and skin health. These formulations typically include active ingredients such as vitamins, peptides, and botanical extracts that provide therapeutic benefits beyond basic skincare. The formulation process involves selecting appropriate emollients, preservatives, and rheological additives to ensure product stability and efficacy. The choice of these ingredients is crucial, as they must not only deliver the active compounds effectively but also maintain the product's sensory attributes, such as texture and fragrance. The development of cosmeceutical products requires a deep understanding of skin biology and the mechanisms by which active ingredients interact with the skin [1]. For instance, antioxidants like vitamins C and E are commonly used to combat oxidative stress and improve skin radiance. Additionally, peptides are incorporated to excite production of collagen, thereby reducing the appearance of wrinkles and fine lines. The formulation must also consider the skin's barrier function, ensuring that the active ingredients can penetrate effectively without causing irritation or damage. Advancements in formulation technologies have led to the creation of more sophisticated cosmeceuticals. Techniques such as encapsulation and controlled-release systems are designed to enhance the stability and bioavailability of active ingredients. These innovations not only improve the efficacy of the products but also extend their shelf life, making them more appealing to consumers seeking long-term skincare solutions [2].

Skin health is a critical aspect of modern cosmeceuticals, as these products aim to provide both aesthetic and therapeutic benefits. Healthy skin serves as a barrier against environmental factors, such as UV radiation and pollutants, which can cause premature aging and other skin conditions. Cosmeceuticals are formulated to support and enhance this barrier function, ensuring that the skin remains resilient and protected. Ingredients like ceramides and hyaluronic acid are commonly used to reinforce the skin's natural moisture barrier, preventing dehydration and maintaining skin elasticity [3]. The growing awareness of skin health has driven consumer demand for products that offer more than just superficial improvements. Modern cosmeceuticals are designed to address underlying skin issues, such as inflammation, hyperpigmentation, and acne. For example, niacinamide is frequently included in formulations for its anti-inflammatory and brightening properties, making it effective in treating conditions like rosacea and melasma. This holistic approach to skincare aligns with the contemporary understanding that maintaining skin health is essential for achieving long-lasting beauty.

Moreover, the emphasis on skin health has led to increased research and development in the cosmeceutical industry. Clinical research is conducted to validate the safety and efficacy of new ingredients and formulations. This scientific rigor ensures that consumers receive products that are not only effective but also safe for prolonged use [4]. As a result, cosmeceuticals have become a trusted category in the skincare market, offering solutions that cater to both cosmetic and health-related needs.

Micro- and nanoemulsions are advanced delivery systems that have revolutionized the field of cosmeceuticals. These emulsions are characterized by their small droplet sizes, typically in the range of 10–200 nm [5], which allow for improved stability and bioavailability of active ingredients. The small size of the droplets enhances the penetration of active compounds into the deeper layers of the skin, ensuring more effective delivery and sustained release. This makes micro- and nanoemulsions particularly suitable for incorporating hydrophobic ingredients that are otherwise difficult to formulate. The preparation of micro- and nanoemulsions involves techniques such as high-pressure homogenization and ultrasonication, which create stable dispersions of oil and water phases. These methods ensure that the emulsions remain stable over time, preventing issues like phase separation and degradation of active ingredients. The use of surfactants and co-surfactants is also crucial in stabilizing these emulsions, as they reduce the interfacial tension between the oil and water phases. This results in a product that is not only effective but also aesthetically pleasing, with a smooth and nongreasy texture [6]. The transformative role of micro- and nanoemulsions in cosmeceuticals extends beyond their enhanced delivery capabilities. These emulsions also offer improved sensory attributes, such as a lightweight feel and rapid absorption, which are highly valued by consumers. Additionally, the ability to incorporate a wide range of active ingredients, from vitamins to peptides, makes these emulsions versatile tools in the formulation of targeted skincare products. As a result, micro- and nanoemulsions have become a cornerstone in the development of innovative cosmeceuticals, driving the industry toward more effective and consumer-friendly solutions [7].

4.2 Micro/nanoemulsions: a game changer in cosmeceuticals

Micro- and nanoemulsions have emerged as revolutionary delivery systems in the field of cosmeceuticals, offering significant advantages over traditional formulations. These emulsions are defined as dispersions of two immiscible liquids (typically oil and water) stabilized by surfactants, with droplet sizes ranging from 10 to 200 nm. The small droplet size of these emulsions enhances their stability and bioavailability, making them ideal for delivering active ingredients deep into the skin [8]. As shown in Figure 4.1. Unlike conventional emulsions, which are prone to phase separation and instability, micro- and nanoemulsions remain stable over extended periods, ensuring consistent performance and efficacy [8, 9]. One of the unique properties of micro- and nanoemulsions is their ability to form clear, thermodynamically stable systems. This stability is achieved through the use of surfactants and co-surfactants, which reduce the interfacial tension between the oil and water phases. The result is a homogeneous mixture that

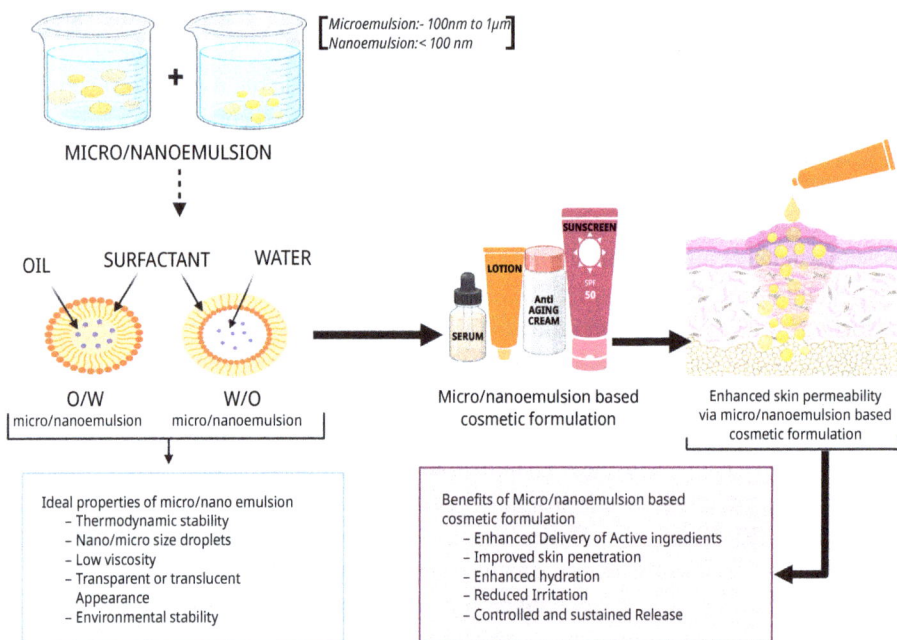

Figure 4.1: Role of micro/nanoemulsions in cosmeceutical formulations.

can effectively encapsulate and protect active ingredients from degradation [10]. Additionally, the small droplet size of these emulsions allows for a larger surface area, enhancing the interaction between the active ingredients and the skin. This property is particularly beneficial for incorporating hydrophobic compounds that are otherwise difficult to formulate [11].

4.2.1 Enhanced stability

Micro- and nanoemulsions offer enhanced stability compared to conventional formulations. The small droplet size of these emulsions reduces the likelihood of phase separation and sedimentation, ensuring that the active ingredients remain evenly distributed throughout the product. This stability is further enhanced by the use of surfactants and co-surfactants, which create a stable interfacial film around the droplets [12]. As a result, micro- and nanoemulsions can maintain their stability over extended periods, even under varying storage conditions. The enhanced stability of micro- and nanoemulsions also translates to improved shelf life and efficacy of cosmeceutical products [13]. Traditional emulsions often suffer from instability issues, such as creaming and coalescence, which can compromise the effectiveness of the active ingredients. In contrast, micro- and nanoemulsions remain stable and homogeneous, ensuring that the active ingredients are consistently delivered to the skin. This stability is particularly important for products containing sensitive or volatile compounds, as it helps to preserve their potency and effectiveness [14].

4.2.2 Improved skin penetration

One of the key advantages of micro- and nanoemulsions is their ability to enhance skin penetration of active ingredients. The small droplet size of these emulsions allows for better absorption and deeper penetration into the skin layers. This is achieved through the increased surface area of the droplets, which facilitates more efficient interaction with the skin's surface. As a result, active ingredients can reach the deeper layers of the skin, where they can exert their therapeutic effects more effectively [15]. Improved skin penetration is particularly beneficial for delivering active ingredients that target specific skin concerns, such as antiaging, pigmentation, and acne. For example, nanoemulsions can enhance the delivery of retinoids and peptides, which are commonly used in antiaging formulations. By improving the penetration of these ingredients, micro- and nanoemulsions can enhance their efficacy and provide more noticeable results. This property makes them a valuable tool in the development of advanced cosmeceutical products that offer targeted and effective skincare solutions.

4.2.3 Controlled and sustained release of active ingredients

Micro- and nanoemulsions also offer the advantage of controlled and sustained release of active ingredients. The small droplet size and stable interfacial film of these emulsions allow for the release of active components over time. This controlled release-type mechanism ensures that the active ingredients are delivered to the skin in a sustained manner, providing prolonged effects. This is mostly beneficial for ingredients that require prolonged exposure to achieve their desired effects, such as antioxidants and anti-inflammatory agents [16]. The controlled release properties of micro- and nanoemulsions also help minimize potential side effects and irritation associated with high concentrations of active ingredients. By delivering the active compounds in a controlled manner, these emulsions can reduce the risk of skin irritation and improve the overall tolerability of the product. This makes them suitable for use in formulations designed for sensitive or reactive skin types. Additionally, the sustained release of active ingredients can enhance the overall efficacy of the product, providing continuous benefits throughout the day [17].

4.3 Formulation strategies for cosmeceuticals

4.3.1 Components of micro/nanoemulsions: oil phase, aqueous phase, and emulsifiers

Micro- and nanoemulsions are advanced delivery systems used in cosmeceuticals to increase both the stability and bioavailability of active substance. These emulsions consist of three main components: the oil phase, the aqueous phase, and emulsifiers. The oil phase typically includes lipophilic substances such as essential oils, fatty acids, and triglycerides, which help in solubilizing lipophilic active ingredients and providing a moisturizing effect on the skin. The aqueous phase, on the other hand, contains hydrophilic substances like water, glycerine, and hydrosols, which aid in hydrating the skin and dissolving water-soluble actives [18, 19]. Emulsifiers are crucial in stabilizing these emulsions by lowering the surface tension between the aqueous and oil phases, thereby preventing phase separation. Common emulsifiers used in micro/nanoemulsions include lecithin, polysorbates, and PEG derivatives. The choice of emulsifier depends on the desired droplet size, stability, and skin compatibility. For instance, lecithin is a natural emulsifier that is well-tolerated by the skin and can form stable nano-emulsions with small droplet sizes, enhancing the penetration of active ingredients. The formulation of micro/nanoemulsions requires careful consideration of the ratio of oil to aqueous phase and the type and concentration of emulsifiers to achieve optimal stability and efficacy [20].

4.3.2 Selection of bioactive ingredients for skin health

The selection of bioactive ingredients in cosmeceuticals is pivotal for addressing various skin concerns and enhancing skin health. Bioactive ingredients are chosen based on their efficacy, safety, and ability to target specific skin issues such as aging, hyperpigmentation, and inflammation. Common bioactive ingredients include antioxidants, peptides, vitamins, and botanical extracts. Antioxidants like vitamin C, vitamin E, and resveratrol are widely used for their ability to neutralize free radicals, thereby preventing oxidative stress and premature aging. Peptides, such as palmitoyl pentapeptide-4, stimulate collagen production and improve skin elasticity, making them effective in antiaging formulations [21].

Botanical extracts, such as green tea, aloe vera, and chamomile, are valued for their soothing, anti-inflammatory, and antimicrobial properties. These extracts not only provide therapeutic benefits but also enhance the overall sensory experience of the product. The selection process involves evaluating the bioavailability, stability, and compatibility of these ingredients with other formulation components [22]. Additionally, the concentration of bioactive ingredients must be optimized to ensure efficacy without causing irritation or adverse reactions. Formulators often rely on scientific research and clinical studies to validate the effectiveness of bioactive ingredients and to develop formulations that deliver tangible skin benefits.

4.3.3 Benefits of plant extracts for skin health

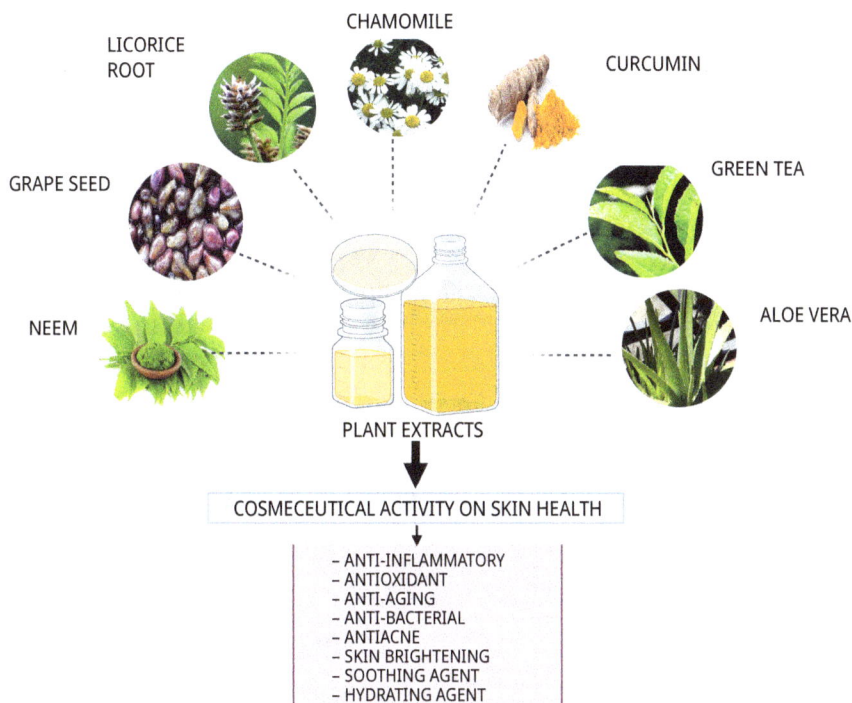

Figure 4.2: Common plants extracts, with their cosmeceutical roles.

Plant-based extracts are integral to the formulation of micro- and nanoemulsions due to their rich composition of bioactive compounds, including antioxidants, anti-inflammatory agents, and antimicrobial substances. These extracts offer numerous benefits for skin health, such as protecting against oxidative stress, reducing inflammation, and preventing microbial infections [23]. As shown in Figure 4.2. The use of micro- and nanoemulsions enhances the bioavailability and stability of these extracts, allowing for better penetration and sustained release of active ingredients into the skin. This advanced delivery system ensures that the beneficial properties of plant extracts are maximized, leading to improved skin hydration, elasticity, and overall health. Moreover, plant extracts in micro- and nanoemulsions can target specific skin concerns more effectively [24]. For instance, antioxidants from plant extracts can neutralize free radicals, thereby preventing premature aging and promoting skin vitality. Anti-inflammatory compounds help soothe irritated skin and reduce redness, while antimicrobial agents can combat acne-causing bacteria and other pathogens [25]. By incorporating plant extracts into micro- and nanoemulsions, skincare formulations can achieve a higher level of efficacy and provide a holistic approach to skin health [26].

4.3.3.1 Green tea extract

Green tea extract is renowned for its high concentration of polyphenols, particularly catechins, which exhibit potent antioxidant properties. These antioxidants help neutralize free radicals, thereby protecting the skin from oxidative stress and premature aging [27]. Additionally, green tea extract has anti-inflammatory properties that can soothe irritated skin and reduce redness, making it beneficial for conditions such as acne and rosacea. The incorporation of green tea extract in micro- and nanoemulsions enhances its stability and bioavailability, ensuring that the skin receives a sustained release of these beneficial compounds. Moreover, green tea extract has been shown to improve skin elasticity and moisture retention. Its ability to inhibit the activity of collagenase, an enzyme that breaks down collagen, helps maintain the skin's structural integrity and firmness [28]. The antimicrobial properties of green tea extract also make it effective in combating acne-causing bacteria, promoting clearer and healthier skin. When formulated in micro- and nanoemulsions, green tea extract can penetrate deeper into the skin layers, providing long-lasting benefits and improving overall skin health [29].

4.3.3.2 Aloe vera

Aloe vera is widely recognized for its soothing and moisturizing properties, making it a staple ingredient in many skincare products. The gel extracted from aloe vera leaves contains vitamins, minerals, enzymes, and amino acids that contribute to its healing and hydrating effects. Aloe vera is particularly effective in treating dry, irritated, and sunburned skin, providing immediate relief and promoting faster healing [30]. The use of aloe vera in micro- and nanoemulsions enhances its absorption and ensures that its active components are delivered efficiently to the skin. In addition to its hydrating properties, aloe vera has anti-inflammatory and antimicrobial effects that can help reduce acne and other skin infections. It also stimulates the production of collagen and elastin, which are essential for maintaining skin elasticity and reducing the appearance of fine lines and wrinkles. The incorporation of aloe vera in advanced delivery systems like micro- and nanoemulsions allows for better penetration and prolonged release of its beneficial compounds, resulting in improved skin texture and overall health [31].

4.3.3.3 Curcumin

Curcumin, the active compound in turmeric, is known for its powerful anti-inflammatory and antioxidant properties. It helps reduce inflammation and oxidative damage, which are common contributors to various skin conditions such as acne, eczema, and psoriasis. Curcumin also has antimicrobial properties that can help prevent and treat bacterial infections on the skin [32]. When formulated in micro- and nanoemulsions, curcumin's bioavailability is significantly enhanced, allowing for better absorption and effectiveness. Furthermore, curcumin has been shown to promote wound healing and reduce the appearance of scars. Its ability to inhibit the production of melanin makes it effective in treating hyperpigmentation and achieving a more even skin tone. The use of curcumin in micro- and nanoemulsions ensures that its active components are delivered directly to the target areas, providing sustained therapeutic effects and improving overall skin health [33].

4.3.3.4 Grape seed extract

Grape seed extract is rich in polyphenols, particularly proanthocyanidins, which are known for their strong antioxidant properties. These antioxidants help protect the skin from free radical damage and reduce the signs of aging, such as fine lines and wrinkles. Grape seed extract also has anti-inflammatory properties that can soothe irritated skin and reduce redness. The incorporation of grape seed extract in micro- and nanoemulsions enhances its stability and bioavailability, ensuring that the skin receives a continuous supply of these beneficial compounds [34]. In addition to its antioxidant and anti-inflammatory effects, grape seed extract promotes collagen synthesis, which is essential for maintaining skin elasticity and firmness. It also helps improve skin hydration and texture, making it a valuable ingredient in antiaging and moisturizing skincare products. The use of grape seed extract in advanced delivery systems like micro- and nanoemulsions allows for better penetration and prolonged release of its active components, resulting in healthier and more youthful-looking skin [35].

4.3.3.5 Licorice root extract

Licorice root extract is well-known for its skin-brightening properties, making it a popular ingredient in products aimed at reducing hyperpigmentation and achieving an even skin tone. The active compound in licorice root extract, glabridin, inhibits the activity of tyrosinase, an enzyme responsible for melanin production. This helps reduce dark spots and discoloration, resulting in a more radiant complexion [36]. The use of licorice root extract in micro- and nanoemulsions enhances its absorption and ensures that its active components are delivered effectively to the skin. Besides its skin-brightening effects, licorice root extract has anti-inflammatory and antioxidant properties that can help soothe irritated skin and protect it from environmental damage. It is also effective in reducing redness and swelling associated with conditions such as rosacea and eczema. The incorporation of licorice root extract in advanced delivery systems like micro- and nanoemulsions allows for better penetration and sustained release of its beneficial compounds, providing long-lasting benefits and improving overall skin health [37].

4.3.3.6 Chamomile extract

Chamomile extract is popular for its calming and anti-inflammatory qualities, making it an excellent substance for irritated and sensitive skin. The active compounds in chamomile, such as bisabolol and chamazulene, help reduce redness, swelling, and discomfort associated with various skin conditions [38]. Chamomile extract also has antioxidant properties that protect the skin from free radical damage and promote a healthy complexion. The use of chamomile extract in micro- and nanoemulsions enhances its stability and bioavailability, ensuring that its active components are delivered efficiently to the skin[12]. In addition to its soothing effects, chamomile extract has antimicrobial properties that can help prevent and treat skin infections [39]. It also promotes wound healing and reduces the appearance of scars, making it a valuable ingredient in skincare products aimed at improving skin texture and overall health. The incorporation of chamomile extract in advanced delivery systems like micro- and nanoemulsions allows for better penetration and prolonged release of its beneficial compounds, resulting in healthier and more resilient skin [40].

4.3.3.7 Neem extract

Neem extract is known for its potent antibacterial, antifungal, and anti-inflammatory properties, making it a valuable ingredient in treating various skin infections and conditions. The active compounds in neem, such as nimbin and azadirachtin, assist in combating acne-causing bacteria and decrease inflammation, promoting clearer and healthier skin. Neem extract also has antioxidant properties that protect the skin from environmental damage and premature aging [41]. The use of neem extract in micro- and nanoemulsions enhances its absorption and ensures that its active components are delivered effectively to the skin. Moreover, neem extract has been shown to improve skin elasticity and hydration, making it beneficial for maintaining a youthful and supple complexion [42]. It also helps soothe irritated skin and reduce redness, making it suitable for sensitive skin types. The incorporation of neem extract in advanced delivery systems like micro- and nanoemulsions allows for better penetration and sustained release of its beneficial compounds, providing long-lasting therapeutic effects and improving overall skin health [43].

4.4 Mechanism of action in skin delivery

4.4.1 Interaction of emulsion with the skin barrier

Wiechers et al. describe the interaction of emulsions with the skin barrier as a complex process influenced by the composition and structure of the emulsion. The stratum corneum, the outermost layer of the skin, acts as a primary barrier to the penetration of substances. Emulsions, particularly those with lipophilic components, can boost the penetration of active substance by interacting with the lipid bilayers of the stratum corneum. This interaction can disrupt the lipid organization, increasing the permeability of the skin and allowing for deeper penetration of the active compounds. The choice of emollients and emulsifiers in the formulation plays a crucial role in determining the extent and rate of skin delivery. Additionally, Wiechers et al. highlight that the partitioning of active ingredients between the emulsion and the skin is a critical factor in skin delivery. The formulation must be designed to optimize the partition coefficient, ensuring that the active ingredients are effectively transferred from the emulsion to the skin. This can be achieved by selecting appropriate emollients that enhance the solubility and diffusion of the active compounds within the skin layers. By carefully formulating emulsions, it is possible to improve the clinical efficacy of topically applied products without increasing the concentration of active ingredients [44].

4.4.2 Role of droplet size and composition in facilitating deeper penetration

Menichetti et al. describe the role of droplet size and composition in facilitating deeper penetration of active ingredients through the skin. Nano-emulsions, with droplet sizes typically ranging from 20 to 200 nm, offer a larger surface area and reduced barrier properties, which enhance their ability to penetrate the skin. The small size of the droplets allows for better interaction with the skin lipids, increasing the permeability of the stratum corneum and facilitating the delivery of active compounds to deeper skin layers. The composition of the droplets, including the type of surfactants and oils used, also plays a significant role in determining the penetration efficiency. Furthermore, Menichetti et al. emphasize that the stability of nano-emulsions is crucial for their effectiveness in skin delivery. Stable nano-emulsions ensure that the active ingredients remain uniformly distributed within the formulation, preventing phase separation and degradation. This stability, combined with the small droplet size, allows for a controlled and sustained release of active compounds, enhancing their therapeutic effects. By optimizing the droplet size and composition, nanoemulsions can significantly improve the bioavailability and efficacy of topically applied products [45].

4.4.3 Synergistic effects between micro/nanoemulsions and bioactive extracts

Nanda et al. [46] describe the synergistic effects between micro/nanoemulsions and bioactive extracts as a promising approach to increase the efficacy of skincare products. The combination of bioactive extracts with micro/nanoemulsions can lead to improved stability, bioavailability, and penetration of the active compounds. The small droplet size and high surface area of nanoemulsions facilitate better interaction with the skin, allowing for a more efficient delivery of bioactive extracts to the target sites. This synergistic interaction can enhance the overall effectiveness of the formulation, providing multiple benefits for skin health. Moreover, Nanda et al. highlight that the encapsulation of bioactive extracts in micro/nanoemulsions can protect them from degradation and enhance their stability. This protection ensures that the active compounds retain their potency and efficacy throughout the shelf life of the product. Additionally, the controlled release properties of micro/nanoemulsions allow for a sustained delivery of bioactive extracts, providing long-lasting therapeutic effects. By leveraging the synergistic effects between micro/nanoemulsions and bioactive extracts, it is possible to develop advanced skincare formulations with enhanced performance and benefits [46].

4.5 Applications of micro/nanoemulsions in cosmeceuticals

4.5.1 Antiaging formulations: role of antioxidants

In their research, Choi et al. emphasize the crucial role of antioxidants in antiaging formulations for protecting the skin from oxidative stress and free radical damage. Antioxidants such as vitamins C and E, green tea polyphenols, and resveratrol are commonly incorporated into micro- and nanoemulsions to enhance their stability and bioavailability. These compounds help neutralize free radicals, which are unstable molecules that can cause cellular damage and accelerate the aging process. By scavenging these free radicals, antioxidants prevent the breakdown of collagen and elastin, thereby maintaining skin firmness and elasticity. The use of micro- and nanoemulsions ensures that these antioxidants are delivered effectively to the deeper layers of the skin, providing long-lasting protection and promoting a youthful appearance [47].

4.5.2 Skin hydration products: hyaluronic acid and ceramide-based emulsion

Patel et al. discuss the benefits of hyaluronic acid and ceramide-based emulsions in skin hydration products. Hyaluronic acid is a potent humectant that can hold up to 1,000 times its weight in water; therefore it is very efficient in maintaining skin moisture levels. Ceramides, on the other hand, are essential lipids that help restore the skin barrier and prevent transepidermal water loss. When formulated in micro- and nanoemulsions, these ingredients can penetrate deeper into the skin, providing intense hydration and improving skin barrier function. This combination not only enhances skin hydration but also improves skin texture and resilience, making it the best choice for dehydrated and dry skin [48, 49].

4.5.3 Acne treatment: neem and tea tree oil microemulsion

Howard et al. highlight the effectiveness of neem and tea tree oil in microemulsions for acne treatment. Neem oil has potent anti-inflammatory and antibacterial qualities that help reduce acne and soothe inflamed skin. Tea tree oil, known for its antimicrobial and anti-inflammatory effects, further enhances the formulation's ability to combat acne. The incorporation of these oils into microemulsions ensures better stability and penetration, allowing the active compounds to spread the deeper layers of the skin where they can show their therapeutic effects. This approach not only helps in reducing acne lesions but also prevents future breakouts, promoting clearer and healthier skin [50].

4.5.4 Skin brightening: licorice root, kojic acid, and arbutin-based products

Sarkar et al. describe the synergistic effects of licorice root, kojic acid, and arbutin in skin brightening formulations. Licorice root extract contains glabridin, which inhibits tyrosinase activity and reduces melanin production, helping to lighten dark spots and even out skin tone. Kojic acid, derived from fungi, also inhibits tyrosinase and has been shown to effectively reduce hyperpigmentation. Arbutin, a natural derivative of hydroquinone, further enhances the skin-brightening effects by preventing melanin formation. When these ingredients are formulated in micro- and nanoemulsions, their stability and bioavailability are significantly improved, allowing for more effective and sustained skin brightening results. This combination provides a comprehensive approach to achieving a radiant and even complexion [51].

4.5.5 Sunscreen formulation: incorporation of green tea polyphenols and grape seed extract

Hegde et al. describe the assimilation of grape seed extract and green tea polyphenols in sunscreen formulations as a novel approach to enhancing UV protection. Grape seed extract is rich in proanthocyanidins, which have strong antioxidant properties that inhibits the skin from UV-induced oxidative damage. Green tea polyphenols, particularly epigallocatechin gallate (EGCG), also provide potent antioxidant effect and anti-inflammatory effect, further enhancing the skin's defense against UV radiation [52]. The use of micro- and nanoemulsions in these formulations ensures that the active compounds are evenly distributed and effectively absorbed by the skin. This not only improves the sunscreen's efficacy but also provides additional skin benefits, such as reducing inflammation and preventing premature aging.

4.6 Challenges in cosmeceuticals

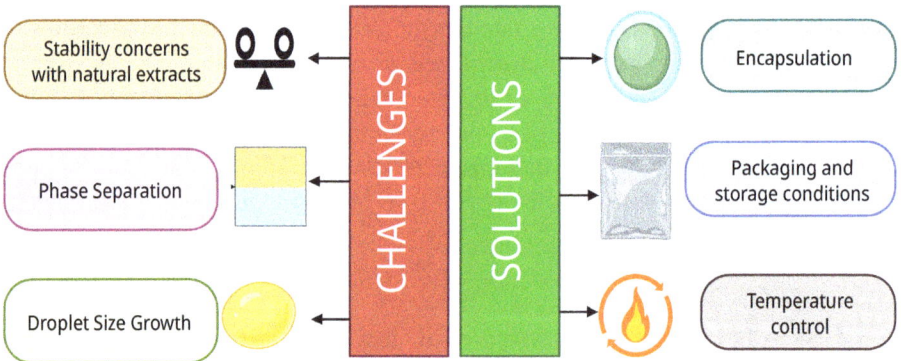

Figure 4.3: Challenges and solutions in cosmeceuticals.

4.6.1 Stability concerns with natural extracts

Natural extracts in cosmeceuticals often face stability issues such as oxidation and degradation, which can compromise their efficacy and shelf life. As illustrated in Figure 4.3. Scheuplein RJ. et al., describe those botanical extracts rich in bioactive compounds that are particularly susceptible to oxidative degradation when exposed to heat, air and light.

This can lead to a loss of potency and effectiveness of the active ingredients [1]. For instance, polyphenols and flavonoids, common in many plant extracts, are prone to oxidation, resulting in reduced antioxidant activity. To address these challenges, researchers have developed various stabilization techniques, including the use of antioxidants, encapsulation, and controlled release systems [53].

4.6.2 Phase separation and droplet size growth

Phase separation and growth in droplet size are significant challenges in the formulation of emulsions. Menichetti et al. explain that emulsions, particularly those containing natural extracts, can experience instability due to the coalescence and Ostwald ripening of droplets. This instability can lead to phase separation, where the oil and water phases separate, resulting in an uneven distribution of active ingredients. The growth of droplet size over time can also affect the texture and appearance of the product, making it less appealing to consumers. To mitigate these issues, formulators often use stabilizers, emulsifiers, and surfactants to balance the integrity of the emulsion and inhibit droplet growth [54].

4.6.3 Techniques to enhance stability and shelf life

Various techniques have been developed to increase the stability and shelf life of cosmeceutical formulations. Nanda et al. describe the use of nanoencapsulation as an effective technique to shield natural extracts from degradation and improve their stability. Nanoencapsulation involves encapsulating active ingredients within nanocarriers, which can shield them from environmental factors e.g. light, heat, and oxygen. This technique not only enhances the stability of the active compounds but also improves their bioavailability and controlled release of active substance. Additionally, the usage of antioxidants and preservatives can further enhance the shelf life of formulations by preventing oxidative degradation and microbial contamination [55, 56].

4.7 Regulatory, safety, and patents

4.7.1 Regulatory guidelines for cosmeceutical formulations

Regulatory guidelines for cosmeceutical formulations are essential to ensure product safety and efficacy. The FDA provides comprehensive regulations for cosmetics, including guidelines on labelling, ingredient safety, and manufacturing practices[1]. In the European Union, the EU Cosmetics Regulation (EC) No. 1223/2009 sets stringent requirements for cosmetic products, including safety assessments, product information files, and notification procedures. These regulations aim to protect consumers by ensuring that cosmeceutical products are safe for use and free from harmful substances. Compliance with these guidelines is crucial for manufacturers to avoid legal issues and ensure market access [57].

4.7.2 Safety evaluation of micro/nanoemulsion-based products: allergenicity and irritation tests

Safety evaluation of micro/nanoemulsion-based products involves rigorous testing to assess their potential for allergenicity and irritation. Wani et al. highlight that the safety of nano-emulsions depends on factors such as composition, particle size, and method of administration. Allergenicity tests, such as the human repeat insult patch test (HRIPT), are organized to evaluate the potential of a product to cause allergic reactions. Irritation tests, including the Draize test, assess the likelihood of skin irritation. These tests are essential to ensure that micro/nanoemulsion-based products are safe for consumer use and do not cause adverse skin reactions [58].

4.8 Key patents in micro/nanoemulsion technology

4.8.1 Innovations in stabilizing emulsions with natural extracts

Recent innovations in stabilizing emulsions with natural extracts have led to the development of more effective and stable formulations. A notable patent by Baek et al. describes a nano-emulsion cosmetic composition that stabilizes high-content oils using a combination of emulsifiers and high-pressure homogenization. This method increased the stability and appearance of the emulsion, making it more suitable for various cosmetic applications. Such innovations are crucial for incorporating natural extracts into cosmeceutical products without compromising their stability and efficacy [59].

4.8.2 Patented delivery systems for enhanced skin absorption

Patented delivery systems for enhanced skin absorption have revolutionized the cosmeceutical industry. The Invisicare® delivery system, for example, is a patented polymer technology that enhances the delivery of active ingredients in topical products. This system forms a protective bond with the skin, allowing for prolonged release and improved absorption of active substance. Such delivery systems are formulated to overcome the skin barrier, ensuring that active ingredients penetrate deeper into the skin for maximum efficacy [60].

4.8.3 Recent trends in green and sustainable cosmeceutical patents

Current development in green and sustainable cosmeceutical patents reflect the industry's shift toward environmentally friendly and sustainable practices. According to a report by JWP Rzecznicy Patentowi, there has been a significant increase in patent applications for green cosmetics, which focus on natural and organic ingredients, biodegradable packaging, and cruelty-free formulations. Companies are investing in sustainable technologies to satisfy consumer demand for environmentally friendly products and reduce their environmental impact. This trend is driving innovation in the cosmeceutical industry, leading to the development of products that are both effective and sustainable [61].

4.9 Conclusion and future outlook

Micro/nanoemulsions have emerged as a transformative innovation in the field of cosmeceuticals, bridging the gap between skincare and pharmaceutical precision. Their ability to enhance stability, improve skin penetration, and provide controlled release of bioactive compounds underscores their potential to revolutionize modern skincare formulations. The inclusion of plant-based extracts, such as green tea, aloe vera, curcumin, and grape seed, has further amplified their efficacy, providing a large range of benefits including antioxidant, antiaging, anti-inflammatory, and skin brightening effects. These attributes make micro/nanoemulsions versatile candidates for applications in antiaging, hydration, acne treatment, and sun protection products. Despite these advantages, challenges such as phase separation, instability of natural extracts, and droplet size growth remain significant hurdles. Addressing these issues will require the development of advanced stabilization techniques and optimized formulation strategies. The future of micro/nanoemulsions in cosmeceuticals is promising, with an increasing focus on sustainability and green technologies. Innovations in eco-friendly emulsifiers, biodegradable materials, and sustainable manufacturing practices will shape the next generation of cosmeceutical products. Additionally, regulatory advancements and comprehensive safety evaluations will ensure consumer confidence and the successful market integration of these sophisticated systems.

References

[1] Scheuplein RJ. Analysis of permeability data for the case of parallel diffusion pathways. Biophys J, 1966;6:1–17.

[2] Scheuplein RJ, Blank IH. Permeability of the skin. Physiol Rev, 1971;51:702–747.

[3] Kasting GB, Barai ND, Wang TF, Nitsche JM. Mobility of water in human stratum corneum. J Pharm Sci, 2003;92:2326–2340.

[4] Illel B, Schaefer H, Wepierre J, Doucet O. Follicles play an important role in percutaneous absorption. J Pharm Sci, 1991;80:424–427.

[5] Singh M, Sharma V, Shrivastav A, Singh P, Verma N. Nanotechnology for novel drug delivery: A systematic review of classification, preparation, characterization, and applications of nanoparticles in drug delivery. Biosci Biotechnol Res Asia, 2023 Dec 31;20(4):1147–1165.

[6] Roberts MS, Mohammed Y, Pastore MN, Namjoshi S, Yousef S, Alinaghi A, et al. Topical and cutaneous delivery using nanosystems. J Contr Release, 2017;247:86–105.

[7] Kircik LH. Effect of skin barrier emulsion cream vs a conventional moisturizer on transepidermal water loss and corneometry in atopic dermatitis: A pilot study. J Drugs Dermatol, 2014 Dec 1;13 (12):1482–1484.

[8] Leite-Silva VR, Sanchez WY, Studier H, Liu DC, Mohammed YH, Holmes AM, et al. Human skin penetration and local effects of topical nano zinc oxide after occlusion and barrier impairment. Eur J Pharm Biopharm, 2016;104:140–147.

[9] Danielsson I, Lindman B. The definition of microemulsion. Colloid Surf, 1981;3:391–392.

[10] Muller RH, Radtke M, Wissing SA. Solid lipid nanoparticles (SLN) and nanostructured lipid carriers (NLC) in cosmetic and dermatological preparations. Adv Drug Deliv Rev, 2002;54(Suppl 1):S131–55.

[11] Boonme P, Junyaprasert VB, Suksawad N, Songkro S. Microemulsions and nanoemulsions: Novel vehicles for whitening cosmeceuticals. J Biomed Nanotechnol, 2009 Aug 1;5(4):373–383.

[12] Touitou E, Dayan N, Bergelson L, Godin B, Eliaz M. Ethosomes – Novel vesicular carriers for enhanced delivery: Characterization and skin penetration properties. J Contr Release, 2000;65:403–418.

[13] Uchegbu IF, Vyas SP. Non-ionic surfactant-based vesicles (niosomes) in drug delivery. Int J Pharm, 1997;172:33–70.

[14] Najafi-Taher R, Ghaemi B, Amani A. Delivery of adapalene using a novel topical gel based on tea tree oil nano-emulsion: Permeation, antibacterial and safety assessments. Eur J Pharm Sci, 2018 Jul 30;120:142–151.

[15] Nastiti CM, Ponto T, Abd E, Grice JE, Benson HA, Roberts MS. Topical nano and microemulsions for skin delivery. Pharmaceutics, 2017 Sep 21;9(4):37.

[16] Mura S, Manconi M, Fadda AM, Sala MC, Perricci J, Pini E, et al. Penetration enhancer-containing vesicles (PEVs) as carriers for cutaneous delivery of minoxidil: In vitro evaluation of drug permeation by infrared spectroscopy. Pharm Dev Technol, 2013;18:1339–1345.

[17] Mariyate J, Bera A. A critical review on selection of microemulsions or nanoemulsions for enhanced oil recovery. J Mol Liq, 2022 May 1;353:118791.

[18] Williams AC, Barry BW. Penetration enhancers. Adv Drug Deliv Rev, 2004;56:603–618. doi: 10.1016/j. addr.2003.10.025

[19] Cevc G, Blume G. Lipid vesicles penetrate into intact skin owing to the transdermal osmotic gradients and hydration force. Biochim Biophys Acta, 1992;1104:226–232. doi: 10.1016/0005-2736(92) 90154-E

[20] Kim S, Shi Y, Kim JY, Park K, Cheng JX. Overcoming the barriers in micellar drug delivery: Loading efficiency, in vivo stability, and micelle-cell interaction. Exp Opin Drug Deliv, 2010;7:49–62. doi: 10.1517/17425240903380446

[21] Borowska K, Wolowiec S, Glowniak K, Sieniawska E, Radej S. Transdermal delivery of 8-methoxypsoralene mediated by polyamidoamine dendrimer G2.5 and G3.5 – In vitro and in vivo study. Int J Pharm, 2012;436:764–770. doi: 10.1016/j.ijpharm.2012.07.067

[22] Hameed A, Fatima GR, Malik K, Muqadas A, Fazalur-Rehman M. Scope of nanotechnology in cosmetics: Dermatology and skin care products. J Med Chem Sci, 2019;2(1):9–16. Ashaolu TJ. Nanoemulsions for health, food, and cosmetics: a review. Environ Chem Lett, 2021 Aug;19 (4):3381–3395.

[23] Menaa F. Emulsions systems for skin care: From macro to nano-formulations. J Pharma Care Health Sys, 2014;1:e104.

[24] Tiwari U, Ganesan NG, Junnarkar J, Rangarajan V. Toward the formulation of bio-cosmetic nanoemulsions: From plant-derived to microbial-derived ingredients. J Dispers Sci Technol, 2020:1–18.

[25] Sonneville-Aubrun O, Yukuyama MN, Pizzino A. Application of nanoemulsions in cosmetics. Nanoemulsions: Elsevier, 2018:435–475.

[26] Sonneville-Aubrun O, Simonnet JT, L'Alloret F. Nanoemulsions: A new vehicle for skincare products. Adv Colloid Interface Sci, 2004;108–109:145–149.

[27] Yilmaz E, Borchert HH. Effect of lipid-containing, positively charged nanoemulsions on skin hydration, elasticity and erythema – An in vivo study. Int J Pharm, 2006;307(2):232–238.

[28] Bernardi DS, Pereira TA, Maciel NR, Bortoloto J, Viera GS, Oliveira GC, et al. Formation and stability of oil-in-water nanoemulsions containing rice bran oil: In vitro and in vivo assessments. J Nanobiotechnol, 2011;9:44.

[29] Trommer H, Neubert R. Overcoming the stratum corneum: The modulation of skin penetration. Skin Pharmacol Physiol, 2006;19(2):106–121.

[30] Aziz ZAA, Mohd-Nasir H, Ahmad A, Peng WL, Chuo SC, Khatoon A, et al. Role of nanotechnology for design and development of cosmeceutical: Application in makeup and skin care. Front Chem, 2019;7:739.

[31] Magrode N, Poomanee W, Kiattisin K, Ampasavate C. Microemulsions and nanoemulsions for topical delivery of tripeptide-3: From design of experiment to anti-sebum efficacy on facial skin. Pharmaceutics, 2024 Apr 19;16(4):554.

[32] Ashaolu TJ. Nanoemulsions for health, food, and cosmetics: A review. Environ Chem Lett, 2021:1–15.

[33] Sonneville-Aubrun O, Simonnet JT, L'alloret F. Nanoemulsions: A new vehicle for skincare products. Adv Colloid Interface Sci, 2004 May 20;108:145–149.

[34] Yousefpoor Y, Amani A, Divsalar A, Mousavi SE, Shakeri A, Sabzevari JT. Anti-rheumatic activity of topical nanoemulsion containing bee venom in rats. Eur J Pharm Biopharm, 2022;172:168–176.

[35] Abbasifard M, Yousefpoor Y, Amani A, Arababadi MK. Topical bee venom nano-emulsion ameliorates serum level of Endothelin-1 in collagen-induced rheumatoid arthritis model. Bionanoscience, 2021;11(3):810–815.

[36] Yousefpoor Y, Amani A, Divsalar A, Vafadar MR. Topical delivery of bee venom through the skin by a water-in-oil nanoemulsion. Nanomed J, 2022;9(2):131–137.

[37] Som I, Bhatia K, Yasir M. Status of surfactants as penetration enhancers in transdermal drug delivery. J Pharm Bioallied Sci, 2012;4(1):2.

[38] Hosmer J, Reed R, Bentley MVL, Nornoo A, Lopes LB. Microemulsions containing medium-chain glycerides as transdermal delivery systems for hydrophilic and hydrophobic drugs. AAPS Pharmscitech, 2009;10(2):589–596.

[39] Resende KX, Corrêa MA, Oliveira AGd, Scarpa MV. Effect of cosurfactant on the supramolecular structure and physicochemical properties of non-ionic biocompatible microemulsions. Ciências Farmacêuticas, 2008;44:35–42.

[40] Souto EB, Cano A, Martins-Gomes C, Coutinho TE, Zielińska A, Silva AM. Microemulsions and nanoemulsions in skin drug delivery. Bioengineering, 2022 Apr 5;9(4):158.

[41] Shaker DS, Ishak RA, Ghoneim A, Elhuoni MA. Nanoemulsion: A review on mechanisms for the transdermal delivery of hydrophobic and hydrophilic drugs. Sci Pharm, 2019;87(3):17.

[42] Grampurohit N, Ravikumar P, Mallya R. Microemulsions for topical use–a review. Ind J Pharm Edu Res, 2011;45(1):100–107.

[43] Menaa F. Emulsions systems for skin care: From macro to nano-formulations. J Pharma Care Health Sys, 2014;1:e104.

[44] Wiechers JW. The influence of emollients on skin penetration. Cosmet Toiletries, 2008;123(1).

[45] Menichetti A, Mordini D, Montalti M. Penetration of microplastics and nanoparticles through skin: Effects of size, shape, and surface chemistry. J Xenobiot, 2024 Dec 31;15(1):6.

[46] Singh N, Verma SM, Singh SK, Verma PRP. Consequences of lipidic nanoemulsions on membrane integrity and ultrastructural morphology of Staphylococcus aureus. Mater Res Express, 2014;1(2):025401.

[47] Choi HY, Lee YJ, Kim CM, Lee YM. Revolutionizing cosmetic ingredients: Harnessing the power of antioxidants, probiotics, plant extracts, and peptides in personal and skin care products. Cosmetics, 2024 Sep 12;11(5):157.

[48] Patel RI, Patel AM, Modi BD. Formulation and evaluation of herbal sunscreen. World J Biol Pharm Health Sci, 2023;13(2):029–40.

[49] Santos-Magalhaes N, Pontes A, Pereira V, Caetano M. Colloidal carriers for benzathine penicillin G: Nanoemulsions and nanocapsules. Int J Pharm, 2000;208(1–2):71–80.

[50] Howard K, et al. How to use neem oil for acne, according to dermatologists. Byrdie, 2022.

[51] Thakur K, Sharma G, Singh B, Jain A, Tyagi R, Chhibber S, et al. Cationic-bilayered nanoemulsion of fusidic acid: An investigation on eradication of methicillin-resistant Staphylococcus aureus 33591 infection in burn wound. Nanomed, 2018;13(8):825–847.

[52] Hegde AR, Kunder MU, Narayanaswamy M, Murugesan S, Furtado SC, Veerabhadraiah BB, Srinivasan B. Advancements in sunscreen formulations: Integrating polyphenolic nanocarriers and nanotechnology for enhanced UV protection. Environ Sci Pollut Res, 2024 May;28:1–22.

[53] Nicolaidou E, Katsambas AD. Pigmentation disorders: Hyperpigmentation and hypopigmentation. Clin Dermatol, 2014;32(1):66–72.

[54] Plensdorf S, Livieratos M, Dada N. Pigmentation disorders: Diagnosis and management. Am Fam Physician, 2017;96(12):797–804.

[55] Engin C, Cayir Y. Pigmentation disorders: A short review. Pigment Disord, 2015;2(189):2376–0427.1000189.

[56] Nautiyal A, Wairkar S. Management of hyperpigmentation: Current treatments and emerging therapies. Pigment Cell Melanoma Res, 2021;34(6):1000–1014.

[57] Hatem S, El Hoffy NM, Elezaby RS, Nasr M, Kamel AO, Elkheshen SA. Background and different treatment modalities for melasma: Conventional and nanotechnology-based approaches. J Drug Deliv Sci Technol, 2020:101984.

[58] Üstündağ Okur N, Çağlar EŞ, Pekcan AN, Okur ME, Ayla Ş. Preparation, optimization and in vivo anti-inflammatory evaluation of hydroquinone loaded microemulsion formulations for melasma treatment. 2019;23(4):662–670.

[59] Sun MC, Xu XL, Lou XF, Du YZ. Recent progress and future directions: The nano-drug delivery system for the treatment of vitiligo. Int J Nanomed, 2020;15:3267.

[60] Alam A, Mustafa G, Agrawal GP, Hashmi S, Khan RA, Alkhayl FF, Ullah Z, Ali MS, Elkirdasy AF, Khan S. A microemulsion-based gel of isotretinoin and erythromycin estolate for the management of acne. J Drug Deliv Sci Technol, 2022 May 1;71:103277.

[61] Patel HK, Barot BS, Parejiya PB, Shelat PK, Shukla A. Topical delivery of clobetasol propionate loaded microemulsion based gel for effective treatment of vitiligo: Ex vivo permeation and skin irritation studies. Colloids Surf B Biointerfaces, 2013;102:86–94.

Sanjay Kumar Gupta*, Aakash Gupta, Debarshi Kar Mahapatra,
Alka Sahu, Sumit Sahu, Priya Komre, and Yamini Sahu

Chapter 5
Accelerating recovery through advanced wound healing applications of micro/ nanoemulsions

Abstract: The restoration of skin integrity and functioning relies on wound healing, a multistep physiological process. It is still very difficult to effectively treat both acute and chronic wounds, even with healthcare innovations. The use of micro/nanoemulsions and other innovative treatment techniques has recently come to the forefront of wound care technology. Promising possibilities for expediting wound healing, these systems are defined by their unique physicochemical features, which allow increased drug solubility, stability, and targeted delivery. The vital function of micro/nanoemulsions in cutting-edge wound healing applications is examined in this chapter. Their anti-inflammatory, antibacterial, and tissue regeneration activities are highlighted, along with their formulation methodologies, important components (such as bioactive compounds, lipids, and surfactants), and modes of action. Chronic, burn, and post-surgical wounds are only a few of the wound types that are covered in detail. More precise wound care is now possible due to nanotechnology-enabled smart and sensitive emulsions. Even while these systems show a lot of promise, they also face problems including reliability, scalability, and regulatory roadblocks. Future directions are discussed in the chapter's last section, with topics such multifunctional emulsions, digital health technology integration, and personalized medicine being highlighted. When it comes to wound healing, micro/nanoemulsions are game changers because they connect fundamental research with clinical application, giving patients fresh hope for better results.

Acknowledgment: The authors acknowledge the help received from Rungta College of Pharmaceutical Sciences and Research, Bhilai.

*Corresponding author: Sanjay Kumar Gupta, Department of Pharmacology, Rungta College of Pharmaceutical Sciences and Research, Kohka, Durg, Bhilai 490024, Chhattisgarh, India,
e-mail: sanjay.gupta0311@gmail.com, ORCID ID: 0000-0003-1084-3530
Aakash Gupta, Priya Komre, Yamini Sahu, Department of Pharmacology, Rungta College of Pharmaceutical Sciences and Research, Durg, Bhilai 490024, Chhattisgarh, India
Debarshi Kar Mahapatra, Chitkara College of Pharmacy, Chitkara University, Rajpura 140401, Punjab, India
Alka Sahu, Sumit Sahu, Department of Pharmaceutics, Rungta College of Pharmaceutical Sciences and Research, Durg, Bhilai 490024, Chhattisgarh, India

https://doi.org/10.1515/9783111593654-006

Researchers, clinicians, and industry stakeholders may benefit from this in-depth review of the potential of micro/nanoemulsions in reshaping wound care practices.

Keywords: Wound healing, microemulsion, nanoemulsion, advanced wound care, nanotechnology

5.1 Introduction

The largest organ in the body is the skin, which consists of the epidermis, dermis, and dermal appendages [1]. A wound is defined as a disruption in the normal structure and function of the skin or underlying tissues, resulting from physical, chemical, thermal, or microbial injury. This breach in the body's protective barrier exposes internal tissues to the external environment, creating a risk of infection, fluid loss, and impaired functionality. Wounds are a common medical concern, affecting individuals across all age groups and demographics. Their management is a critical aspect of healthcare, as improper healing can lead to complications such as infections, chronic pain, and disability. Understanding the nature of wounds, their classification, underlying pathophysiology, and the biological processes involved in healing is essential for effective treatment and prevention of complications [2].

Wounds can be classified based on various criteria, including their cause, depth, duration, and healing trajectory. The two primary categories are acute wounds and chronic wounds:

- **Acute wounds**: These wounds result from sudden trauma, such as cuts, abrasions, lacerations, surgical incisions, or burns. They typically follow a predictable and timely healing process, progressing through the well-defined stages of hemostasis, inflammation, proliferation, and remodeling. Acute wounds generally heal within 8–12 weeks, depending on the severity and location of the injury [3].
- **Chronic wounds**: Chronic wounds fail to progress through the normal stages of healing and often remain stuck in the inflammatory phase for extended periods. They are frequently associated with underlying medical conditions such as diabetes, venous insufficiency, or arterial disease. Examples include diabetic foot ulcers (DFUs), venous leg ulcers, and pressure sores. Chronic wounds are characterized by prolonged healing times, high risk of infection, and significant impact on the patient's quality of life [4].

Wounds can also be classified based on their depth (superficial, partial-thickness, or full-thickness) and cause (traumatic, surgical, or pathological). Understanding these classifications helps healthcare providers tailor treatment strategies to meet the specific needs of the wound and the patient [5].

5.1.1 Pathophysiology of wound healing

Wound healing is a complex and dynamic process that involves the coordinated inter-action of cells, extracellular matrix components, and biochemical signaling pathways. The process is typically divided into four overlapping phases (Figure 5.1) [6]:

i.   **Hemostasis**: Immediately after injury, the body initiates hemostasis to stop bleed-ing. Platelets aggregate at the injury site, forming a clot that seals the wound and provides a temporary barrier against infection.
ii. **Inflammation**: The inflammatory phase is characterized by recruiting immune cells, such as neutrophils and macrophages, to the wound site. These cells clear debris, prevent infection, and release growth factors that stimulate the next phase of healing.
iii. **Proliferation**: During this phase, fibroblasts and endothelial cells proliferate to form new tissue. Collagen is deposited, and new blood vessels are formed (angio-genesis) to supply oxygen and nutrients to the healing tissue. Epithelial cells mi-grate across the wound bed to restore the skin barrier.
iv. **Remodeling**: The final phase involves the maturation and reorganization of the newly formed tissue. Collagen fibers are realigned to increase tensile strength, and excess cells are removed through apoptosis. This phase can last for months or even years, depending on the severity of the wound.

5.1.2 Significance of wound care in healthcare

Wound care is a critical component of healthcare, with significant implications for patient outcomes, quality of life, and healthcare costs. Chronic wounds, in particular, pose a substantial burden on healthcare systems worldwide, requiring prolonged treatment, frequent hospital visits, and specialized care. Effective wound manage-ment not only accelerates healing but also prevents complications, reduces hospital readmissions, and lowers overall healthcare costs. Advancements in wound care, such as the use of bioactive dressings, growth factor therapies, and micro/nanoemul-sions, have revolutionized the field, offering more effective and personalized treat-ment options. These innovations address the limitations of traditional wound care methods and improve outcomes for patients with complex or hard-to-heal wounds.

The primary objective of wound healing is to restore the integrity of damaged tis-sues and ensure their proper functioning. When the skin or underlying tissues are compromised due to trauma, surgery, or chronic conditions, the body initiates a cas-cade of biological processes to repair the injury. These processes include hemostasis to stop bleeding, inflammation to clear debris and pathogens, proliferation to regener-ate tissue, and remodeling to strengthen and mature the new tissue. Effective wound healing ensures that the structural barrier of the skin is restored, preventing further damage and enabling the affected area to regain its normal function. Without this process, even minor injuries could lead to severe complications, such as chronic non-healing wounds, loss of mobility, or permanent tissue damage.

5.1.3 Advancements in wound healing technologies

The field of wound healing has witnessed remarkable advancements in recent years, driven by innovations in biotechnology, materials science, and pharmacology. Techniques such as the use of stem cells, growth factors, and micro/nanoemulsions have revolutionized wound care by accelerating healing, reducing scarring, and addressing the underlying causes of chronic wounds. These technologies not only improve patient outcomes but also pave the way for personalized wound care strategies tailored to individual needs.

5.1.4 Challenges in traditional wound care

Traditional wound care has long been the cornerstone of managing acute and chronic wounds, relying on established methods such as gauze dressings, antiseptics, and basic wound cleaning techniques. While these approaches have been effective to some extent, they are increasingly being scrutinized for their limitations in addressing the complex and multifaceted nature of modern wound care needs. The challenges associated with traditional wound care are multifaceted, encompassing clinical, economic, and patient-centered concerns. These challenges highlight the need for innovative and advanced wound care solutions to improve outcomes and reduce the burden on healthcare systems.

5.1.4.1 Ineffectiveness in managing chronic wounds

One of the most significant challenges of traditional wound care is its limited efficacy in managing chronic wounds, such as DFUs, venous leg ulcers, and pressure sores. Chronic wounds are characterized by prolonged inflammation, impaired healing mechanisms, and a high risk of infection. Traditional methods, which often focus on passive wound coverage and basic infection control, fail to address the underlying biological and physiological factors that impede healing. For instance, gauze dressings, while inexpensive and widely available, do not provide the moist wound environment necessary for optimal healing. Additionally, they can adhere to the wound bed, causing trauma during dressing changes and further delaying recovery. This inadequacy often results in prolonged healing times, increased patient discomfort, and a higher likelihood of complications.

5.1.4.2 Risk of infection and poor infection control

Infection is a major complication in wound care, particularly in chronic and nonhealing wounds. Traditional wound care methods, such as the use of antiseptics like hydrogen peroxide or iodine, can be overly harsh, damaging healthy tissue and impeding the healing process. Moreover, these methods often lack the sustained antimicrobial activity required to prevent recurrent infections. The frequent need for dressing changes in traditional wound care also increases the risk of introducing pathogens into the wound, further exacerbating the problem. Inadequate infection control not only delays healing but can also lead to severe systemic infections, such as sepsis, which pose significant risks to patient health and increase healthcare costs.

5.1.4.3 Lack of personalized and targeted therapies

Traditional wound care approaches are often generic and do not account for the unique needs of individual patients or specific wound types. For example, a diabetic patient with poor circulation and neuropathy requires a fundamentally different approach to wound care compared to a patient with a surgical wound. Traditional methods lack the precision and customization needed to address the diverse etiologies and healing requirements of different wounds. This one-size-fits-all approach often results in suboptimal outcomes, particularly for patients with complex medical conditions or those at high risk of complications.

5.1.4.4 Patient discomfort and poor quality of life

Traditional wound care methods can be uncomfortable and distressing for patients. Frequent dressing changes, the use of harsh antiseptics, and the physical trauma caused by adhesive dressings can lead to significant pain and discomfort. For patients with chronic wounds, this discomfort is often prolonged, negatively impacting their quality of life. Additionally, the visibility of wounds and the need for ongoing care can lead to psychological distress, social isolation, and a diminished sense of well-being. These factors underscore the need for wound care solutions that prioritize patient comfort and holistic care.

5.1.4.5 Economic burden on healthcare systems

The inefficiencies of traditional wound care contribute to a substantial economic burden on healthcare systems. Chronic wounds, in particular, require prolonged treatment, frequent hospital visits, and specialized care, all of which drive up healthcare costs. The high rates of complications, such as infections and amputations, further exacerbate this burden. Traditional methods often necessitate more frequent interventions and longer healing times, increasing the overall cost of care. In resource-limited settings, the reliance on outdated and inefficient wound care practices can strain healthcare infrastructure and limit access to effective treatments.

5.1.4.6 Limited focus on advanced healing mechanisms

Traditional wound care methods primarily focus on passive wound coverage and basic infection control, neglecting the advanced biological mechanisms involved in wound healing. Modern research has highlighted the importance of factors such as angiogenesis, extracellular matrix remodeling, and the role of growth factors and stem cells in promoting healing. Traditional approaches do not leverage these insights, resulting in missed opportunities to accelerate healing and improve outcomes. For example, the use of gauze dressings does not facilitate the delivery of bioactive compounds or the creation of an optimal healing environment, which are critical for addressing the complex pathophysiology of chronic wounds.

5.1.5 Role of advanced therapeutics

Advanced therapeutics has revolutionized the field of wound care, offering innovative solutions to address the limitations of traditional methods and improve patient outcomes. These cutting-edge approaches leverage advancements in biotechnology, materials science, and pharmacology to target the complex biological processes involved in wound healing. By focusing on precision, personalization, and enhanced efficacy, advanced therapeutics are transforming the way wounds are managed, particularly in the context of chronic and hard-to-heal wounds. Below is a comprehensive exploration of the role and impact of advanced therapeutics in modern wound care.

5.1.5.1 Targeting the underlying pathophysiology of wounds

One of the most significant contributions of advanced therapeutics is its ability to address the underlying pathophysiology of wounds. Chronic wounds, such as DFUs, venous leg ulcers, and pressure sores, are characterized by prolonged inflammation, impaired angiogenesis, and disrupted extracellular matrix remodeling. Advanced therapeutics, including growth factors, stem cell therapies, and bioactive molecules, target these specific mechanisms to promote healing. For example, recombinant human platelet-derived growth factor (rhPDGF) has been shown to stimulate cell proliferation and tissue regeneration, while stem cell therapies enhance angiogenesis and modulate the inflammatory response. By addressing the root causes of impaired healing, these therapies offer a more effective and sustainable approach to wound management.

5.1.5.2 Enhanced drug delivery systems

Advanced therapeutics has also introduced innovative drug delivery systems that improve the efficacy and precision of wound treatments. Microemulsions and nanoemulsions, for instance, enable the controlled release of active ingredients, such as antimicrobial agents, anti-inflammatory drugs, and growth factors, directly to the wound site. These systems enhance the bioavailability of therapeutic agents, reduce systemic side effects, and ensure sustained action over time. Additionally, nanotechnology-based delivery systems, such as nanofibers and hydrogels, provide a moist wound environment, promote cell migration, and facilitate the gradual release of bioactive compounds [7, 8]. These advancements not only accelerate healing but also minimize the risk of complications, such as infections and scarring.

5.1.5.3 Personalized and precision medicine

The advent of advanced therapeutics has paved the way for personalized and precision medicine in wound care. By leveraging genetic, molecular, and clinical data, healthcare providers can tailor treatments to the unique needs of individual patients. For instance, gene therapy and RNA-based interventions, such as siRNA, can modulate specific genes involved in wound healing, offering targeted solutions for patients with genetic predispositions to impaired healing. Similarly, bioengineered skin substitutes and autologous cell therapies can be customized to match the patient's tissue type, reducing the risk of rejection and improving outcomes. This personalized approach ensures that treatments are not only more effective but also aligned with the patient's specific healing requirements.

5.1.5.4 Promotion of regenerative healing

Advanced therapeutics emphasizes regenerative healing, which aims to restore the structure and function of damaged tissues rather than merely closing the wound. Techniques such as tissue engineering, 3D bioprinting, and the use of extracellular matrix (ECM) scaffolds facilitate the regeneration of skin, blood vessels, and other tissues. These approaches mimic the natural healing process, promoting the formation of functional tissue with minimal scarring. For example, ECM-based scaffolds provide a framework for cell migration and proliferation, while 3D-printed skin grafts offer precise reconstruction of complex wounds. By focusing on regeneration, advanced therapeutics not only improves cosmetic outcomes but also restores the functional integrity of the affected area.

5.1.5.5 Combating antimicrobial resistance

The rise of antimicrobial resistance poses a significant challenge in wound care, particularly in the management of infected wounds. Advanced therapeutics offers novel solutions to combat this issue through the use of antimicrobial peptides, phage therapy, and smart dressings with built-in antimicrobial properties. These approaches provide targeted and sustained antimicrobial activity without contributing to resistance. For instance, antimicrobial peptides disrupt bacterial cell membranes, offering broad-spectrum activity against multidrug-resistant pathogens. Similarly, smart dressings release antimicrobial agents in response to changes in wound pH or temperature, ensuring timely and effective infection control. These innovations are critical for addressing the growing threat of antimicrobial resistance in wound care.

5.1.5.6 Improving patient comfort and quality of life

Advanced therapeutics prioritizes patient comfort and quality of life by reducing the frequency of dressing changes, minimizing pain, and accelerating healing. For example, hydrogels and foam dressings provide a moist wound environment, reduce friction, and alleviate discomfort [9]. Similarly, bioactive dressings that release growth factors or anti-inflammatory agents promote faster healing, reducing the duration of treatment and the associated psychological stress. By enhancing patient comfort and reducing the burden of care, advanced therapeutics contributes to improved mental and emotional well-being, particularly for patients with chronic wounds.

5.1.5.7 Economic and healthcare system benefits

While advanced therapeutics often involves higher upfront costs, they offer significant long-term economic benefits by reducing the overall cost of wound care. Faster healing times, fewer complications, and lower rates of hospital readmissions translate into cost savings for healthcare systems. Additionally, the prevention of chronic wound-related complications, such as amputations and systemic infections, reduces the economic burden on patients and society. By improving outcomes and optimizing resource utilization, advanced therapeutics represents a cost-effective solution for modern wound care.

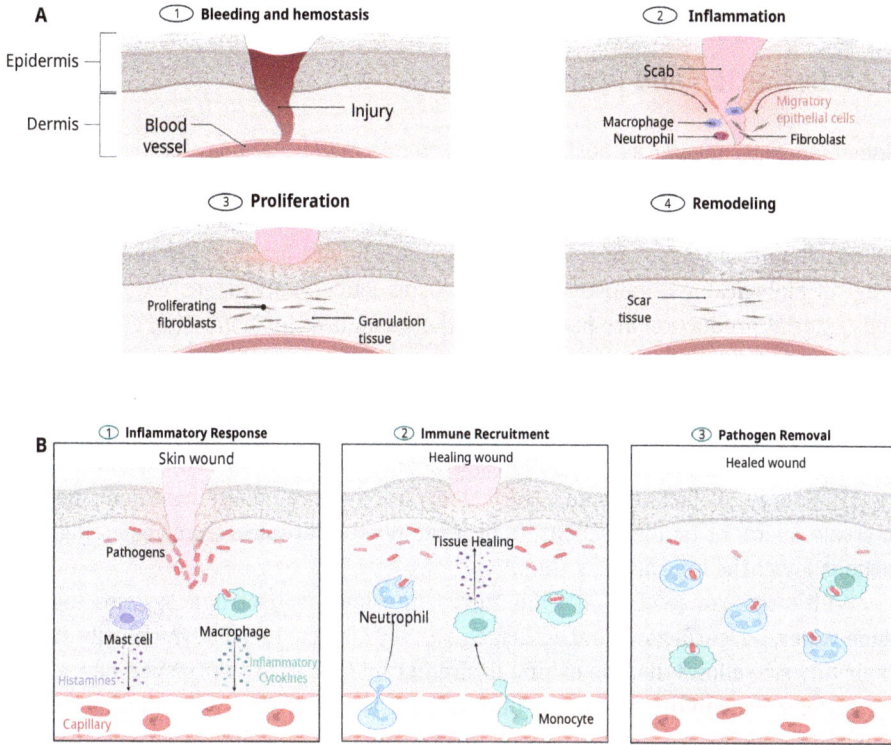

Figure 5.1: The wound healing process and immune response.

(A) The four stages of wound healing: **(1)** bleeding and hemostasis, where a vascular injury leads to clot formation; **(2)** inflammation, characterized by immune cell recruitment and scab formation; **(3)** proliferation, involving fibroblast activity and granulation tissue formation; and **(4)** remodeling, where scar tissue develops as the wound matures.

(B) The immune response to a wound: **(1)** the inflammatory response, where mast cells release histamines and macrophages initiate cytokine signaling to combat pathogens; **(2)** immune recruitment, with neutrophils and monocytes facilitating tissue healing; and **(3)** pathogen removal, leading to a fully healed wound with resolved immune activity.

5.2 Micro/nanoemulsions

5.2.1 Definition and properties

Under some circumstances, surfactants, oils, water, and/or nanoparticles may be combined to form a liquid-in-liquid colloidal dispersion system called a micro- or nanoemulsion. With their tiny droplet sizes and distinctive thermodynamic features, colloidal dispersions of surfactant-stabilized immiscible liquids are indispensable in a wide range of biological and industrial contexts. Drug delivery, medicines, cosmetics, food, and material production are just a few of the many uses for both. They may be used for entrapping drugs that are hydrophobic as well as hydrophilic [10].

With a size range of 20–200 nm, nanoemulsions exhibit kinetic stability. Their tiny size allows them to exhibit great stability, enhanced drug administration, controllable rheology, a high surface area-to-volume ratio, and optical clarity or minor opacity [11]. Energy, surfactants, and cosurfactants are the three main ingredients needed to create micro- or nanoemulsions. In order to prevent droplets from coalescing, surfactants lower the interfacial tension.

With a size range of 10–100 nm, microemulsions are isotropic systems that include water, oil, surfactant, and cosurfactant; they are also thermodynamically stable. Their tiny size allows them to exhibit desired properties such as transparency, stability, better permeability, improved bioavailability, and rheology, in addition to improved drug delivery and stability [12]. Microemulsions are able to develop on their own without the need for high-energy sources, unlike conventional emulsions. Due to their unique physicochemical properties, microemulsions have found utility in biomedical applications, food processing, medicines, and cosmetics [13].

5.2.2 Formulation techniques

5.2.2.1 High-energy methods

High-shear homogenizers, microfluidizers, and ultrasonicators are high-energy mechanical devices that reduce size. This process makes it simple to transform oil into an emulsion. This method was mostly used to prepare nanoemulsions because they required a lot of energy to prepare. High-energy methods like microfluidization, high-shear homogenization, and sonication facilitate nanoemulsion formation and are sometimes also used in microemulsion method [14 –16].

5.2.2.2 Microfluidization

Devices called microfluidizers are used to mix materials at the microscale. Before passing through a microfluidizer, water, oil, surfactant, and cosurfactant are combined as macroemulsions. There are various kinds of microchannels in microfluidizers that aid in mixing at the microsize level. This technique produced a micro/nano with increased stability and a smaller, uniform droplet size [17].

5.2.2.3 High-shear homogenization

The method's most popular approach uses a high-energy source to create the tiniest micro- or nanoemulsion particles. The macroemulsion undergoes a series of transformations in a high-shear homogenizer, including cavitation, turbulence, and hydraulic shear, to provide a fine-grained emulsion. The kind of homogenizer used, the structure of the sample, and the parameters of duration, temperature, and energy intensity are all critical for micro/nano-size emulsions. Particle size, for instance, is inversely proportional to the intensity of the applied energy. In order to produce a better emulsion, high-shear homogenizers employ small-molecule emulsifiers such as hexadecyl-trimethyl-ammonium, sodium dodecyl sulfate, and Tween 20 [17].

5.2.2.4 Ultrasonication

Ultrasonication is the best of the high-energy techniques. Ultrasonic waves with frequency of 20 kHz or higher are used in the creation of emulsions. Cavitation force is used to disassemble microemulsions in order to create nanoemulsions and shrink their sizes. When fluids cause microbubbles to develop, expand, and eventually collapse, this phenomenon is known as cavitation. By manipulating the ultrasonic energy and duration using piezoelectric probes, the desired particle size and stability may be achieved by the generation of ultrasonic waves [17].

5.2.2.5 Low-energy methods

Since microemulsions require less energy to form, this method is typically used to prepare them. They are also occasionally employed in the preparation of nanoemulsions. The cumulative behavior of water, oil, surfactant, and cosurfactant was examined for this method. Phase inversion and spontaneous emulsification are the two categories of low-energy techniques [14–16].

5.2.2.6 Spontaneous emulsification

When it comes to making polymeric nanoparticles, the nanoprecipitation method is quite similar to spontaneous emulsification. Microemulsions may be produced using the phase titration method, also known as spontaneous emulsification [18]. Microemulsions may be classified as W/O or O/W based on their composition alone, which indicates whether they are rich in water or oil. A thorough analysis must be conducted to exclude metastable systems. But instead of polymer, oil is used. Making a drug-containing hydrophilic surfactant like mygliol with an organic or oil phase, an oil-soluble surfactant like Span, and a somewhat water-miscible organic solvent like acetone or ethyl acetate are all part of the process. While the process may also include adding oil to water in the case of W/O emulsions, the organic phase is often introduced drop by drop to the agitated water phase to create tiny nanoscale emulsions [19].

5.2.2.7 Phase inversion technique

The phase inversion temperature (PIT) technique creates nanoemulsions and micro-emulsions by capitalizing on the temperature-dependent variations in surfactant solubility between water and oil. An orderly transformation from a W/O to O/W emulsion, or vice versa, is achieved by the use of a bi-continuous intermediary phase. There are two main categories of phase inversion techniques: transitional and catastrophic. It is common practice to quickly quench a mixture of surfactant, water, and oil after heating it to a certain temperature (PIT, or formulation blend-specific). A phase inversion occurs when the interfacial structure opens and reverses as a consequence of a change in temperature from low to high. The subsequent rapid quenching seals the interfacial structure, trapping any remaining oil or water. The young droplets stay there for a long time because of the thick surfactant coating; this is a bottom-up approach. Because heat is delivered via PIT techniques, heat-sensitive medications may not be able to be used. The medication, oil, water, and surfactant must all be solubilized with each other for the phase change to go off without a hitch. The only factor that controls destabilization is Ostwald ripening. A catastrophic phase inversion occurs when the fractional volume of the scattered phase changes, leading to a phase inversion. A more stable emulsion may be formed from an emulsion that has been prepared in a way that deviates from Bancroft's guidelines [20].

5.2.3 Advantage of micro/nanoemulsion

5.2.3.1 Pharmaceutical and drug delivery

- Micro/nanoemulsions are used for different pharmaceutical drug delivery such as oral, topical, intravenous, and pulmonary routes to enhance therapeutic use. Nanoemulsions increase the solubility and bioavailability of water-insoluble drugs. They facilitate targeted drug delivery and sustained release, which renders them applicable in cancer treatment, vaccine delivery, and transdermal drug delivery [21–23].
- Micro/nanoemulsions are suitable for both lipophilic and hydrophilic drugs. Drug absorption is increased by the increased surface area caused by the smaller droplet size (10–200 nm) [23, 24].
- Micro/nanoemulsions decrease dosing interval, dosing frequency, and side effects [21, 23].
- Micro/nanoemulsions are mostly safe because they are prepared by using oil, water and surfactants, which are biocompatible and biodegradable [23, 25].
- They protect drugs from light, oxidation, and hydrolysis and prolong their shelf life [23, 24].

5.2.3.2 Cosmeceuticals

Micro/nanoemulsions are used in cosmetics to prepare a variety of creams, including moisturizers, sunscreen, and antiaging products. Additionally, cosmetic products are also used to deliver active pharmaceutical product with high penetration into the skin that gives nongreasy and light feel to the skin [26, 27].

5.2.3.3 Food and beverage industry

Applied to encapsulate bioactive substances such as vitamins, antioxidants, and essential oils, micro/nanoemulsions enhance the stability and absorption of nutrients in functional foods and beverages [27, 28].

5.2.3.4 Agriculture

They boost the effectiveness of pesticides and lessen environmental contamination by enhancing agrochemical solubility and penetration into plant tissues [25].

5.2.3.5 Biomedical applications

Micro/nanoemulsions are utilized in disinfectants, wound healing, and as carriers of imaging agents for medical diagnostics due to their antimicrobial qualities [27, 29, 30].

5.3 Mechanisms of action in wound healing

The unusual physicochemical features and multifunctional capacities of micro/nano-emulsions have made them advanced therapeutic systems in wound healing. The anti-inflammatory effects, antibacterial activity, increased moisture retention, and enhanced cellular regeneration are some of the main ways in which these systems achieve their therapeutic benefits [31]. In order to optimize the design and use of micro/nanoemulsions in wound care, it is vital to understand these processes (Figure 5.2). An in-depth analysis of these processes and their role in facilitating efficient wound healing is shown below.

5.3.1 Anti-inflammatory effects

While some degree of inflammation is necessary for wound healing to occur, chronic wounds may develop when inflammation is either too severe or lasts too long. Micro/nanoemulsions are crucial for regulating the inflammatory response and facilitating effective and balanced healing.

- *Targeted delivery of anti-inflammatory agents:* Micro/nanoemulsions can encapsulate and deliver anti-inflammatory drugs, such as corticosteroids or nonsteroidal anti-inflammatory drugs (NSAIDs), directly to the wound site. This targeted delivery reduces systemic side effects and ensures high local concentrations of the drug, effectively controlling inflammation.
- *Natural bioactive compounds:* Many micro/nanoemulsions incorporate natural bioactive compounds, such as curcumin, aloe vera, or omega-3 fatty acids, which possess inherent anti-inflammatory properties. These compounds inhibit the production of pro-inflammatory cytokines (e.g., TNF-α and IL-6) and promote the resolution of inflammation, creating a conducive environment for healing.
- *Reduction of oxidative stress:* Chronic inflammation is often associated with elevated levels of reactive oxygen species (ROS), which can damage tissues and delay healing. Micro/nanoemulsions containing antioxidants, such as vitamin E or polyphenols, neutralize ROS and reduce oxidative stress, further supporting the anti-inflammatory effects [32].

5.3.2 Antimicrobial activity

Infections are a major complication in wound healing, particularly in chronic and open wounds. Micro/nanoemulsions provide effective antimicrobial activity, preventing infections and promoting a sterile environment for healing:

- *Encapsulation of antimicrobial agents:* Micro/nanoemulsions can encapsulate a wide range of antimicrobial agents, including antibiotics (e.g., ciprofloxacin, gentamicin), silver nanoparticles, and essential oils (e.g., tea tree oil, eucalyptus oil). These agents are released in a controlled manner, ensuring sustained antimicrobial activity at the wound site.
- *Broad-spectrum activity:* The combination of multiple antimicrobial agents within a single emulsion can provide broad-spectrum activity against bacteria, fungi, and viruses. This is particularly important for wounds colonized by multidrug-resistant pathogens, which are challenging to treat with conventional therapies [26].
- *Enhanced penetration:* The small droplet size of micro/nanoemulsions allows for deep penetration into the wound bed, ensuring that antimicrobial agents reach the site of infection effectively. This is especially beneficial for biofilms, which are often resistant to traditional treatments [33].

5.3.3 Enhanced cellular regeneration

Cellular regeneration is a cornerstone of wound healing, involving the proliferation and migration of cells to rebuild damaged tissues. Micro/nanoemulsions enhance this process by delivering growth factors, stem cells, and other regenerative agents directly to the wound site:

- *Delivery of growth factors:* Micro/nanoemulsions can encapsulate and deliver growth factors, such as epidermal growth factor (EGF), vascular endothelial growth factor (VEGF), and fibroblast growth factor (FGF). These growth factors stimulate cell proliferation, angiogenesis, and extracellular matrix formation, accelerating tissue regeneration.
- *Stem cell therapy:* The incorporation of stem cells or stem cell-derived exosomes into micro/nanoemulsions has shown promise in promoting tissue regeneration. Stem cells modulate the wound environment, enhance angiogenesis, and promote the differentiation of cells into functional tissues.
- *Collagen synthesis:* Micro/nanoemulsions can deliver agents that promote collagen synthesis, such as vitamin C or proline. Collagen is essential for providing structural support to the newly formed tissue and improving the tensile strength of the wound [34].

5.3.4 Improved moisture retention

A moist wound environment is essential for optimal healing, as it promotes cell migration, reduces scab formation, and minimizes tissue dehydration. Micro/nanoemulsions excel in maintaining moisture balance at the wound site:

- *Occlusive properties:* The lipid components of micro/nanoemulsions create an occlusive barrier on the wound surface, preventing moisture loss and maintaining a hydrated environment. This is particularly important for dry or necrotic wounds, which require rehydration to support healing.
- *Humectants and moisturizers:* Micro/nanoemulsions can incorporate humectants, such as glycerin or hyaluronic acid, which attract and retain water in the wound bed. These agents ensure a consistently moist environment, facilitating cell migration and tissue repair.
- *Reduction of exudate buildup:* While maintaining moisture is important, excessive exudate can impede healing. Micro/nanoemulsions can be designed to absorb excess exudate while retaining essential moisture, creating a balanced environment that supports healing [35].

5.4 Key components of micro/nanoemulsions for wound healing

Because of their tiny droplet size, excellent stability, and higher bioavailability, micro-emulsions and nanoemulsions have become particularly useful delivery methods in wound healing [36]. Carefully chosen components of these systems enhance tissue regeneration, modify the wound environment, and transport therapeutic substances directly to the wound site to hasten healing [37]. Advanced encapsulation techniques, natural bioactive chemicals, lipids and surfactants, active pharmaceutical ingredients (APIs), and micro/nanoemulsions are the main components of wound healing solutions. Improving the performance and utility of these emulsions depends on each individual component [38].

5.4.1 Lipids and surfactants

Micro- and nanoemulsions rely on lipids and surfactants (Table 5.1) as their building blocks to provide the stability and structural integrity needed for efficient medication delivery. The oil phase of the emulsion is composed of lipids, which may encapsulate bioactive molecules and lipophilic medicines. These lipids include medium-chain tri-glycerides (MCTs), phospholipids, and fatty acids. In addition to facilitating cell migration and tissue regeneration, these lipids help moisten the wound bed, which is cru-

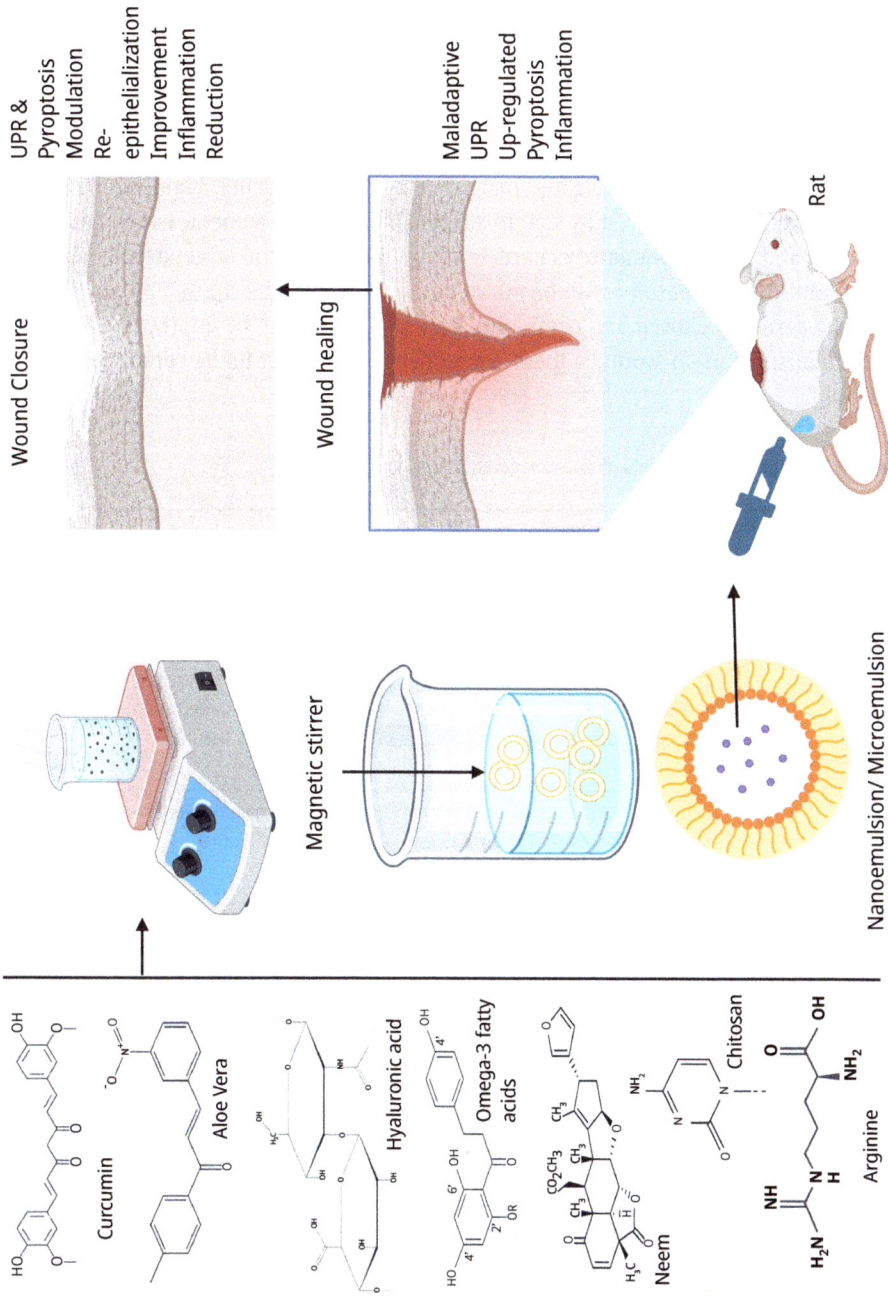

Figure 5.2: Fabricated micro/nanoemulsion as wound healing accelerator.

cial for a successful healing process. To stabilize the emulsion, surfactants lower the interfacial tension between the oil and water phases; they are amphiphilic molecules. Wound healing emulsions often include surfactants such as lecithin, sorbitan esters (e.g., Span 80), and polysorbates (e.g., Tween 80). Not only do these surfactants make the emulsion more stable and prolong its shelf life, but they also make medications that are not very water-soluble and increase their bioavailability. Emulsion therapeutic effectiveness is enhanced by the presence of surfactants, some of which have antibacterial and anti-inflammatory characteristics. To achieve the target droplet size, stability, and drug release profile of micro/nanoemulsions, the lipids and surfactants must be carefully chosen and combined. For the emulsion to be appropriate for topical administration on wounds, these components also affect its biocompatibility and safety [39].

Table 5.1: Lipids and surfactants of micro/nanoemulsions.

Category	Lipid/ surfactant	Properties	Role in wound healing	References
Lipids	Medium-chain triglycerides (MCTs)	High solubility for lipophilic drugs, biocompatible, and easily metabolized	Enhances drug delivery and provides a moist wound environment	[67]
	Phospholipids (e.g., lecithin)	Amphiphilic, forms stable bilayers, and improves emulsion stability	Enhances skin penetration and promotes tissue regeneration	[68]
	Oleic acid	Monounsaturated fatty acid, enhances drug solubility, and biocompatible	Improves drug delivery and has anti-inflammatory properties	[69]
	Squalene	Natural lipid, antioxidant, and enhances skin absorption	Protects against oxidative stress and improves wound hydration	[70]
	Glyceryl monostearate	Nonionic, stabilizes emulsions, and biocompatible	Enhances drug delivery and provides a protective barrier on the wound	[71]
Surfactants	Polysorbates (e.g., Tween 80)	Nonionic, reduces interfacial tension, and improves emulsion stability	Enhances drug solubility and bioavailability, and stabilizes the emulsion.	[72]
	Sorbitan esters (e.g., Span 80)	Nonionic, forms stable emulsions, and biocompatible	Improves drug encapsulation and delivery to the wound site	[73]
	Lecithin	Natural surfactant, amphiphilic, and biocompatible	Enhances skin penetration and promotes tissue regeneration	[74]

Table 5.1 (continued)

Category	Lipid/ surfactant	Properties	Role in wound healing	References
	Sodium lauryl sulfate (SLS)	Anionic, reduces surface tension, and enhances drug solubility	Improves drug delivery but may cause irritation; used in low concentrations	[75]
	Cetyl alcohol	Nonionic, stabilizes emulsions, and provides a smooth texture	Enhances emulsion stability and improves wound hydration	[76]
Cosurfactants	Ethanol	Enhances solubility of surfactants and lipids, and improves emulsion formation	Facilitates the formation of stable micro/ nanoemulsions	[77]
	Propylene glycol	Enhances drug solubility and stabilizes emulsions	Improves drug delivery and provides a moist wound environment	[78]
	Polyethylene glycol (PEG)	Nonionic, improves solubility, and stabilizes emulsions	Enhances drug delivery and reduces immunogenicity	[79]
Natural surfactants	Saponins	Natural, biodegradable, and forms stable emulsions	Provides antimicrobial and anti-inflammatory effects	[80]
	Quillaja saponin	Natural, forms stable nanoemulsions, and biocompatible	Enhances drug delivery and provides antimicrobial properties	[81]
	Glycyrrhizin	Natural, anti-inflammatory, and stabilizes emulsions	Reduces inflammation and promotes wound healing	[82]

5.4.2 Active pharmaceutical ingredients (APIs)

To target the specific biological mechanisms at work during wound healing, micro/ nanoemulsions include APIs. Antibiotics, anti-inflammatory drugs, growth factors, and analgesics are all examples of APIs that may aid in wound healing. For instance, dexamethasone and antibiotics like ciprofloxacin and gentamicin alleviate inflammation and discomfort, respectively, and are used for the prevention and treatment of wound infections. Particularly crucial in the healing of chronic wounds are growth factors including VEGF and EGF, which promote cell proliferation, angiogenesis, and tissue regeneration. To maximize their therapeutic potential, these APIs are enhanced by including them into micro/nanoemulsions, which prolongs their release, guarantees targeted distribution to the wound site, and boosts their stability. Table 5.2 shows

Table 5.2: Recent reports on active substance-loaded microemulsions and nanoemulsions with multifarious applications.

Study focus	Key findings	Type of emulsion	Active ingredient(s)	Wound type	Outcome metrics	References
Nanoemulsions for drug delivery in wound healing	Enhanced penetration of antimicrobial agents, accelerated wound closure, and reduced infection rates	Nanoemulsion	Antimicrobial agents	Infected wounds	Wound closure rate, infection reduction	[83]
Microemulsions for topical delivery of curcumin in chronic wounds	Significant anti-inflammatory and antioxidant effects, promoting faster tissue regeneration.	Microemulsion	Curcumin	Chronic wounds	Tissue regeneration, inflammation reduction	[84]
Dual-drug loaded nanoemulsions for diabetic wound healing	Improved angiogenesis and collagen deposition, enhancing diabetic wound healing	Nanoemulsion	Insulin, metformin	Diabetic wounds	Angiogenesis, collagen deposition	[85]
Essential oil-based nanoemulsions for infected wound management	Potent antimicrobial activity against multidrug-resistant bacteria, reducing wound infection	Nanoemulsion	Tea tree oil	Infected wounds	Antimicrobial activity, infection reduction	[86]
Chitosan-based nanoemulsions for burn wound healing	Accelerated re-epithelialization and reduced scar formation in burn wounds	Nanoemulsion	Chitosan	Burn wounds	Re-epithelialization, scar reduction	[87]
Microemulsions for delivery of growth factors in chronic wounds	Improved cell proliferation and migration, enhancing chronic wound closure	Microemulsion	VEGF, EGF	Chronic wounds	Cell proliferation, wound closure rate	[88]
Nanoemulsions for transdermal delivery of silver nanoparticles in wound care	Enhanced antimicrobial activity and reduced cytotoxicity, promoting wound healing	Nanoemulsion	Silver nanoparticles	General wounds	Antimicrobial activity, cytotoxicity reduction	[89]

Description	Effect	Formulation	Active agent	Wound type	Outcome	Ref.
Hyaluronic acid-based nanoemulsions for scarless wound healing	Improved hydration and collagen alignment, reducing scar formation	Nanoemulsion	Hyaluronic acid	Surgical wounds	Scar reduction, hydration levels	[90]
Microemulsions for delivery of herbal extracts in wound healing	Accelerated wound closure and reduced inflammation	Microemulsion	Aloe vera, calendula	General wounds	Wound closure rate, inflammation reduction	[91]
Nanoemulsions for co-delivery of antibiotics and anti-inflammatory agents	Reduced infection and inflammation, promoting faster healing	Nanoemulsion	Ciprofloxacin, dexamethasone	Infected wounds	Infection reduction, inflammation reduction	[92]
Microemulsions for delivery of omega-3 fatty acids in wound healing	Enhanced anti-inflammatory effects and improved tissue regeneration	Microemulsion	Omega-3 fatty acids	Chronic wounds	Tissue regeneration, inflammation reduction	[93]
Nanoemulsions for delivery of siRNA in wound healing	Effectively silenced pro-inflammatory genes, reducing chronic inflammation and promoting healing	Nanoemulsion	siRNA	Chronic wounds	Gene silencing, inflammation reduction	[94]
Microemulsions for delivery of probiotics in infected wounds	Reduced bacterial load and promoted tissue repair in infected wounds	Microemulsion	Probiotics	Infected wounds	Bacterial load reduction, tissue repair	[95]
Nanoemulsions for delivery of stem cell-derived exosomes in wound healing	Enhanced angiogenesis and collagen synthesis, accelerating wound closure	Nanoemulsion	Stem cell-derived exosomes	Chronic wounds	Angiogenesis, collagen synthesis	[96]
Microemulsions for delivery of resveratrol in diabetic wound healing	Improved antioxidant activity and reduced oxidative stress, enhancing diabetic wound healing	Microemulsion	Resveratrol	Diabetic wounds	Antioxidant activity, oxidative stress reduction	[97]

(continued)

Table 5.2 (continued)

Study focus	Key findings	Type of emulsion	Active ingredient(s)	Wound type	Outcome metrics	References
Nanoemulsions for delivery of nitric oxide in chronic wounds	Improved blood flow and tissue regeneration in chronic wounds	Nanoemulsion	Nitric oxide	Chronic wounds	Blood flow improvement, tissue regeneration	[98]
Microemulsions for delivery of vitamin E in burn wound healing	Reduced oxidative damage and promoted faster re-epithelialization in burn wounds	Microemulsion	Vitamin E	Burn wounds	Oxidative damage reduction, re-epithelialization	[99]
Nanoemulsions for delivery of antimicrobial peptides in wound healing	Broad-spectrum antimicrobial activity and enhanced wound closure	Nanoemulsion	Antimicrobial peptides	Infected wounds	Antimicrobial activity, wound closure rate	[100]
Microemulsions for delivery of collagen in wound healing	Improved extracellular matrix formation and tissue regeneration	Microemulsion	Collagen	General wounds	Extracellular matrix formation, tissue regeneration	[101]
Nanoemulsions for delivery of growth hormone in chronic wounds	Enhanced cell proliferation and tissue repair in chronic wounds	Nanoemulsion	Growth hormone	Chronic wounds	Cell proliferation, tissue repair	[102]

that the wound type, wound severity, and patient-specific healing needs determine which APIs are most appropriate. Emulsifying several APIs allows for the simultaneous treatment of wounds in a number of ways, including infection prevention, tissue regeneration, and scar reduction [40].

5.4.3 Natural bioactive compounds

Because of their biocompatibility, low toxicity, and multifunctional qualities, natural bioactive substances produced from microbes, animals, or plants have garnered a lot of interest in wound healing (Table 5.3) [41, 42]. Ideal candidates for integration into micro/nanoemulsions include substances with antibacterial [43], anti-inflammatory, antioxidant, and wound-healing characteristics, such as curcumin, honey, aloe vera, and essential oils. For example, turmeric's curcumin, a polyphenol, may speed up wound healing, decrease inflammation, and increase collagen production [44]. In a similar vein, aloe vera calms the skin and speeds up tissue regeneration; honey kills germs and creates a moist wound environment. Lavender and tea tree oils, among others, have anti-inflammatory and antibacterial properties. Improving the therapeutic efficiency of micro/nanoemulsions with natural bioactive substances lessens the need for manufactured medications, which might cause adverse effects or add to the problem of antimicrobial resistance [42]. Wound healing results may be further improved by using these chemicals alone or in conjunction with APIs to achieve synergistic effects [45].

Table 5.3: Natural bioactive compounds as wound healing agents.

Bioactive compound/ sources	Source	Properties	Role in wound healing	References
Curcumin	Turmeric (*Curcuma longa*)	Anti-inflammatory, antioxidant, antimicrobial, and promotes collagen synthesis	Reduces inflammation, prevents infections, and accelerates wound closure	[103]
Honey	Bees (*Apis mellifera*)	Antimicrobial, anti-inflammatory, and promotes moist wound healing	Prevents infections, reduces inflammation, and enhances tissue regeneration.	[104]
Aloe vera	*Aloe barbadensis* Miller	Anti-inflammatory, moisturizing, and promotes cell proliferation	Soothes the skin, reduces inflammation, and accelerates re-epithelialization	[105]

Table 5.3 (continued)

Bioactive compound/ sources	Source	Properties	Role in wound healing	References
Tea tree oil	*Melaleuca alternifolia*	Antimicrobial, antifungal, and anti-inflammatory.	Effective against multidrug-resistant pathogens and reduces wound infections	[106]
Chitosan	Crustacean shells	Antimicrobial, biodegradable, and promotes tissue regeneration	Enhances wound closure, reduces infections, and improves collagen deposition	[107]
Omega-3 fatty acids	Fish oil and flaxseed	Anti-inflammatory and promotes angiogenesis	Reduces chronic inflammation and enhances blood vessel formation in wounds	[108]
Resveratrol	Grapes and berries	Antioxidant, anti-inflammatory, and promotes cell proliferation	Reduces oxidative stress, accelerates wound closure, and prevents scarring	[109]
Calendula extract	*Calendula officinalis*	Anti-inflammatory, antimicrobial, and promotes tissue regeneration	Reduces inflammation, prevents infections, and enhances wound healing	[110]
Vitamin E	Plant oils and nuts	Antioxidant, moisturizing, and promotes collagen synthesis	Protects against oxidative damage, improves skin hydration, and enhances tissue repair	[111]
Lavender oil	*Lavandula angustifolia*	Antimicrobial, anti-inflammatory, and soothing	Reduces pain, prevents infections, and promotes relaxation of wound tissues	[112]
Neem extract	*Azadirachta indica*	Antimicrobial, antifungal, and anti-inflammatory	Effective against a wide range of pathogens and reduces inflammation in chronic wounds	[113]
Green tea extract	*Camellia sinensis*	Antioxidant, anti-inflammatory, and promotes collagen synthesis	Reduces oxidative stress, accelerates wound closure, and improves skin elasticity	[114]
Chamomile extract	*Matricaria chamomilla*	Anti-inflammatory, antimicrobial, and soothing	Reduces inflammation, prevents infections, and promotes relaxation of wound tissues	[115]

Table 5.3 (continued)

Bioactive compound/ sources	Source	Properties	Role in wound healing	References
Oregano oil	*Origanum vulgare*	Antimicrobial, antifungal, and antioxidant	Effective against multidrug-resistant pathogens and reduces oxidative stress in wounds	[116]
Manuka honey	*Leptospermum scoparium*	Antimicrobial, anti-inflammatory, and promotes tissue regeneration	Highly effective against resistant bacteria and enhances wound closure	[117]
Centella asiatica	*Centella asiatica*	Promotes collagen synthesis, anti-inflammatory, and antioxidant	Enhances tissue regeneration, reduces scarring, and improves wound strength	[118]
Pomegranate extract	*Punica granatum*	Antioxidant, antimicrobial, and promotes angiogenesis	Reduces oxidative stress, prevents infections, and enhances blood vessel formation	[119]
Rosemary oil	*Rosmarinus officinalis*	Antimicrobial, anti-inflammatory, and antioxidant.	Reduces inflammation, prevents infections, and protects against oxidative damage	[120]
Propolis	Bees (*Apis mellifera*)	Antimicrobial, anti-inflammatory, and promotes tissue regeneration	Prevents infections, reduces inflammation, and accelerates wound closure	[121]

5.4.4 Encapsulation strategies

For micro/nanoemulsion therapeutic agent distribution and release optimization, encapsulation methods are crucial. To prevent degradation and provide regulated release at the wound site, these solutions employ innovative techniques to encapsulate APIs and bioactive chemicals inside the emulsion droplets. Some common techniques for encapsulation include ultrasonication, high-pressure homogenization, and solvent evaporation. Incorporating nanocarriers, including liposomes, solid-lipid nanoparticles (SLNs), and polymeric nanoparticles into the emulsion is a promising encapsulation strategy. The encapsulating medicines are further protected, made more stable, and may be delivered to particular cell types or tissues with the help of these nanocarriers. For instance, SLNs provide prolonged release and enhanced skin penetration, while liposomes may encapsulate hydrophilic and hydrophobic medications concurrently. Using stimuli-responsive encapsulation systems is another novel method; these systems release their payload in reaction to certain triggers, such as changes in wound site temperature, pH, or enzyme activity. In order to minimize unwanted effects and maximize therapeutic agent effectiveness, these systems control their release to specific locations and times. To make sure that therapeutic drugs reach the wound site in enough quantities to work, encapsulation techniques are crucial for enhancing their bioavailability and pharmacokinetics. Enhancing their overall effectiveness in wound healing, micro/nanoemulsions may deliver a broad spectrum of therapeutic agents with accuracy and efficiency by optimizing encapsulation [46].

5.5 Applications in wound care

5.5.1 Chronic wounds

The complicated underlying pathophysiology, protracted healing periods, and high infection risk of chronic wounds (e.g., pressure sores, venous leg ulcers, and DFUs) make them a formidable obstacle in wound management. One promising approach to wound management is the use of micro/nanoemulsions, which may deliver therapeutic chemicals that target particular healing obstacles. For example, emulsions containing growth factors such as VEGF or EGF may promote cell proliferation and angiogenesis, leading to tissue regeneration in diabetic ulcers [47, 48]. Similarly, emulsions laden with antimicrobials help fight infections caused by MRSA, a kind of bacteria that often causes persistent wounds. Reducing the frequency of dressing changes, micro/nanoemulsions are able to permeate the wound bed and release their active components over time, maintaining therapeutic concentrations. To speed up the healing process even further, natural bioactive substances like honey or curcumin may be used to boost the anti-inflammatory and antioxidant benefits. Micro/nanoemulsions provide a thorough and efficient method for healing chronic wounds by treating the fundamental reasons for these difficult situations, including poor circulation, persistent inflammation, and microbial colonization [49].

5.5.2 Acute injuries

To promote the efficient healing of cuts, scrapes, and lacerations while minimizing complications, it is essential to provide prompt and effective treatment. Because of their capacity to transport medicinal substances straight to the site of injury, micro/nanoemulsions work wonders in this setting. A typical consequence of acute injuries is infections; however, emulsions containing antibiotics or silver nanoparticles may prevent these infections. Patients might experience less edema and discomfort when they use anti-inflammatory medications like dexamethasone or natural ingredients like aloe vera. To ensure that therapeutic drugs reach the afflicted tissues effectively, micro/nanoemulsions are able to penetrate deep into the wound bed due to their tiny droplet size. Imperative for cell migration and tissue regeneration, the moisturizing characteristics of these emulsions also aid in keeping the wound area wet. When time is of the essence, micro/nanoemulsions are a lifesaver for treating acute injuries since they shorten the healing period and lessen the likelihood of problems [50].

5.5.3 Burn wounds

There is a considerable chance of infection, scarring, and substantial tissue damage in burn wounds, making them among the most difficult injuries to cure. Therapeutic drugs that stimulate tissue regeneration, decrease inflammation, and prevent infections may be delivered via micro/nanoemulsions, which provide various benefits in burn wound care. Emulsions containing growth factors, including FGF, might hasten re-epithelialization by, for example, stimulating cell proliferation in the skin. In a similar vein, antimicrobial medicines like essential oils or silver sulfadiazine help safeguard burn victims against infections, a leading cause of morbidity. Burn victims have rapid alleviation from pain and suffering because of the cooling and moisturizing properties of micro/nanoemulsions. The anti-inflammatory and antioxidant properties may be further enhanced by adding natural bioactive ingredients like curcumin or honey, which further promotes healing. Micro/nanoemulsions provide a holistic and efficient method of burn treatment by enhancing functional and aesthetic results while tackling the specific problems associated with burn wounds, such as significant tissue damage and a high risk of infection [51].

5.5.4 Post-surgical healing

Infections, dehiscence (reopening of the wound), and severe scarring are consequences of post-surgical wounds that must be carefully managed to promote efficient and quick healing. Here, micro/nanoemulsions really shine since they can transport medicinal substances tailored to the unique requirements of surgical wounds. As an example, one frequent consequence of post-surgical wounds is surgical site infections; emulsions containing antibiotics or silver nanoparticles help prevent these infections. Anti-inflammatory medications help alleviate swelling and discomfort, leading to improved patient comfort. These agents can be synthetic or natural, such as chamomile extract. The wound site is protected throughout the healing process using micro/nanoemulsions because they release therapeutic agents gradually, minimizing the frequency of dressing changes. Furthermore, growth factors like EGF or VEGF may speed up the healing process by stimulating angiogenesis and tissue regeneration. In order to improve results and shorten recovery periods, micro/nanoemulsions provide a thorough and efficient method of post-surgical wound care by tackling specific issues including infection risk and the need for quick tissue regeneration [52].

5.6 Advances in technology and research

Thanks to new research and technological developments, wound healing has come a long way in the last several years. For instance, micro/nano-emulsions have become potent instruments for administering medicinal substances, influencing the wound milieu, and encouraging tissue repair. These innovations are marked by the creation of sensitive and intelligent emulsions, the incorporation of nanotechnology to allow personalization, and the use of complementary treatments to improve healing results. What follows is an in-depth analysis of these recent innovations and how they affect wound care.

5.6.1 Smart and responsive emulsions

Smart and responsive emulsions represent a significant leap forward in wound care, offering the ability to adapt to the dynamic conditions of the wound environment. These emulsions are designed to release their payload in response to specific triggers, such as changes in pH, temperature, or the presence of enzymes, ensuring that therapeutic agents are delivered precisely when and where they are needed:

– *pH-Responsive emulsions:* Chronic wounds often exhibit an alkaline pH due to bacterial colonization and prolonged inflammation. pH-responsive emulsions can release antimicrobial or anti-inflammatory agents in response to these alkaline conditions, targeting infection and inflammation more effectively.
– *Temperature-responsive emulsions:* In burn wounds or inflamed tissues, localized temperature increases can trigger the release of cooling agents, analgesics, or anti-inflammatory drugs, providing immediate relief and promoting healing.
– *Enzyme-responsive emulsions:* Certain enzymes, such as matrix metalloproteinases (MMPs), are overexpressed in chronic wounds. Enzyme-responsive emulsions can release therapeutic agents, such as MMP inhibitors or growth factors, in the presence of these enzymes, addressing the underlying causes of impaired healing [53].

5.6.2 Nanotechnology in customizing emulsions

The advancements in nanotechnology have completely changed the way micro/nano-emulsions may be customized. Now, delivery systems can be designed with pinpoint accuracy to meet the unique requirements of various types of wounds. Researchers have shown that emulsions may be made more stable, bioavailable, and therapeutically effective, by using the specific characteristics of nanomaterials:

– *Nanocarriers:* The incorporation of nanocarriers, such as liposomes, SLNs, and polymeric nanoparticles, into emulsions allows for the encapsulation and controlled release of therapeutic agents. These nanocarriers protect the active ingredients from degradation, enhance their penetration into the wound bed, and provide sustained release over time.
– *Surface modifications:* Nanotechnology enables the surface modification of emulsion droplets to improve their targeting and adhesion to specific tissues or cells. For example, the addition of ligands or antibodies to the surface of emulsion droplets can enhance their binding to receptors on fibroblasts or endothelial cells, promoting tissue regeneration and angiogenesis.
– *Multifunctional emulsions:* Nanotechnology allows for the development of multifunctional emulsions that combine multiple therapeutic agents or properties in a single system. For instance, an emulsion can be designed to deliver both antimicrobial agents and growth factors, addressing infection and promoting tissue regeneration simultaneously [54].

5.6.3 Synergistic therapies

To improve the overall effectiveness of wound healing, synergistic treatments combine several therapeutic substances or techniques. Synergistic therapies, which use the synergistic effects of many treatments, are more successful than single-agent therapies in addressing the complex wound healing process:

– *Combination of antimicrobial and anti-inflammatory agents:* Chronic wounds often involve both infection and inflammation. Combining antimicrobial agents, such as silver nanoparticles or antibiotics, with anti-inflammatory drugs, such as corticosteroids or natural compounds like curcumin, can address both issues simultaneously, promoting faster and more effective healing [44].
– *Growth factors and stem cells:* The combination of growth factors, such as EGF or VEGF, with stem cell therapies can enhance tissue regeneration and angiogenesis. Stem cells can modulate the wound environment and promote the formation of new blood vessels, while growth factors stimulate cell proliferation and differentiation.

– *Natural and synthetic compounds:* The integration of natural bioactive compounds, such as honey or aloe vera, with synthetic drugs can provide a balanced approach to wound healing. Natural compounds often possess antimicrobial, anti-inflammatory, and antioxidant properties, while synthetic drugs offer targeted and potent therapeutic effects [55].

5.7 Challenges and limitations

There are several obstacles to overcome before fully realizing the potential of micro/nanoemulsions to transform wound care. In order to fully use these sophisticated treatment systems, it is essential to resolve issues, which include the technological, regulatory, and economic spheres. Regulatory and safety concerns, cost-effectiveness, stability, and scalability are only a few of the important constraints and issues linked with micro/nanoemulsions.

5.7.1 Stability and scalability

The creation of stable and scalable micro/nanoemulsions during manufacture and storage is a major problem. When we talk about an emulsion's stability, we're referring to how well it holds its physicochemical characteristics, such as the size of its droplets, how uniform it is, and how well it encapsulates drugs, over time. Conversely, emulsions must be scalable if their quality and performance can be mass-produced without degradation:

– *Physical instability:* Micro/nanoemulsions are prone to physical instability, such as coalescence (merging of droplets), flocculation (clumping of droplets), or phase separation. These issues can lead to inconsistent drug delivery and reduced therapeutic efficacy. Stabilizing agents, such as surfactants and cosurfactants, are often required to maintain the integrity of the emulsion, but their selection and optimization can be complex.

– *Chemical instability:* The APIs or bioactive compounds encapsulated in the emulsion may degrade over time due to exposure to light, heat, or oxygen. This degradation can reduce the therapeutic efficacy of the emulsion and limit its shelf life.

– *Scalability challenges:* Translating laboratory-scale formulations to industrial-scale production is a significant hurdle. Factors such as the choice of equipment, production conditions, and quality control measures must be carefully optimized to ensure consistency and reproducibility. Scaling up production while maintaining the stability and efficacy of the emulsion requires significant investment in technology and expertise [56].

5.7.2 Regulatory and safety considerations

There are many safety and regulatory concerns with developing and selling micro/
nanoemulsions for wound healing. To guarantee their quality, safety, and effectiveness, these emulsions must undergo thorough examination before they can be approved as innovative drug delivery systems:
- *Biocompatibility and toxicity:* The components of micro/nanoemulsions, such as
  lipids, surfactants, and encapsulated drugs, must be biocompatible and nontoxic.
  However, certain surfactants or nanomaterials may pose risks of cytotoxicity,
  immunogenicity, or long-term accumulation in the body. Comprehensive preclinical and clinical studies are required to assess these risks and establish safe usage
  guidelines.
- *Regulatory approval:* Regulatory agencies, such as the U.S. Food and Drug Administration (FDA) and the European Medicines Agency (EMA), require extensive
  data on the formulation, manufacturing process, stability, and therapeutic efficacy of micro/nanoemulsions. The approval process can be time-consuming and
  costly, particularly for novel or complex formulations.
- *Standardization and quality control:* Ensuring the consistency and quality of
  micro/nanoemulsions is critical for regulatory approval and clinical success. Standardized protocols for manufacturing, characterization, and testing are essential
  to meet regulatory requirements and ensure patient safety [57].

5.7.3 Cost-effectiveness

Micro- and nanoemulsions have been widely used in wound treatment because of
their cost-effectiveness. The development, manufacture, and implementation of these
sophisticated systems may be costly, which raises issues regarding affordability and
accessibility:
- *High development costs:* The research and development of micro/nanoemulsions
  involves significant investment in materials, technology, and expertise. The need
  for extensive preclinical and clinical studies further adds to the cost.
- *Manufacturing expenses:* The production of micro/nanoemulsions requires specialized equipment and processes, such as high-pressure homogenization or ultrasonication, which can be costly. Additionally, the use of high-quality ingredients,
  such as purified lipids and surfactants, increases production expenses.
- *Economic burden on healthcare systems:* While micro/nanoemulsions may reduce
  long-term healthcare costs by improving healing outcomes and preventing complications, their high upfront costs can be a barrier to adoption, particularly in
  resource-limited settings. Ensuring cost-effectiveness requires careful consideration of the balance between therapeutic benefits and economic feasibility [58].

5.8 Future perspectives

Prompted by developments in formulation science, digital health technologies, and personalized medicine, the area of micro/nanoemulsions for wound healing is undergoing fast evolution. These new developments have the potential to revolutionize wound care by making treatments more effective, precise, and accessible. Personalized medicine, new developments in formulation, and the use of micro/nanoemulsions are the main topics of this in-depth look at the field's future prospects.

5.8.1 Emerging trends in formulation

The cutting edge of innovation in micro/nanoemulsions for wound healing is the creation of improved formulations. To rectify the shortcomings of existing formulations and enhance treatment results, researchers are investigating new components, delivery mechanisms, and functional characteristics:

- *Multifunctional emulsions:* Future formulations are likely to incorporate multiple therapeutic agents, such as antimicrobials, anti-inflammatory drugs, and growth factors, within a single emulsion. These multifunctional emulsions can address the complex and multifaceted nature of wound healing, providing a comprehensive approach to treatment.
- *Natural and sustainable ingredients:* There is a growing emphasis on the use of natural and sustainable ingredients, such as plant-based oils, biodegradable polymers, and bioactive compounds derived from natural sources. These ingredients not only enhance the biocompatibility and safety of emulsions but also align with the global trend toward sustainable and eco-friendly healthcare solutions.
- *Stimuli-responsive systems*: Advances in stimuli-responsive emulsions, which release their payload in response to specific triggers such as pH, temperature, or enzyme activity, are expected to play a key role in future formulations. These systems ensure targeted and controlled delivery of therapeutic agents, minimizing side effects and maximizing efficacy [59].
- *Nanotechnology-driven innovations:* The integration of nanotechnology into emulsion formulations is expected to continue, with the development of advanced nanocarriers, such as lipid nanoparticles, polymeric nanoparticles, and hybrid systems. These nanocarriers can enhance the stability, bioavailability, and therapeutic efficacy of emulsions, paving the way for more effective wound care solutions [60].

5.8.2 Integration with digital health technologies

There is a great potential for digital health technologies in improvement of wound care by combining micro/nanoemulsions with real-time monitoring, data-driven decision-making, and patient interaction:
-   *Smart dressings:* The development of smart dressings embedded with sensors and micro/nanoemulsions is a promising trend. These dressings can monitor wound parameters, such as pH, temperature, and moisture levels, in real time and release therapeutic agents as needed. This approach ensures timely and precise treatment, improving healing outcomes.
-   *Telemedicine and remote monitoring:* Digital health platforms, such as telemedicine and mobile health applications, can facilitate remote monitoring of wound healing progress. Patients can share data on wound conditions with healthcare providers, who can then adjust treatment plans accordingly. This integration enhances accessibility to care, particularly for patients in remote or underserved areas.
-   *Artificial intelligence (AI) and machine learning:* AI and machine learning algorithms can analyze large datasets on wound healing to identify patterns, predict outcomes, and optimize treatment strategies. These technologies can guide the development of personalized micro/nanoemulsion formulations and improve clinical decision-making [61].

5.8.3 Role in personalized medicine

Micro- and nanoemulsions are in a prime position to contribute significantly to the paradigm change towards personalized medicine, which is a major movement in the healthcare industry. Personalized medicine seeks to enhance treatment results while decreasing adverse effects by customizing formulations to each patient's specific requirements:
-   *Patient-specific formulations:* Advances in genomics, proteomics, and metabolomics are enabling the development of patient-specific formulations. For example, genetic profiling can identify patients who are more likely to benefit from specific growth factors or antimicrobial agents, allowing for the customization of micro/nanoemulsions to their unique needs.
-   *Precision delivery systems:* Micro/nanoemulsions can be designed to target specific cell types or tissues within the wound, enhancing the precision of drug delivery. For instance, surface modifications with ligands or antibodies can enable targeted delivery to fibroblasts or endothelial cells, promoting tissue regeneration and angiogenesis.

– *Adaptive therapies:* The development of adaptive therapies, which adjust their activity based on real-time feedback from the wound environment, is another promising trend. For example, emulsions that release therapeutic agents in response to changes in wound pH or enzyme activity can provide personalized and dynamic treatment [62].

5.9 Case studies and clinical evidence

An increasing amount of clinical data and case studies supports the effectiveness and potential of micro/nanoemulsions in wound healing. The advantages, disadvantages, and practical aspects of using micro/nanoemulsions in wound care may be better understood via these comparative research and real-world applications. In order to emphasize their importance in progressing the area, the following is an exhaustive examination of effective applications and comparative effectiveness studies.

5.9.1 Successful applications

Acute injuries, chronic wounds, burns, and wounds caused by surgery are just some of the many wound types that have benefited from the use of micro/nanoemulsions in clinical settings. The efficacy and adaptability of micro/nanoemulsions in enhancing healing and patient outcomes are shown by these fruitful applications:

– *DFUs:* They are a common and challenging complication of diabetes, often leading to infections and amputations. A case study involving the use of a nanoemulsion loaded with curcumin and silver nanoparticles demonstrated significant improvements in wound closure rates and reduced infection rates in patients with DFUs. The nanoemulsion provided sustained release of curcumin, which reduced inflammation, and silver nanoparticles, which prevented bacterial colonization [63].

– *Burn wounds:* In a clinical trial, a microemulsion containing honey and aloe vera was used to treat second-degree burn wounds. The emulsion promoted rapid re-epithelialization, reduced pain, and minimized scarring compared to traditional dressings. The natural bioactive compounds in the emulsion provided antimicrobial, anti-inflammatory, and moisturizing effects, creating an optimal environment for healing [64].

– *Chronic venous leg ulcers:* A study involving a nanoemulsion loaded with growth factors (EGF and VEGF) and an antimicrobial agent (ciprofloxacin) showed promising results in treating chronic venous leg ulcers. The emulsion accelerated wound closure, improved angiogenesis, and reduced bacterial load, addressing the key challenges associated with chronic wounds [65].

– *Post-surgical wounds:* In a post-surgical wound care study, a microemulsion containing dexamethasone and hyaluronic acid was applied to surgical incisions. The emulsion reduced inflammation, promoted collagen deposition, and minimized scarring, leading to faster recovery and improved cosmetic outcomes [33].

5.9.2 Comparative effectiveness studies

When compared to more conventional wound care practices and other cutting-edge treatments, the efficacy of micro/nanoemulsions may be better understood via comparative effectiveness studies. Clinical and economic value of micro/nanoemulsions may be better understood and optimized with the aid of these investigations:

– *Micro/nanoemulsions versus traditional dressings:* A comparative study evaluated the effectiveness of a nanoemulsion containing tea tree oil and silver nanoparticles against traditional gauze dressings in treating infected wounds. The nanoemulsion demonstrated superior antimicrobial activity, faster wound closure, and reduced inflammation compared to gauze dressings. The study concluded that nanoemulsions offer a more effective and efficient alternative to traditional methods.

– *Micro/nanoemulsions versus hydrogels:* Another study compared the performance of a microemulsion loaded with growth factors and a commercially available hydrogel in treating DFUs. The microemulsion showed significantly higher rates of wound closure and improved angiogenesis, which can be attributed to its ability to provide sustained release of growth factors and maintain a moist wound environment.

– *Cost-effectiveness analysis:* A cost-effectiveness analysis compared the long-term economic impact of using micro/nanoemulsions versus standard wound care methods for chronic wounds. The study found that while the upfront costs of micro/nanoemulsions were higher, their ability to reduce healing times, prevent complications, and lower hospital readmissions resulted in significant cost savings over time.

– *Patient-reported outcomes:* Comparative studies have also examined patient-reported outcomes, such as pain levels, comfort, and quality of life, in individuals treated with micro/nanoemulsions versus traditional methods. Patients using micro/nanoemulsions reported lower pain levels, greater comfort, and higher satisfaction with their treatment, underscoring the patient-centered benefits of these advanced therapies [66].

5.10 Conclusion

A major step forward in medical research has been creating and using micro/nanoemulsions for wound healing. These cutting-edge delivery methods have shown incredible promise in providing focused, efficient, and patient-centered solutions to the complicated and multifaceted problems of wound healing. A multipurpose vehicle for therapeutic agent delivery, wound environment modulation, and tissue regeneration promotion, micro/nanoemulsions make use of their distinct physicochemical features. Their therapeutic effectiveness and versatility are highlighted by their capacity to reduce inflammation, fight bacteria, promote cell regeneration, and keep moisture in.

An increasing amount of clinical data and positive case studies demonstrate the revolutionary effects of micro/nanoemulsions in the treatment of various wound types, such as burns, chronic wounds, acute injuries, and wounds that have occurred after surgery. By showing better healing results, fewer complications, and more patient satisfaction, comparative effectiveness studies further highlight their superiority over conventional wound care treatments. In addition, digital health tools, nanotechnology, and stimuli-responsive systems have all come together to pave the way for more precise and personalized care of wounds, meeting the specific requirements of each patient.

Stability, scalability, regulatory clearance, and cost-effectiveness are still issues, despite the encouraging developments. Researchers, physicians, and business leaders must work together to find solutions to these problems. If these challenges can be overcome, micro/nanoemulsions can be commercialized and used by patients in a variety of healthcare contexts.

When digital health technology, personalized medicine, and sophisticated therapies all come together, it will revolutionize wound care. Innovative solutions that improve healing results, lower healthcare costs, and increase patients' quality of life are offered by micro/nanoemulsions, which are positioned to play a vital role in this paradigm shift. The area is always evolving, so there is a lot of room for future advancements in wound healing. This will lead to therapies that are more effective, efficient, and sustainable.

As a conclusion, micro/nanoemulsions provide a revolutionary strategy for wound healing by overcoming the shortcomings of conventional approaches and resolving all of the intricate problems associated with wound care. Healthcare practitioners have the opportunity to transform wound treatment by using these cutting-edge solutions. This will provide hope and healing to millions of patients throughout the globe. With the ongoing progress in research and invention, micro/nanoemulsions have great potential to aid in wound healing. Their status as a foundational tool in contemporary medicine is only going to strengthen.

Conflict of interest: No conflict of interest declared.

Funding: No agency provided any funds.

References

[1]   Fu W, et al. Opportunities and challenges of nanomaterials in wound healing: Advances, mechanisms, and perspectives. Chem Eng J, Sep. 2024;495, 153640, doi: 10.1016/j.cej.2024.153640

[2]   Saroj S, et al. Plant extracellular nanovesicle-loaded hydrogel for topical antibacterial wound healing *in vivo*. ACS Appl Bio Mater, Jan. 2025;8(1), 1–11, doi: 10.1021/acsabm.4c00992

[3]   Raziyeva K, Kim Y, Zharkinbekov Z, Kassymbek K, Jimi S, Saparov A. Immunology of acute and chronic wound healing. Biomolecules, May 2021;11(5), 700, doi: 10.3390/biom11050700

[4]   Demidova-Rice TN, Hamblin MR, Herman IM. Acute and impaired wound healing. Adv Skin Wound Care, Jul. 2012;25(7), 304–314, doi: 10.1097/01.ASW.0000416006.55218.d0

[5]   Patel Y, et al. Integrated image and location analysis for wound classification: A deep learning approach. Sci Rep, Mar. 2024;14(1), 7043, doi: 10.1038/s41598-024-56626-w

[6]   Wilkinson HN, Hardman MJ. Wound healing: Cellular mechanisms and pathological outcomes. Open Biol, Sep. 2020;10(9), doi: 10.1098/rsob.200223

[7]   Kumar D, Gupta SK, Verma A, Ajazuddin. Transferosome as nanocarrier for drug penetration enhancement across skin: A comparison with liposome and ethosome. Current Nanomedicine, Nov. 2024;15, doi: 10.2174/0124681873333341241030065955

[8]   Ali A, Saroj S, Saha S, Gupta SK, Rakshit T, Pal S. Glucose-responsive chitosan nanoparticle/poly(vinyl alcohol) hydrogels for sustained insulin release *in vivo*. ACS Appl Mater Interfaces, Jul. 2023;15(27), 32240–32250, doi: 10.1021/acsami.3c05031

[9]   Gupta SK, Verma A Magnetic Hydrogel and Its Biomedical Applications. In *Magnetic Polymer Composites and Their Emerging Applications*, 1st ed. S Ganguly, S Margels, P Das Eds., Boca Raton: CRC Press, 2024 ch. 8, pp. 138–159. doi: 10.1201/9781003454236

[10]   Nikolaev B, Yakovleva L, Fedorov V, Li H, Gao H, Shevtsov M. Nano- and microemulsions in biomedicine: From theory to practice. Pharmaceutics, Jul. 2023;15(7), 1989, doi: 10.3390/pharmaceutics15071989

[11]   Jadhav ST, Salunkhe VR, Bhinge SD. Nanoemulsion drug delivery system loaded with imiquimod: A QbD-based strategy for augmenting anti-cancer effects. Futur J Pharm Sci, Dec. 2023;9(1), 120, doi: 10.1186/s43094-023-00568-z

[12]   Szumała P, Kaplińska J, Makurat-Kasprolewicz B, Mania S. Microemulsion delivery systems with low surfactant concentrations: Optimization of structure and properties by glycol cosurfactants. Mol Pharm, Jan. 2023;20(1), 232–240, doi: 10.1021/acs.molpharmaceut.2c00599

[13]   Kazemi M, Mohammadifar M, Aghadavoud E, Vakili Z, Aarabi MH, Talaei SA. Deep skin wound healing potential of lavender essential oil and licorice extract in a nanoemulsion form: Biochemical, histopathological and gene expression evidences. J Tissue Viability, May 2020;29(2), 116–124, doi: 10.1016/j.jtv.2020.03.004

[14]   Sneha K, Kumar A. Nanoemulsions: Techniques for the preparation and the recent advances in their food applications. Innovative Food Sci Emerg Technol, Mar. 2022; 76, 102914, doi: 10.1016/j.ifset.2021.102914

[15]   Singh Y, et al. Nanoemulsion: Concepts, development and applications in drug delivery. J Control Release, Apr. 2017;252, 28–49, doi: 10.1016/j.jconrel.2017.03.008

[16]   Solans C, Izquierdo P, Nolla J, Azemar N, Garcia-Celma MJ. Nano-emulsions. Curr Opin Colloid Interface Sci, Oct. 2005;10(3–4), 102–110, doi: 10.1016/j.cocis.2005.06.004

[17]   Kumar M, Bishnoi RS, Shukla AK, Jain CP. Techniques for formulation of nanoemulsion drug delivery system: A review. Prev Nutr Food Sci, Sep. 2019;24(3), 225–234, doi: 10.3746/pnf.2019.24.3.225

[18]   Li Z, et al. Advances of spontaneous emulsification and its important applications in enhanced oil recovery process. Adv Colloid Interface Sci, Mar. 2020;277, 102119, doi: 10.1016/j.cis.2020.102119

[19]   Ryu K-A, Park PJ, Kim S-B, Bin B-H, Jang D-J, Kim ST. Topical delivery of coenzyme Q10-loaded microemulsion for skin regeneration. Pharmaceutics, Apr. 2020;12(4), 332, doi: 10.3390/pharmaceutics12040332

[20]   Alam MS, et al. Evaluation of wound healing potential and biochemical estimation of sage oil nanoemulsion on animal model. Pak J Pharm Sci, Jul. 2021;34:(4), 1385–1392.

[21]   Shakeel F, Baboota S, Ahuja A, Ali J, Aqil M, Shafiq S. Nanoemulsions as vehicles for transdermal delivery of aceclofenac. AAPS PharmSciTech, Oct. 2007;8(4), 191, doi: 10.1208/pt0804104

[22]   Date AA, Desai N, Dixit R, Nagarsenker M. Self-nanoemulsifying drug delivery systems: Formulation insights, applications and advances. Nanomedicine, Dec. 2010;5(10), 1595–1616, doi: 10.2217/nnm.10.126

[23]   Tenjarla S Microemulsions: An overview and pharmaceutical applications. Crit Rev Ther Drug Carrier Syst, 1999;16(5), 461–521.

[24]   Jaiswal M, Dudhe R, Sharma PK. Nanoemulsion: An advanced mode of drug delivery system. 3 Biotech, Apr. 2015;5(2), 123–127, doi: 10.1007/s13205-014-0214-0

[25]   Salvia-Trujillo L, Soliva-Fortuny R, Rojas-Graü MA, McClements DJ, Martín-Belloso O. Edible nanoemulsions as carriers of active ingredients: A review. nnu Rev Food Sci Technol, Feb. 2017;8(1), 439–466, doi: 10.1146/annurev-food-030216-025908

[26]   Faria-Silva AC, et al. Nanoemulsions for cosmetic products. In *Nanocosmetics*, Arun Nanda, Sanju Nanda, Tuan Anh Nguyen, Susai Rajendran, Yassine Slimani (Eds.), Elsevier, 2020, pp. 59–77, doi: https://doi.org/10.1016/B978-0-12-822286-7.00004-8

[27]   Mitra D. Microemulsion and its application: An inside story. Mater Today Proc, 2023;83, 75–82, doi: 10.1016/j.matpr.2023.01.149

[28]   Ozogul Y, et al. Recent developments in industrial applications of nanoemulsions. Adv Colloid Interface Sci, Jun. 2022;304, 102685, doi: 10.1016/j.cis.2022.102685

[29]   Verma A, Gupta SK. History and Techniques of Bioimaging. In *Magnetic Quantum Dots for Bioimaging*, 1st ed. vol. 1, AR Rajabzadeh, S Srinivasan, P Das, S Ganguly (Eds.), New York: CRC Press, 2023, pp. 22. doi: 10.1201/9781003319870

[30]   Shah P, Bhalodia D, Shelat P. Nanoemulsion: A pharmaceutical review. Syst Rev Pharm, 2010;1(1), 24, doi: 10.4103/0975-8453.59509

[31]   Akbari S, Nour AH. Emulsion types, stability mechanisms and rheology: A review. Int J Innov Res Sci Stud , Sep. 2018;1(1), 11–17, doi: 10.53894/ijirss.v1i1.4

[32]   Dolgachev VA, et al. Dermal nanoemulsion treatment reduces burn wound conversion and improves skin healing in a porcine model of thermal burn injury. J Burn Care Res, Nov. 2021;42(6), 1232–1242, doi: 10.1093/jbcr/irab118

[33]   Alhakamy N, et al. RETRACTED: Fluoxetine ecofriendly nanoemulsion enhances wound healing in diabetic rats: In vivo efficacy assessment. Pharmaceutics, May 2022;14(6), 1133, doi: 10.3390/pharmaceutics14061133

[34]   Farahani H, Barati A, Arjomandzadegan M, Vatankhah E. Nanofibrous cellulose acetate/gelatin wound dressing endowed with antibacterial and healing efficacy using nanoemulsion of Zataria multiflora. Int J Biol Macromol, Nov. 2020;162, 762–773, doi: 10.1016/j.ijbiomac.2020.06.175

[35]   Almukainzi M, et al. Gentiopicroside PLGA nanospheres: Fabrication, in vitro characterization, antimicrobial action, and in vivo effect for enhancing wound healing in diabetic rats. Int J Nanomed, Mar. 2022;17, 1203–1225, doi: 10.2147/IJN.S358606

[36]   Kumar V, et al. An updated review on nanoemulsion: Factory for food and drug delivery. Curr Pharma Biotechnol, Dec. 2024;25(17), 2218–2252, doi: 10.2174/0113892010267771240211124950

[37]   Chavda VP, et al. Nanoemulsions: Summary of a decade of research and recent advances. Nanomedicine, Mar. 2024;19(6), 519–536, doi: 10.2217/nnm-2023-0199

[38]   Rizg WY, et al. Tailoring of geranium oil-based nanoemulsion loaded with pravastatin as a nanoplatform for wound healing. Polymers (Basel), May 2022;14(9), 1912, doi: 10.3390/polym14091912

[39]   Back PI, et al. Hydrogels containing soybean isoflavone aglycones-rich fraction-loaded nanoemulsions for wound healing treatment – In vitro and in vivo studies. Colloids Surf B Biointerfaces, Dec. 2020;196, 111301, doi: 10.1016/j.colsurfb.2020.111301

[40]   Alam P, Shakeel F, Anwer MK, Foudah AI, Alqarni MH. Wound healing study of eucalyptus essential oil containing nanoemulsion in rat model. Journal of oleo science, 2018;67(8), 957–968, doi: 10.5650/jos.ess18005

[41]   Choudhary MK, Bodakhe SH, Gupta SK. Assessment of the antiulcer potential of moringa oleifera root-bark extract in rats. J Acupunct Meridian Stud, Aug. 2013;6(4), 214–220, doi: 10.1016/j.jams.2013.07.003

[42]   Sinha S, Gupta SK, Sahu P, Gupta A, Ajazuddin. Capsaicin: unveiling its therapeutic potential and pharmacological actions. Curr Bioact Compd, Dec. 2024;21, doi: 10.2174/0115734072338291241129095347

[43]   Gupta SK, Verma A Toxicity and Safety Aspects of Inulin. In *Inulin for Pharmaceutical Applications*, W Akram, N Mishra, T Haider Eds., Singapore: Springer Nature Singapore, 2025, 169–187, doi: 10.1007/978-981-97-9056-2_9

[44]   Shinde N, Chauhan AS, Gupta SK, Bodakhe SH, Pandey DP. Antifertility studies of curcumin and andrographolide combination in female rats. Asian Pac J Reprod, Sep. 2015;4(3), 188–194, doi: 10.1016/j.apjr.2015.06.012

[45]   Pollock A, Berge E. How to do a systematic review. International Journal of Stroke, Feb. 2018;13(2), 138–156, doi: 10.1177/1747493017743796

[46]   Okur ME, et al. Evaluation of burn wound healing activity of novel fusidic acid loaded microemulsion based gel in male Wistar albino rats. Saudi Pharmaceut J, Mar. 2020;28(3), 338–348, doi: 10.1016/j.jsps.2020.01.015

[47]   Liu Y, Liu Y, Deng J, Li W, Nie X. Fibroblast growth factor in diabetic foot ulcer: Progress and therapeutic prospects. Front Endocrinol (Lausanne), Oct. 2021;12, doi: 10.3389/fendo.2021.744868

[48]   Zheng S-Y, et al. Therapeutic role of growth factors in treating diabetic wound. World J Diabetes, Apr. 2023;14(4), 364–395, doi: 10.4239/wjd.v14.i4.364

[49]   Raya SA, Mohd Saaid I, Abbas Ahmed A, Abubakar Umar A. A critical review of development and demulsification mechanisms of crude oil emulsion in the petroleum industry. J Pet Explor Prod Technol, Apr. 2020;10(4), 1711–1728, doi: 10.1007/s13202-020-00830-7

[50]   Maatouk B, et al. Sulfated alginate/polycaprolactone double-emulsion nanoparticles for enhanced delivery of heparin-binding growth factors in wound healing applications. Colloids Surf B Biointerfaces, Dec. 2021;208, 112105, doi: 10.1016/j.colsurfb.2021.112105

[51]   Alam P, Ansari MJ, Anwer MK, Raish M, Kamal YKT, Shakeel F. Wound healing effects of nanoemulsion containing clove essential oil. Artif Cells Nanomed Biotechnol, Apr. 2017;45(3), 591–597, doi: 10.3109/21691401.2016.1163716

[52]   De Assis KMA, et al. Bicontinuous microemulsions containing Melaleuca alternifolia essential oil as a therapeutic agent for cutaneous wound healing. Drug Deliv Transl Res, Dec. 2020;10(6), 1748–1763, doi: 10.1007/s13346-020-00850-0

[53]   Shanmugapriya K, Kim H, Saravana PS, Chun B-S, Kang HW. Astaxanthin-alpha tocopherol nanoemulsion formulation by emulsification methods: Investigation on anticancer, wound healing, and antibacterial effects. Colloids Surf B Biointerfaces, Dec. 2018;172, 170–179, doi: 10.1016/j.colsurfb.2018.08.042

[54]   Vater C, et al. Lecithin-based nanoemulsions of traditional herbal wound healing agents and their effect on human skin cells. Eur J Pharm Biopharm, Jan. 2022;170, 1–9, doi: 10.1016/j.ejpb.2021.11.004

[55]   Page MJ, et al. The PRISMA 2020 statement: An updated guideline for reporting systematic reviews. BMJ, Mar. 2021;71, doi: 10.1136/bmj.n71

[56]   Ahmad N, et al. Ultrasonication techniques used for the preparation of novel Eugenol-Nanoemulsion in the treatment of wounds healings and anti-inflammatory. J Drug Deliv Sci Technol, Aug. 2018;46, 461–473, doi: 10.1016/j.jddst.2018.06.003

[57]   Abdellatif MM, Elakkad YE, Elwakeel AA, Allam RM, Mousa MR. Formulation and characterization of propolis and tea tree oil nanoemulsion loaded with clindamycin hydrochloride for wound healing: In-vitro and in-vivo wound healing assessment. Saudi Pharmaceut J, Nov. 2021;29(11), 1238–1249, doi: 10.1016/j.jsps.2021.10.004

[58]   Zain MSC, Edirisinghe SL, Kim C-H, De Zoysa M, Shaari K. Nanoemulsion of flavonoid-enriched oil palm (Elaeis guineensis Jacq.) leaf extract enhances wound healing in zebrafish. Phytomedicine Plus, Nov. 2021;1(4), 100124, doi: 10.1016/j.phyplu.2021.100124

[59]   Rajgopal B, et al. Emerging trends in hydrogels for the treatment of vaginal candidiasis: A comprehensive review. Feb. 2025;Recent Adv Anti-Infect. Drug Discov, 20, doi: 10.2174/0127724344348928250220063431

[60]   Guliani A, Pooja MV, Kumari A, Acharya A. Retaining the 'essence' of essential oil: Nanoemulsions of citral and carvone reduced oil loss and enhanced antibacterial efficacy via bacterial membrane perturbation. J Drug Deliv Sci Technol, Feb. 2021 61, 102243, doi: 10.1016/j.jddst.2020.102243

[61]   Chen M-Y. Progress in the application of artificial intelligence in skin wound assessment and prediction of healing time. Am J Transl Res, 2024;16(7), 2765–2776, doi: 10.62347/MYHE3488

[62]   Yousry C, Saber MM, Abd-Elsalam WH. A cosmeceutical topical water-in-oil nanoemulsion of natural bioactives: design of experiment, in vitro characterization, and in vivo skin performance against uvb irradiation-induced skin damages. Int J Nanomed, Jul. 2022;17, 2995–3012, doi: 10.2147/IJN.S363779

[63]   Sethuram L, Thomas J, Mukherjee A, Chandrasekaran N. A review on contemporary nanomaterial-based therapeutics for the treatment of diabetic foot ulcers (DFUs) with special reference to the Indian scenario. Nanoscale Adv, 2022;4(11), 2367–2398, doi: 10.1039/D1NA00859E

[64]   Gul H, et al. The therapeutic application of tamarix aphylla extract loaded nanoemulsion cream for acid-burn wound healing and skin regeneration. Medicina (B Aires), Dec. 2022;59(1), 34, doi: 10.3390/medicina59010034

[65]   Zhang Y, et al. Prevention and repair of ultraviolet b-induced skin damage in hairless mice via transdermal delivery of growth factors immobilized in a gel-in-oil nanoemulsion. ACS Omega, Mar. 2023;8(10), 9239–9249, doi: 10.1021/acsomega.2c07343

[66]   Chakraborty T, Gupta S, Nair A, Chauhan S, Saini V. Wound healing potential of insulin-loaded nanoemulsion with aloe vera gel in diabetic rats. J Drug Deliv Sci Technol, Aug. 2021;64, 102601, doi: 10.1016/j.jddst.2021.102601

[67]   Zulfakar MH, Pubadi H, Ibrahim SI, Hairul NM. Medium-chain triacylglycerols (MCTs) and their fractions in drug delivery systems : A systematic review. J Oleo Sci, 2024;73(3), 23204, doi: 10.5650/jos.ess23204

[68]   Luo Y, Vivaldi Marrero E, Choudhary V, Bollag WB. Phosphatidylglycerol to Treat Chronic Skin Wounds in Diabetes. Pharmaceutics, May 2023;15(5), 1497, doi: 10.3390/pharmaceutics15051497

[69]   Atef B, Ishak RAH, Badawy SS, Osman R. Exploring the potential of oleic acid in nanotechnology-mediated dermal drug delivery: An up-to-date review. J Drug Deliv Sci Technol, Jan. 2022 67, 103032, doi: 10.1016/j.jddst.2021.103032

[70]   Shanmugarajan TS, Selvan NK, Uppuluri VNVA. Development and characterization of squalene-loaded topical agar-based emulgel scaffold: wound healing potential in full-thickness burn model. Int J Low Extrem Wounds, Dec 2021;20(4), 364–373, doi: 10.1177/1534734620921629

[71]   Ferro AC, De Souza Paglarini C, Rodrigues Pollonio MA, Lopes Cunha R. Glyceryl monostearate-based oleogels as a new fat substitute in meat emulsion. Meat Sci, Apr 2021;174, 108424, doi: 10.1016/j.meatsci.2020.108424

[72]   Kerwin BA. Polysorbates 20 and 80 used in the formulation of protein biotherapeutics: structure and degradation pathways. J Pharm Sci, Aug 2008;97(8), 2924–2935, doi: 10.1002/jps.21190

[73] Fiume MM, et al. Safety assessment of sorbitan esters as used in cosmetics. Int J Toxicol, Sep 2019;38(2_suppl), 60S–80S, doi: 10.1177/1091581819871877

[74] Gutiérrez-Méndez N, Chavez-Garay DR, Leal-Ramos MY. Lecithins: A comprehensive review of their properties and their use in formulating microemulsions. J Food Biochem, Jul. 2022;46(7), doi: 10.1111/jfbc.14157

[75] Asio JRG, Garcia JS, Antonatos C, Sevilla-Nastor JB, Trinidad LC. Sodium lauryl sulfate and its potential impacts on organisms and the environment: A thematic analysis. Emerg Contam, Mar. 2023;9(1), 100205, doi: 10.1016/j.emcon.2023.100205

[76] Flavia D, et al. Polymorphism, crystallinity and hydrophilic-lipophilic balance (HLB) of cetearyl alcohol and cetyl alcohol as raw materials for solid lipid nanoparticles (SLN). Aspects of Nanotechnology, Nov. 2018;1(1), doi: 10.36959/758/540

[77] Le Daré B, Gicquel T. Therapeutic applications of ethanol: A review. J Pharm Pharm Sci, Oct. 2019;22, 525–535, doi: 10.18433/jpps30572

[78] Fowles JR, Banton MI, Pottenger LH. A toxicological review of the propylene glycols. Crit Rev Toxicol, Apr. 2013;43(4), 363–390, doi: 10.3109/10408444.2013.792328

[79] D'souza AA, Shegokar R. Polyethylene glycol (PEG): A versatile polymer for pharmaceutical applications. Expert Opin Drug Deliv, Sep. 2016;13(9), 1257–1275, doi: 10.1080/17425247.2016.1182485

[80] Sharma K, et al. Saponins: A concise review on food related aspects, applications and health implications. Food Chemistry Advances, Oct. 2023;2, 100191, doi: 10.1016/j.focha.2023.100191

[81] Reichert CL, Salminen H, Weiss J. *Quillaja* Saponin characteristics and functional properties. Annu Rev Food Sci Technol, Mar. 2019;10(1), 43–73, doi: 10.1146/annurev-food-032818-122010

[82] Wahab S, et al. Glycyrrhiza glabra (Licorice): A comprehensive review on its phytochemistry, biological activities, clinical evidence and toxicology. Plants, Dec. 2021;10(12), 2751, doi: 10.3390/plants10122751

[83] Souto EB, Cano A, Martins-Gomes C, Coutinho TE, Zielińska A, Silva AM. Microemulsions and nanoemulsions in skin drug delivery. Bioengineering, Apr. 2022;9(4), 158, doi: 10.3390/bioengineering9040158

[84] Guo JW, Pu C-M, Liu C-Y, Lo S-L, Yen Y-H. Curcumin-loaded self-microemulsifying gel for enhancing wound closure. Skin Pharmacol Physiol, 2020;33(6), 300–308, doi: 10.1159/000512122

[85] Mirrezaei N, Yazdian-Robati R, Oroojalian F, Sahebkar A, Hashemi M. Recent developments in nano-drug delivery systems loaded by phytochemicals for wound healing. Mini Rev Med Chem, Dec 2020;20(18), 1867–1878, doi: 10.2174/1389557520666200807133022

[86] Garcia CR, et al. Nanoemulsion delivery systems for enhanced efficacy of antimicrobials and essential oils. Biomater Sci, 2022;10(3), 633–653, doi: 10.1039/D1BM01537K

[87] Malik MR, et al. Formulation and characterization of chitosan-decorated multiple nanoemulsion for topical delivery in vitro and ex vivo. Molecules, May 2022;27(10), 3183, doi: 10.3390/molecules27103183

[88] Çelebi N, Türkyilmaz A, Gönül B, Özogul C. Effects of epidermal growth factor microemulsion formulation on the healing of stress-induced gastric ulcers in rats. J Control Release, Oct. 2002;83(2), 197–210, doi: 10.1016/S0168-3659(02)00198-0

[89] Sosa AM, Igartúa DE, Martínez LM, Del V. Alonso S, Prieto MJ, Martinez CS. Multifunctional emulsion for potential wound-healing applications: A development for co-administration of essential vitamins, silver nanoparticles, and potent active ingredients. J Drug Deliv Sci Technol, Nov. 2024;101, 106128, doi: 10.1016/j.jddst.2024.106128

[90] Vigani B, Rossi S, Sandri G, Bonferoni MC, Caramella CM, Ferrari F. Hyaluronic acid and chitosan-based nanosystems: A new dressing generation for wound care. Expert Opin Drug Deliv, Jul. 2019;16(7), 715–740, doi: 10.1080/17425247.2019.1634051

[91] Leanpolchareanchai J, Teeranachaideekul V. Topical microemulsions: Skin irritation potential and anti-inflammatory effects of herbal substances. Pharmaceuticals, Jul. 2023;16(7), 999, doi: 10.3390/ph16070999

[92] Weimer P, Kirsten CN, De Araújo Lock G, Nunes KAA, Rossi RC, Koester LS. Co-delivery of beta-caryophyllene and indomethacin in the oily core of nanoemulsions potentiates the anti-inflammatory effect in LPS-stimulated macrophage model. Eur J Pharm Biopharm, Oct. 2023;191, 114–123, doi: 10.1016/j.ejpb.2023.08.020

[93] Emad NA, Sultana Y, Aqil M, Saleh A, Kamaly OA, Nasr FA. Omega-3 fatty acid-based self-microemulsifying drug delivery system (SMEDDS) of pioglitazone: Optimization, in vitro and in vivo studies. Saudi J Biol Sci, Sep. 2023;30(9), 103778, doi: 10.1016/j.sjbs.2023.103778

[94] Padilla MS, Tangsangasaksri M, Chang -C-C, Mecozzi S. MCT nanoemulsions for the efficient delivery of siRNA. J Pharm Sci, Mar. 2024;113(3), 764–771, doi: 10.1016/j.xphs.2023.11.013

[95] Song Y, et al. Advancements in the transdermal drug delivery systems utilizing microemulsion-based gels. Curr Pharm Des, Oct. 2024;30(35), 2753–2764, doi: 10.2174/0113816128305190240718112945

[96] Liu M, et al. Recent advances in nano-drug delivery systems for the treatment of diabetic wound healing. Int J Nanomed, Mar. 2023;18, 1537–1560, doi: 10.2147/IJN.S395438

[97] Hu C, Wang Q, Ma C, Xia Q. Non-aqueous self-double-emulsifying drug delivery system: A new approach to enhance resveratrol solubility for effective transdermal delivery. Colloids Surf A Physicochem Eng Asp, Jan. 2016;489, 360–369, doi: 10.1016/j.colsurfa.2015.11.017

[98] Ruan L, et al. Dual-delivery temperature-sensitive hydrogel with antimicrobial and anti-inflammatory brevilin a and nitric oxide for wound healing in bacterial infection. Gels, Mar. 2024;10 (4), 219, doi: 10.3390/gels10040219

[99] Ali M, et al. novel curcumin-encapsulated α -tocopherol nanoemulsion system and its potential application for wound healing in diabetic animals. Biomed Res Int, Jan. 2022;2022(1), doi: 10.1155/2022/7669255

[100] Kumari M, Nanda DK. Potential of Curcumin nanoemulsion as antimicrobial and wound healing agent in burn wound infection. Burns, Aug. 2023;49(5), 1003–1016, doi: 10.1016/j.burns.2022.10.008

[101] Ding Y, et al. Microemulsion-thermosensitive gel composites as *in situ* -forming drug reservoir for periodontitis tissue repair through alveolar bone and collagen regeneration strategy. Pharm Dev Technol, Jan. 2023;28(1), 30–39, doi: 10.1080/10837450.2022.2161574

[102] Zhang Y, et al. Prevention and repair of ultraviolet b-induced skin damage in hairless mice via transdermal delivery of growth factors immobilized in a gel-in-oil nanoemulsion. ACS Omega, Mar. 2023;8(10), 9239–9249, doi: 10.1021/acsomega.2c07343

[103] Kumari A, et al. Wound-healing effects of curcumin and its nanoformulations: A comprehensive review. Pharmaceutics, Oct. 2022;14(11), 2288, doi: 10.3390/pharmaceutics14112288

[104] Tashkandi H. Honey in wound healing: An updated review. Open Life Sci, Oct. 2021;16(1), 1091–1100, doi: 10.1515/biol-2021-0084

[105] Oryan A, Mohammadalipour A, Moshiri A, Tabandeh MR. Topical application of aloe vera accelerated wound healing, modeling, and remodeling. Ann Plast Surg, Jul. 2016;77(1), 37–46, doi: 10.1097/SAP.0000000000000239

[106] Chin KB, Cordell B. The effect of tea tree oil (*Melaleuca alternifolia*) on wound healing using a dressing model. J Altern Complement Med, Dec. 2013;19(12), 942–945, doi: 10.1089/acm.2012.0787

[107] Rajinikanth B, Rajkumar DSR, Keerthika K, Vijayaragavan V. Chitosan-based biomaterial in wound healing: A review. Cureus, Feb. 2024;doi: 10.7759/cureus.55193

[108] McDaniel JC, Belury M, Ahijevych K, Blakely W. Omega-3 fatty acids effect on wound healing. Wound Repair Regen, May 2008;16(3), 337–345, doi: 10.1111/j.1524-475X.2008.00388.x

[109] Jia Y, et al. Emerging effects of resveratrol on wound healing: A comprehensive review. Molecules, Oct. 2022;27(19), 6736, doi: 10.3390/molecules27196736

[110] Givol O, Kornhaber R, Visentin D, Cleary M, Haik J, Harats M. A systematic review of *Calendula officinalis* extract for wound healing. Wound Repair Regen, Sep. 2019;27(5), 548–561, doi: 10.1111/wrr.12737

[111] Hobson R. Vitamin E and wound healing: An evidence-based review. Int Wound J, Jun. 2016;13(3), 331–335, doi: 10.1111/iwj.12295

[112] Samuelson R, Lobl M, Higgins S, Clarey D, Wysong A. The effects of lavender essential oil on wound healing: A review of the current evidence. J Altern Complement Med, Aug. 2020;26(8), 680–690, doi: 10.1089/acm.2019.0286

[113] Nasrine A, et al. Neem (Azadirachta Indica) and silk fibroin associated hydrogel: Boon for wound healing treatment regimen. Saudi Pharmaceut J, Oct. 2023;31(10), 101749, doi: 10.1016/j.jsps.2023.101749

[114] Asadi SY, et al. Effect of green tea (Camellia sinensis) extract on healing process of surgical wounds in rat. Int J Surg, May 2013;11(4), 332–337, doi: 10.1016/j.ijsu.2013.02.014

[115] Niknam S, et al. Polyherbal combination for wound healing: Matricaria chamomilla L. and Punica granatum L. DARU J Pharm Sci, May 2021;29(1), 133–145, doi: 10.1007/s40199-021-00392-x

[116] Avola R, Granata G, Geraci C, Napoli E, Graziano ACE, Cardile V. Oregano (Origanum vulgare L.) essential oil provides anti-inflammatory activity and facilitates wound healing in a human keratinocytes cell model. Food Chem Toxicol, Oct. 2020;144, 111586, doi: 10.1016/j.fct.2020.111586

[117] Kapoor N, Yadav R. Manuka honey. Natl J Maxillofac Surg, May 2021;12(2), 233–237, doi: 10.4103/njms.NJMS_154_20

[118] Witkowska K, Paczkowska-Walendowska M, Garbiec E, Cielecka-Piontek J. Topical Application of centella asiatica in wound healing: Recent insights into mechanisms and clinical efficacy. Pharmaceutics, Sep. 2024;16(10), 1252, doi: 10.3390/pharmaceutics16101252

[119] Bahadoram M, Hassanzadeh S, Bahadoram S, Mowla K. Effects of pomegranate on wound repair and regeneration. World J Plast Surg, Mar. 2022;11(1), 157–159, doi: 10.52547/wjps.11.1.157

[120] Khezri K, Farahpour MR, Mounesi Rad S. Accelerated infected wound healing by topical application of encapsulated Rosemary essential oil into nanostructured lipid carriers. Artif Cells Nanomed Biotechnol, Dec. 2019;47(1), 980–988, doi: 10.1080/21691401.2019.1582539

[121] El-Sakhawy M, Salama A, Tohamy H-AS. Applications of propolis-based materials in wound healing. Arch Dermatol Res, Dec. 2023;316(1), 61, doi: 10.1007/s00403-023-02789-x

Amrita Thakur* and Gunjan Jeswani
Chapter 6
Precision oncology: nanoemulsions as pioneers in the delivery of anticancer agents, immunomodulators, and vaccines

Abstract: Precision oncology represents a paradigm shift in cancer treatment, emphasizing tailored therapeutic strategies based on individual patient profiles. Among the innovative drug delivery systems revolutionizing this field, micro- and nanoemulsions have emerged as promising carriers for anticancer agents, immunomodulators, and vaccines. These colloidal systems offer several advantages, including enhanced solubility of poorly water-soluble drugs, targeted delivery, controlled release, and reduced systemic toxicity. This chapter delves into the design, formulation, and optimization of nanoemulsions, highlighting their physicochemical properties and how these contribute to improved therapeutic outcomes in oncology. Emphasis is placed on their role in delivering chemotherapeutic agents with precision, modulating immune responses for immunotherapy, and serving as platforms for cancer vaccines. The integration of these systems with advanced targeting ligands, imaging agents, and personalized medicine approaches is also explored, underscoring their transformative potential in cancer care. The chapter concludes with a discussion of current challenges, regulatory perspectives, and future directions for clinical translation.

Keywords: Precision oncology, nanoemulsions, targeted drug delivery, immunomodulators, cancer vaccines

6.1 Introduction

Nanoemulsions are kinetically stable colloidal dispersions of two immiscible liquids, stabilized by surfactants or co-surfactants, with droplet sizes typically ranging between 20 and 200 nm. They are designed to overcome the limitations of conventional drug delivery systems, such as low bioavailability, poor aqueous solubility, and inefficient targeting of therapeutic agents [1, 2]. The nanoscale size of the droplets confers unique properties, including enhanced surface area, high kinetic stability, and the ability to solubilize hydrophobic drugs, making nanoemulsions an attractive platform

*Corresponding author: Amrita Thakur, School of Pharmacy, Vishwakarma University, Survey No. 2, 3,4, Kondhwa Main Rd, Laxmi Nagar, Betal Nagar, Kondhwa, Pune 411048, Maharashtra, India
Gunjan Jeswani, School of Pharmacy Shri Shankaracharya Professional University, Bhilai 490020, Chhattisgarh, India

https://doi.org/10.1515/9783111593654-007

for pharmaceutical applications [3–5]. Nanoemulsions can be classified into oil-in-water (O/W), water-in-oil (W/O), and bi-continuous systems based on the phase composition. These formulations have gained significant attention due to their tenable properties and ability to encapsulate a wide range of drugs, including small molecules, peptides, and nucleic acids [6, 7]. Advances in preparation techniques, such as high-pressure homogenization, ultrasonication, and low-energy emulsification methods, have enabled the scalable production of nanoemulsions, while maintaining their stability and reproducibility [8–10].

6.1.1 Significance in cancer therapy, immunomodulation, and vaccines

Nanoemulsions have demonstrated substantial potential in cancer therapy due to their ability to selectively deliver chemotherapeutic agents to tumor tissues via enhanced permeability and retention (EPR) effects, thereby reducing systemic toxicity [11, 12]. Their surface can be modified with ligands like folic acid or antibodies to improve active targeting of cancer cells, further enhancing therapeutic efficacy [13, 14]. For instance, paclitaxel-loaded nanoemulsions have shown improved cytotoxicity against cancer cells, while minimizing side effects in preclinical studies [15, 16]. In the field of immunomodulation, nanoemulsions facilitate the delivery of immunotherapeutics, including cytokines and monoclonal antibodies, to specific immune cells, thereby improving immune response modulation [17, 18]. Recent advancements in vaccine development have also leveraged nanoemulsion technology to create more effective adjuvant systems. Nanoemulsions can enhance antigen uptake by antigen-presenting cells, improve cross-presentation, and stimulate robust cellular and humoral immune responses [19, 20]. For example, squalene-based nanoemulsions have been utilized as adjuvants in influenza vaccines, demonstrating improved immunogenicity and safety profiles [21, 22].

6.1.2 Advantages of nanoemulsions over conventional systems

Nanoemulsions offer several advantages over conventional drug delivery systems, including enhanced solubilization of poorly water-soluble drugs, improved drug stability, and prolonged systemic circulation time [23, 24]. Their nanoscale size facilitates passive targeting of diseased tissues via the EPR effect, enabling site-specific drug delivery [25, 26]. Additionally, their high surface area allows for controlled drug release, minimizing peak plasma concentrations and associated side effects [27]. Another significant advantage is their ability to bypass physiological barriers, such as the blood–brain barrier (BBB), which restricts the entry of many therapeutic agents into the central nervous system [28]. Nanoemulsions have been successfully used to deliver drugs for the treatment of neurological disorders like Alzheimer's and Parkinson's disease [29]. Furthermore, nanoemulsions exhibit high biocompatibility and biodegradability, making them safe and effective for clinical use [30]. Table 6.1 provides a comparative overview of various drug delivery systems over the conventional drug delivery system, explaining the advantages and applications of these systems.

Table 6.1: Comparative overview of drug delivery systems.

Drug delivery system	Advantages	Limitations	Applications	Reference
Nanoemulsions	Enhanced solubilization, targeted drug delivery, and improved bioavailability	Potential stability issues and scalability challenges	Cancer therapy, immunotherapy, and vaccine delivery	[31, 32]
Liposomes	Biocompatibility, reduced toxicity, and prolonged circulation time	Prone to leakage and expensive production	Antifungal drugs and gene therapy	[33, 34]
Polymeric micelles	Good solubilization of hydrophobic drugs and enhanced stability	Limited loading capacity and possible cytotoxicity	Chemotherapy and protein drug delivery	[35, 36]
Solid lipid nanoparticles (SLNs)	Controlled drug release and good biocompatibility	Risk of drug expulsion and low drug loading efficiency	Anti-inflammatory drugs and neurodegenerative disease treatment	[37, 38]
Nanocrystals	Enhanced dissolution rate and high drug loading	Requires surfactants for stability	Poorly water-soluble drugs	[39, 40]
Microemulsions	Thermodynamically stable and easy to scale-up	High surfactant concentration required	Topical formulations and transdermal delivery	[41, 42]
Self-nanoemulsifying drug delivery systems (SNEDDS)	Self-emulsifies in GI tract and improved absorption	Requires careful formulation optimization	Oral drug delivery	[43, 44]
Dendrimers	Precise molecular architecture and high drug-loading capacity	Potential toxicity and complex synthesis	Targeted drug delivery and gene therapy	[45, 46]
Inorganic nanoparticles	High stability and unique surface properties	Biocompatibility concerns and regulatory challenges	Imaging, diagnostics and photothermal therapy	[47, 48]

Table 6.1 (continued)

Drug delivery system	Advantages	Limitations	Applications	Reference
Carbon nanotubes	Large surface area and electrical properties	Toxicity and long-term retention in the body	Cancer drug delivery and biosensors	[49, 50]
Gold nanoparticles	High biocompatibility and ease of surface modification	Costly production and potential accumulation in tissues	Photothermal therapy and diagnostics	[51, 52]
Iron oxide nanoparticles	Magnetic properties enable targeted delivery	Potential oxidative stress and clearance issues	MRI contrast agents and drug targeting	[53, 54]
Nanosponges	High porosity and controlled drug release	Limited large-scale production	Drug detoxification and sustained release formulations	[55, 56]
Hydrogel nanoparticles	Biocompatible and hydrophilic nature	Swelling behavior affects drug release	Wound healing and tissue engineering	[57, 58]
Chitosan nanoparticles	Mucoadhesive and biodegradable	pH-Sensitive degradation	Ocular drug delivery and vaccine carriers	[59, 60]
Silica nanoparticles	High stability and large surface area	Biodegradability concerns	Drug carriers and gene delivery	[61, 62]
Protein nanoparticles	Biodegradable and nontoxic	Limited drug loading	Enzyme replacement therapy and vaccine delivery	[63, 64]
Peptide nanoparticles	High specificity and immunogenicity	Short shelf life	Cancer immunotherapy and peptide vaccines	[65, 66]
Quantum dots	Fluorescence properties enable tracking	Heavy metal toxicity	Bioimaging and theranostics	[67, 68]
Lipid nanoparticles (LNPs)	Protects mRNA and efficient delivery	Stability issues and potential immune response	mRNA vaccines (e.g., COVID-19 vaccines)	[69, 70]
Hybrid nanocarriers	Combines benefits of multiple systems	Complex synthesis and regulatory hurdles	Multimodal therapy and personalized medicine	[71, 72]
Ceramic nanoparticles	High stability and tunable properties	Brittle nature and limited biodegradability	Bone regeneration and dental applications	[73, 74]

Table 6.1 (continued)

Drug delivery system	Advantages	Limitations	Applications	Reference
Extracellular vesicles (EVs)	Natural origin and excellent biocompatibility	Isolation and scalability issues	Regenerative medicine and RNA delivery	[75, 76]
Metal-organic frameworks (MOFs)	High porosity and tunable structure	Potential toxicity and complex synthesis	Gas delivery and sustained drug release	[77, 78]

6.1.3 Limitations in the delivery of anticancer agents, immunomodulators, and vaccines

Despite significant advancements in drug delivery technologies, several limitations persist, particularly in the delivery of anticancer agents, immunomodulators, and vaccines. One primary challenge is poor bioavailability caused by rapid degradation, metabolism, and clearance of therapeutic agents before they reach their target site [79, 80]. Hydrophobic drugs like paclitaxel and docetaxel, commonly used in cancer therapy, exhibit low solubility, necessitating the use of toxic excipients or high doses to achieve therapeutic effects [81, 82]. This not only reduces efficacy but also increases systemic toxicity, leading to severe side effects such as myelosuppression, neuropathy, and gastrointestinal disturbances [83]. Target specificity remains another hurdle in cancer therapy. While conventional drugs indiscriminately affect cancerous and normal tissues, targeted delivery systems still face challenges such as off-target effects and difficulty penetrating the tumor microenvironment [84, 85]. Tumors often possess dense extracellular matrices and abnormal vasculature, hindering the effective transport and accumulation of therapeutic agents [86]. Similarly, immunomodulators, such as cytokines and monoclonal antibodies, encounter limitations due to systemic toxicity and insufficient activation of immune responses at the desired site [87, 88]. Vaccine delivery also faces numerous challenges, including the instability of antigens during formulation, storage, and administration, which compromises their immunogenicity [89, 90]. Conventional vaccine formulations require cold-chain storage and distribution, limiting access in resource-poor settings [91]. Additionally, traditional adjuvants often fail to induce robust, long-lasting immunity without causing undesirable inflammatory responses [92].

6.1.4 Opportunities for nanoemulsion-based systems

Nanoemulsion-based systems have emerged as a promising solution to address the challenges associated with traditional drug delivery methods. These systems offer enhanced solubility and bioavailability for hydrophobic drugs, enabling lower doses and reducing systemic toxicity [93, 94]. Their nanoscale size facilitates deeper penetration into tumor tissues, while surface modifications, such as ligand conjugation, enable targeted delivery to cancer cells, minimizing off-target effects [95, 96]. For immunomodulators, nanoemulsions provide a sustained release profile and protect sensitive therapeutic agents from enzymatic degradation, improving therapeutic efficacy [97, 98]. Furthermore, nanoemulsions can act as adjuvants in vaccine delivery, enhancing antigen uptake by antigen-presenting cells and promoting robust immune responses [99, 100]. Their ability to encapsulate both hydrophilic and hydrophobic compounds enables the co-delivery of antigens and adjuvants in a single formulation, improving vaccine stability and efficacy [101, 102]. Nanoemulsions also address the logistical challenges of vaccine delivery by enhancing the stability of antigens, reducing the dependency on cold-chain storage [103, 104]. This makes them particularly beneficial for improving vaccination coverage in remote and resource-constrained regions [105].

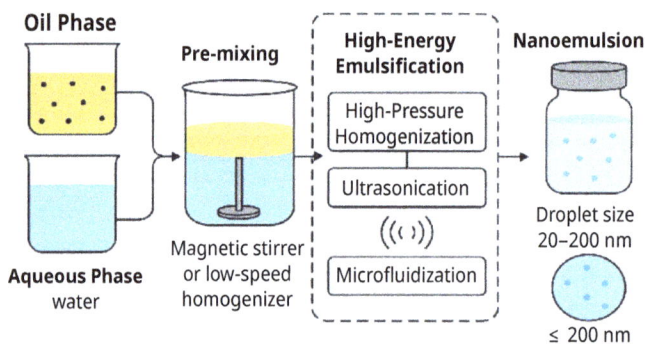

Figure 6.1: Schematic representation of nanoemulsion prepared using high-emulsification method.

6.2 Nanoemulsion technology: composition and mechanism

Nanoemulsions are submicron-sized colloidal systems, composed of distinct phases, including a lipid phase, an aqueous phase, surfactants, and co-surfactants. The lipid phase typically comprises oils such as medium-chain triglycerides, castor oil, or essential oils, which act as carriers for hydrophobic drugs [106, 107]. These oils improve drug solubility and stability by encapsulating hydrophobic molecules within their core. The aqueous phase constitutes water or water-miscible solvents that provide a continuous medium for dispersing lipid droplets [108]. Surfactants, such as polysorbates, lecithins, or sodium dodecyl sulfate, lower the interfacial tension between the lipid and aqueous phases, enabling the formation of stable nano-sized droplets [109, 110]. Co-surfactants, including alcohols or glycols, enhance the surfactant's efficacy by further reducing interfacial tension and stabilizing the system [111]. The balance between these components is critical to achieving a stable nanoemulsion with desirable physicochemical properties, such as droplet size, surface charge, and stability [112]. Figure 6.1 describes the process of preparation of nanoemulsions using high-energy emulsification. The preparation of a stable nanoemulsion involves mixing of aqueous and oil phases in the presence of a suitable surfactant and co-surfactants. This pre-emulsified mixture is then subjected to high-pressure homogenization to reduce the droplet size and ensure uniform dispersion. The resulting nano-emulsion typically exhibits droplet sizes ranging from 20 to 200 nm.

6.2.1 Formulation methods

6.2.1.1 High-energy methods

High-energy methods utilize mechanical energy to break down larger droplets into nano-sized particles. Ultrasonication employs high-frequency sound waves to create intense cavitation, resulting in droplet disruption and size reduction [113, 114]. High-pressure homogenization is another widely used method, where a coarse emulsion is passed through a narrow orifice under high pressure, creating intense shear and turbulence that produce nano-sized droplets [115]. These methods are particularly effective in producing small, uniform droplets, making them suitable for scalable production [116].

6.2.1.2 Low-energy methods

Low-energy methods rely on the spontaneous formation of nanoemulsions by exploiting the physicochemical properties of the system. Phase inversion involves altering the temperature or composition to achieve a phase transition that reduces interfacial tension, leading to the self-assembly of nano-sized droplets [117]. Spontaneous emulsification, on the other hand, occurs when an organic phase containing oil, surfactants, and co-surfactants is introduced into an aqueous phase under specific conditions, resulting in the formation of nanoemulsions without the need for external energy [118,

119]. These methods are cost-effective and suitable for thermosensitive drugs or compounds [120].

6.2.1.3 Mechanism of drug delivery

6.2.1.3.1 Drug encapsulation, absorption, and controlled release

Nanoemulsions serve as versatile carriers for both hydrophilic and hydrophobic drugs, encapsulating therapeutic molecules within their oil core or surfactant layers [121]. Encapsulation protects drugs from enzymatic degradation, enhancing their bioavailability and therapeutic efficacy [122]. Upon administration, nanoemulsions release drugs in a controlled manner, enabling sustained therapeutic levels over an extended period [123].

6.2.1.3.2 Role of size, charge, and surface properties in targeting tissues

The nanoscale size of droplets enhances drug penetration into tissues, especially in poorly vascularized regions like tumors [124]. The small size also facilitates endocytosis, allowing drugs to bypass biological barriers such as the blood–brain barrier [125]. Surface charge and properties influence biodistribution and cellular uptake. For instance, positively charged nanoemulsions exhibit enhanced interactions with negatively charged cell membranes, improving internalization [126]. Additionally, surface modifications, such as PEGylation or ligand conjugation, enable active targeting to specific tissues or receptors, further improving therapeutic outcomes [127, 128].

6.3 Delivery of anticancer agents using nanoemulsions

Recent advancements in nanoemulsion technology have shown promise in delivering key anticancer agents effectively. Nanoemulsions have been successfully used to deliver hydrophobic chemotherapeutic drugs such as doxorubicin, paclitaxel, and curcumin, improving their solubility, bioavailability, and therapeutic efficacy [129, 130]. Doxorubicin-loaded nanoemulsions have demonstrated improved pharmacokinetics and reduced cardiotoxicity in preclinical models, making them a safer alternative to conventional formulations [131]. Paclitaxel, a poorly water-soluble drug, has shown enhanced tumor accumulation and reduced systemic toxicity when delivered via nanoemulsions [132]. Similarly, curcumin, a natural compound with potent anticancer properties, has exhibited improved stability and bioavailability in nanoemulsion systems, leading to significant anticancer effects in vitro and in vivo [133, 134]. Synergistic drug combinations encapsulated in nanoemulsions have gained attention for their potential to overcome drug resistance and enhance therapeutic efficacy. For instance, combining doxorubicin and curcumin in a single nanoemulsion formulation has shown superior cytotoxicity against multidrug-resistant cancer cells, compared to individual drugs [135]. Other studies have explored dual encapsulation of paclitaxel and cisplatin, demonstrating synergistic effects and reduced side effects in preclinical tumor models [136, 137].

6.3.1 Targeting tumor microenvironment

Nanoemulsions are uniquely suited to target the tumor microenvironment (TME) due to their nanoscale size and modifiable surface properties. One of the primary mechanisms employed by nanoemulsions is the EPR effect, which allows nano-sized particles to passively accumulate in tumor tissues due to leaky vasculature and poor lymphatic drainage [138, 139]. This passive targeting increases drug concentration in the tumor site, reducing off-target effects and systemic toxicity [140]. Active targeting strategies further enhance the therapeutic potential of nanoemulsions. Surface modifications with ligands, antibodies, or peptides enable specific binding to overexpressed receptors on cancer cells or in the TME [141]. For example, folate-conjugated nanoemulsions have shown increased uptake in tumors expressing folate receptors, leading to enhanced anticancer activity [142]. Similarly, antibody-coated nanoemulsions targeting HER2 receptors have demonstrated efficacy in HER2-positive breast cancer models [143]. These targeted approaches improve drug delivery precision, minimize side effects, and enhance therapeutic outcomes.

6.3.2 Case studies and clinical trials

Recent preclinical and clinical research highlights the potential of nanoemulsions in oncology. Several studies have demonstrated the efficacy of nanoemulsion-based drug delivery systems in improving the therapeutic index of anticancer agents. For example, a preclinical study on doxorubicin-loaded nanoemulsions reported significant tumor growth inhibition in breast cancer xenograft models, with reduced cardiotoxicity, compared to conventional formulations [144]. Similarly, paclitaxel and doxorubicin nanoemulsions have shown superior efficacy in ovarian and lung cancer models, demonstrating increased tumor penetration and reduced adverse effects [145–147]. In clinical settings, several nanoemulsion-based formulations are under investigation. A notable example is curcumin-loaded nanoemulsions, which have been evaluated in randomized, double-blinded phase 1 controlled research for breast cancer patients to alleviate their joint pain, demonstrating improved patient outcomes [148]. A randomized, prospective, double-blinded pilot study compared the efficacy of three photosensitizers – hexylaminolevulinate, aminolevulinic acid nanoemulsion,

and methylaminolevulinate – in photodynamic therapy for superficial basal cell carci-
nomas [149]. Similarly, a randomized pilot study evaluated the effectiveness, safety,
and adherence of nanoemulsion curcumin at 50 mg and 100 mg doses in obese women
at high risk for breast cancer [150]. The research focused on curcumin's potential to reduce inflammation in breast and adipose tissue by modulating pro-inflammatory biomarkers. Reducing such inflammation may help lower the risk of developing breast
cancer. While true nanoemulsion formulations are still emerging, their safety and efficacy profiles in trials indicate their potential to revolutionize cancer therapy. Safety remains a critical consideration in the clinical translation of nanoemulsions. Recent trials
have reported favorable safety profiles, with minimal systemic toxicity and immunogenicity These findings underscore the potential of nanoemulsion-based systems to address longstanding challenges in cancer drug delivery, paving the way for future innovations.

6.4 Nanoemulsions in immunomodulation

Immunomodulators play a critical role in the treatment of various conditions, including cancer, autoimmune diseases, and infections. These agents are designed to either
stimulate or suppress the immune system to achieve therapeutic outcomes. In cancer
therapy, immunomodulators such as cytokines, checkpoint inhibitors, and monoclonal antibodies aim to enhance the body's immune response against tumors [151, 152].
In autoimmune diseases, these agents help regulate the immune system to prevent
attacks on the body's own tissues, while in infectious diseases, immunomodulators
boost the immune response to fight pathogens [153, 154]. Despite the potential of immunomodulators, their clinical efficacy is often limited by factors such as poor bioavailability, rapid clearance from the body, and instability. These limitations underscore the
need for advanced drug delivery systems that can improve the pharmacokinetics, stability, and targeting of immunomodulators. Nanoemulsions have emerged as a promising platform to address these challenges, offering enhanced delivery of immunomodulatory agents.

6.4.1 Nanoemulsions as immunomodulator carriers

Recent research has demonstrated the potential of nanoemulsions to serve as effective carriers for a wide range of immunomodulatory agents. These include cytokines,
monoclonal antibodies, and small-molecule immunomodulators, all of which require
controlled delivery to optimize their therapeutic effects. Nanoemulsions offer several
advantages, including the ability to encapsulate both hydrophobic and hydrophilic
compounds, provide sustained release, and improve the stability of sensitive mole-

cules [155, 156]. For example, cytokines such as interleukin-2 (IL-2) and interferons (IFNs) have been successfully delivered via nanoemulsion formulations, enhancing their efficacy in cancer immunotherapy by increasing their stability and reducing side effects [157]. Similarly, nanoemulsions have been utilized to deliver monoclonal antibodies, such as trastuzumab, improving the pharmacokinetics and reducing off-target effects [158, 159]. Furthermore, small-molecule immunomodulators like rapamycin and cyclosporine have been encapsulated in nanoemulsions to enhance their bioavailability and therapeutic potential in conditions like autoimmune diseases and organ transplantation [160, 161].

6.4.2 Challenges and solutions

One of the major challenges in immunomodulation therapy is overcoming immune system evasion, which can limit the effectiveness of treatments. The immune system often recognizes and clears therapeutic agents before they can exert their full effects. Nanoemulsions can address this issue by protecting immunomodulators from immune surveillance through their small size and by evading recognition by the immune system [162]. Additionally, surface modifications of nanoemulsions, such as PE-Gylation, can further improve circulation time and prevent premature elimination by the reticuloendothelial system [163]. Another significant challenge in the use of immunomodulators is improving their bioavailability and stability. Many immunomodulatory agents, particularly cytokines, are prone to degradation and require frequent dosing due to their short half-life. Nanoemulsions have been shown to enhance the stability of these agents by providing a protective environment within the formulation, thus increasing their shelf life and reducing the need for frequent administration [164, 165]. Moreover, nanoemulsions can be designed for controlled release, providing sustained drug delivery and reducing the need for high doses, thereby minimizing side effects [166].

6.4.3 Recent research

Recent studies have shown promising results in the use of nanoemulsions for the delivery of immunomodulators. For instance, a study demonstrated that IL-2-loaded nanoemulsions enhanced tumor growth inhibition and prolonged survival in a murine model of melanoma [167]. Similarly, the use of nanoemulsions for the delivery of monoclonal antibodies like trastuzumab has shown increased tumor targeting and reduced systemic toxicity, providing a more effective treatment option for HER2-positive breast cancer [159]. Additionally, nanoemulsions have been explored in the treatment of autoimmune diseases, with studies showing enhanced stability and bioavailability of immunosuppressive drugs like cyclosporine A when delivered in nanoemulsion formulations [168]. Clinical trials investigating the use of nanoemulsions in immunotherapy are also underway, with promising early results showing improved therapeutic outcomes and fewer side effects compared to conventional delivery systems [169, 170].

6.5 Nanoemulsions for vaccine delivery

Vaccine development has evolved significantly over the past few decades, with traditional methods focusing on inactivated or live-attenuated pathogens, subunit vaccines, and adjuvants to boost immune responses. However, these approaches often face challenges related to stability, efficiency, and the need for frequent booster doses. With the rise of new infectious diseases and the increasing demand for more effective vaccines, innovative strategies, including the use of advanced vaccine delivery systems, have gained considerable attention [171, 172]. Nanoemulsion-based vaccine delivery has emerged as a promising alternative to conventional systems. Nanoemulsions, due to their small size, biocompatibility, and ability to encapsulate both hydrophilic and hydrophobic antigens, offer several advantages over traditional delivery systems. These include improved antigen presentation, enhanced stability, and the potential for sustained release, making them ideal candidates for developing vaccines against various diseases, including infectious diseases and cancer [173, 174].

6.5.1 Nanoemulsion-based vaccines: recent innovations

The application of nanoemulsions in vaccine delivery has led to notable advancements, particularly in the development of vaccines for COVID-19, influenza, and cancer. In the case of COVID-19, several nanoemulsion-based vaccines have been developed, which enhance the immunogenicity of mRNA or protein-based antigens. These nanoemulsions provide an effective means of delivering antigens and adjuvants, leading to improved immune responses and reducing the need for higher doses or frequent boosters [175, 176]. Similarly, for influenza, nanoemulsion-based vaccine formulations have been explored to improve immune responses and ensure better protection against seasonal variants [177, 178]. In the realm of cancer immunotherapy, nanoemulsion-based vaccines are being studied to deliver tumor-associated antigens and adjuvants. These vaccines aim to activate the immune system to recognize and attack cancer cells. The ability of nanoemulsions to encapsulate and protect the antigens from degradation, while promoting efficient delivery to the target site, has been shown to significantly enhance the immunogenicity of cancer vaccines [179, 180]. The advantages of nanoemulsions in vaccine formulations also extend to their role as adjuvants. The small size and surface properties of nanoemulsions allow for increased interaction with immune cells, enhancing the uptake of antigens by antigen-presenting cells (APCs) and thereby stimulating stronger immune responses [181]. Additionally, nanoemulsions can be tailored to improve the solubility and stability of antigens, leading to more consistent and long-lasting immunity [182]. Figure 6.2 demonstrates the mechanism of nanoemulsions in boosting immune response in the form of vaccines. The figure also describes that the nanoemulsion in the form of vaccines increases the antigen uptake and stimulates humoral and cellular responses. This leads to a stronger and longer immune response.

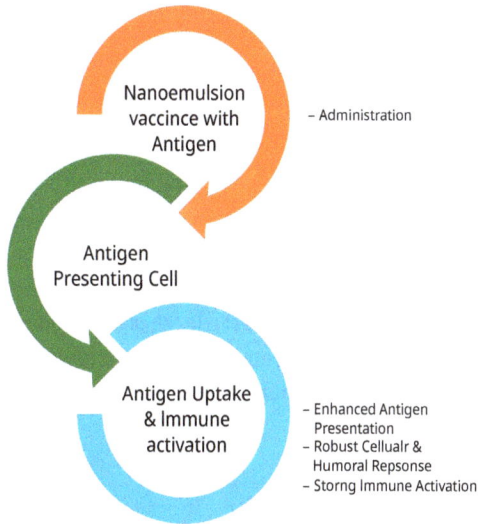

Figure 6.2: Mechanism of nanoemulsion in form of vaccine to enhance immune response.

6.5.2 Vaccine stability and immunogenic response

One of the key challenges in vaccine development is ensuring stability, particularly for vaccines that contain sensitive proteins or nucleic acids. Nanoemulsions have been found to improve the stability of these antigens by providing a protective environment within their lipid core, preventing degradation due to environmental factors such as temperature, pH, and enzymatic activity [183, 184]. Furthermore, nanoemulsions can facilitate controlled release of antigens, leading to more efficient presentation to the immune system and enhancing both humoral and cellular immune responses [185]. Studies have shown that nanoemulsions enhance the immunogenicity of various antigens by promoting their uptake by dendritic cells and other APCs. This leads to the activation of both innate and adaptive immune responses, resulting in a more robust and long-lasting immune protection. The controlled release properties of nanoemulsions also ensure that antigens are delivered in a sustained manner, reducing the frequency of booster doses and improving overall vaccine efficacy [186, 187].

6.5.3 Case studies

A growing body of clinical and preclinical research supports the use of nanoemulsion-based vaccines for a range of diseases. For example, clinical trials evaluating the use of nanoemulsion-based adjuvants in the COVID-19 vaccine development process have demonstrated enhanced immune responses and the potential for faster, more efficient vaccine production [188, 189]. Similarly, ongoing trials with nanoemulsion-based influenza vaccines are showing promising results in terms of enhanced protection against seasonal and pandemic strains [190, 191]. Cancer vaccines utilizing nanoemulsion delivery systems are also undergoing clinical evaluation. Nanoemulsion-based cancer vaccines have been shown to promote stronger immune responses and improved tumor targeting in preclinical models, with several studies moving toward clinical trials [192, 193]. In addition to these infectious disease and cancer vaccines, nanoemulsion-based systems are also being explored for their potential in vaccine delivery against diseases such as malaria, HIV, and tuberculosis [194, 195]. The commercialization of nanoemulsion-based vaccines is progressing, with several vaccine formulations already reaching advanced stages of clinical trials. These vaccines show promise not only in terms of efficacy but also in their potential for large-scale production, stability, and cost-effectiveness. As more research is conducted, nanoemulsion-based vaccines are expected to play an increasingly important role in global vaccination programs and disease prevention [196, 197].

6.5.4 Mechanisms of cellular uptake and targeting

The efficient delivery of nanoemulsion-based drug carriers to target cells and tissues largely depends on their ability to be taken up by cells. Nanoemulsions, due to their small size, can exploit several cellular uptake mechanisms, such as endocytosis, transcytosis, and receptor-mediated uptake, to facilitate the internalization of encapsulated drugs:

- **Endocytosis** is the most common pathway for nanoemulsion uptake, where the cell membrane engulfs the nanoemulsion particles and forms vesicles to internalize them into the cytoplasm. This process can occur through several forms, such as clathrin-mediated, caveolae-mediated, and macropinocytosis, depending on the size, charge, and surface properties of the nanoemulsion [198, 199].
- **Transcytosis** allows nanoemulsions to cross cellular barriers, such as the blood-brain barrier (BBB) or the intestinal epithelium. This process involves the transport of nanoemulsion particles across polarized cells by vesicle-mediated translocation, ensuring that therapeutic agents can reach otherwise hard-to-access tissues [200].
- **Receptor-mediated uptake** involves the recognition and binding of nanoemulsions to specific cell surface receptors, followed by internalization. This pathway enables the targeted delivery of drugs to specific cells or tissues overexpressing

the corresponding receptors, enhancing the therapeutic efficacy and minimizing off-target effects [201, 202].

6.5.5 Strategies for improved targeting

To further enhance the targeting specificity and cellular uptake of nanoemulsions, various strategies are employed. One prominent approach is the functionalization of nanoemulsions with targeting moieties, such as antibodies, aptamers, peptides, or small molecules. These targeting ligands can selectively bind to specific receptors or antigens present on the surface of target cells, such as tumor cells or immune cells, thereby improving the efficiency of drug delivery to the desired site of action.

- **Antibodies** and aptamers are commonly used targeting ligands due to their high specificity and affinity for their corresponding targets. Functionalizing nanoemulsions with antibodies or aptamers enables the delivery of drugs to cancer cells, specific immune cells, or other disease-related tissues [203, 204].
- **Peptides**, such as RGD (arginine-glycine-aspartic acid) or TAT peptides, can also be used to target specific cell types or tissues. For instance, RGD peptides target integrin receptors overexpressed on cancer cells, while TAT peptides enhance cellular uptake by interacting with the cell membrane [205, 206].

Examples of receptor-targeted nanoemulsion formulations include nanoemulsions functionalized with anti-HER2 antibodies for targeting HER2-positive breast cancer cells, or aptamers targeting the epidermal growth factor receptor (EGFR) in tumor therapy [207, 208]. These targeted nanoemulsions have shown improved therapeutic outcomes, including enhanced tumor accumulation and reduced systemic toxicity.

6.6 Recent research in nanotoxicology

As the clinical application of nanoemulsions continues to expand, understanding their safety profile and potential toxicity is crucial. Nanotoxicology is the study of the toxic effects of nanoparticles, including nanoemulsions, on biological systems, and aims to identify strategies to mitigate risks associated with their use:

- **Toxicity considerations** include evaluating the potential for cellular damage, inflammation, and oxidative stress caused by nanoemulsions. The size, surface charge, and composition of nanoemulsions are key factors influencing their interaction with cells and tissues. For instance, positively charged nanoemulsions may induce more significant toxicity due to their increased interaction with negatively charged cell membranes, whereas larger particles may be cleared more effectively by the immune system [209, 210].

– **Mitigation strategies** focus on designing nanoemulsions with improved biocompatibility. Surface modifications with hydrophilic polymers, such as polyethylene glycol (PEG), can reduce the immunogenicity and toxicity of nanoemulsions by preventing their aggregation and enhancing their stability in biological fluids. Additionally, the use of biocompatible lipids and nontoxic surfactants can further reduce adverse effects [211, 212]. Recent studies have also highlighted the importance of in vitro and in vivo testing to assess the safety of nanoemulsion formulations, ensuring that they do not cause long-term toxicity or adverse immune responses. Moreover, regulatory frameworks are being developed to ensure that nanoemulsion-based drug delivery systems meet safety standards before clinical use [213, 214].

6.7 Combination therapies

Recent advancements in nanoemulsion-based drug delivery have enabled the development of multimodal therapies that integrate chemotherapy, immunotherapy, and vaccines within a single platform. This approach enhances therapeutic efficacy by targeting different aspects of disease progression simultaneously, improving patient outcomes and overcoming resistance mechanisms [215, 216]. For instance, co-delivery of chemotherapeutic agents and immune checkpoint inhibitors in nanoemulsions has demonstrated synergistic effects – boosting tumor regression, while reactivating the immune system [217, 218]. A notable example includes a nanoemulsion formulation combining paclitaxel with anti-PD-1 antibodies, enabling targeted delivery to tumors and enhancing both cytotoxic and immune-mediated responses [219, 220]. Similarly, integrating cancer vaccines with chemotherapeutics in a single nanoemulsion, such as peptide-based vaccines with doxorubicin, has shown promising preclinical results, including heightened anti-tumor immunity and tumor regression [221, 222].

6.8 Challenges and future prospects

6.8.1 Regulatory challenges

As the use of nanoemulsions in drug delivery continues to expand, regulatory challenges remain a significant barrier to their widespread adoption in clinical settings. Regulatory authorities, such as the U.S. Food and Drug Administration (FDA) and the European Medicines Agency (EMA), have stringent guidelines for the approval of novel drug delivery systems. One of the primary concerns is the lack of standardized protocols for the evaluation of nanoemulsion products, which complicates their regulatory approval process. Despite the advantages of nanoemulsions, their complexity in composition, stability, and performance presents difficulties for regulators in ensuring product safety and efficacy. Moreover, the long-term toxicity and the potential environmental impact of nanomaterials are critical considerations that need to be thoroughly addressed before they can be approved for widespread clinical use [223, 224]. FDA has initiated the Nanotechnology Task Force to provide guidance on the regulation of nanomaterials, including nanoemulsions. However, regulatory uncertainty still exists regarding the quantification and characterization of nanoparticles, especially in terms of their size distribution, surface properties, and release profiles. Similarly, the EMA has been slow to establish uniform guidelines, and its regulatory approach remains fragmented, with varying requirements across member states. To address these challenges, both agencies are working toward creating clear, harmonized guidelines to streamline the approval of nanoemulsion-based drug delivery systems [225].

6.8.2 Challenges in large-scale production

Scaling up the production of nanoemulsions for commercial applications presents another set of challenges, primarily related to scalability, reproducibility, and stability. The methods used for preparing nano-emulsions, such as ultrasonication, high-pressure homogenization, and phase inversion, work well in laboratory-scale formulations but often face difficulties when adapted to large-scale production:

- **Scalability**: While techniques like ultrasonication can produce nanoemulsions with high stability and small droplet sizes at the laboratory scale, these processes may not be easily scalable due to energy consumption **and** high operational costs. Furthermore, scaling up these processes can lead to issues with uniformity and reproducibility, which are critical for ensuring consistent product quality [226, 227].
- **Stability**: Nanoemulsions are inherently susceptible to physical instability, including phase separation, creaming, and coalescence. Achieving long-term stability, while maintaining their therapeutic efficacy, requires the optimization of both formulation and processing conditions. Factors such as temperature, ionic strength, and pH can affect the physical properties of nanoemulsions, making it challenging to produce large quantities with consistent quality [228]. Additionally, ensuring the stability of bioactive compounds encapsulated in nanoemulsions during large-scale manufacturing processes is an ongoing issue that needs to be addressed for their commercialization.

6.9 Future directions

As nanoemulsion technology continues to evolve, several future directions in drug delivery, targeting, and formulation techniques hold great promise:
– **Innovations in drug delivery and targeting**: One of the most exciting prospects for nanoemulsions is their potential for targeted drug delivery. Recent advances in surface functionalization techniques allow for the attachment of specific targeting ligands, such as antibodies, aptamers, and peptides, to the surface of nanoemulsions. This enables the selective delivery of therapeutic agents to diseased tissues, such as tumors, while minimizing systemic side effects. Additionally, stimuli-responsive nanoemulsions that can release drugs in response to specific environmental triggers, such as pH, temperature, or enzymes, are also gaining attention for their potential to further enhance therapeutic efficacy [229, 230].
– **Personalized medicine**: Nanoemulsions hold great potential for personalized medicine, where therapies are tailored to individual patients based on their genetic and physiological characteristics. By incorporating patient-specific biomarkers and leveraging nanoemulsion formulations that can adapt to specific disease profiles, the precision of treatments can be significantly improved. Nanoemulsions can also enable the co-delivery of multiple drugs in personalized combinations, enhancing therapeutic outcomes for complex diseases like cancer, autoimmune disorders, and infections [231].
– **Advances in AI and computational modeling**: The integration of artificial intelligence (AI) and computational modeling in the design of nanoemulsions is an emerging area of research. AI can be used to predict the optimal formulation parameters for nanoemulsions, including the choice of lipids, surfactants, and drug candidates, to achieve desired properties such as size, stability, and drug release profiles. Additionally, computational models can simulate the interaction of nanoemulsions with biological systems, providing insights into their pharmacokinetics, toxicity profiles, and mechanisms of action. This approach could expedite the development of novel nanoemulsion-based therapies by reducing the time and cost associated with experimental trials [232]. Table 6.2 describes emerging technologies in nanoemulsions.

Table 6.2: Emerging technologies in nanoemulsion formulation.

Technology	Principle	Benefits	Potential applications	Reference
AI-assisted formulation design	Uses machine learning algorithms to optimize nanoemulsion parameters	Faster formulation development and improved reproducibility	Personalized medicine and targeted therapy	[233]
Hybrid nanoemulsions	Combines multiple nanocarriers (e.g., liposomes and nanoemulsions)	Enhanced stability and dual drug loading	Analgesic property and multifunctional vaccines	[234, 235]
Microfluidics-based nanoemulsion production	Uses microscale fluid dynamics to generate uniform droplets	Highly controlled size distribution and scalable production	Precision medicine and advanced drug formulations	[236, 237]
Ultrasound-assisted nanoemulsion synthesis	Uses sound waves for emulsification	Energy-efficient and high droplet uniformity	Drug delivery and cosmetic formulations	[238, 239]
Bioinspired nanoemulsions	Mimics natural biological systems	High biocompatibility and enhanced cell uptake	Food and cosmetic formulation	[240]
Stimuli-responsive nanoemulsions	Releases drugs in response to pH, temperature, or light	Controlled release and reduced side effects	Drug delivery	[241]
PEGylated nanoemulsions	Surface modification with PEG to improve circulation	Longer half-life, reduced immune response	Anticancer drug delivery	[242, 243]
Charge-reversible nanoemulsions	pH-dependent charge shifts for targeted delivery	Site-specific drug release and better absorption	Drug delivery	[244, 245]
Edible nanoemulsions	Food-grade formulations for oral drug delivery	Improved taste-masking and enhanced absorption	Nutraceuticals, functional foods	[246, 247]
Lipid-based nanoemulsions	Natural lipid components enhance bioavailability	High biocompatibility and reduced toxicity	Cancer therapy and drug delivery system	[248, 249]

6.10 Conclusion

Recent research into nanoemulsion-based drug delivery systems has underscored their significant potential in overcoming many of the challenges faced by conventional drug delivery methods. The advancements in the formulation and composition of nanoemulsions have shown promise in enhancing the solubility, stability, and bioavailability of various therapeutic agents, particularly those that are poorly soluble in water. Nanoemulsions have proven to be effective carriers for a wide range of bioactive compounds, including anticancer drugs, immunomodulators, and vaccine antigens, allowing for improved therapeutic outcomes across diverse disease areas. Key findings from recent studies have highlighted the versatility of nanoemulsions in addressing complex delivery challenges. For anticancer therapies, nanoemulsions have been successful in delivering drugs like doxorubicin and paclitaxel, improving their stability and targeting capabilities. Additionally, their potential for combination therapies – merging chemotherapy, immunotherapy, and vaccines – has opened new avenues for synergistic treatments, offering enhanced efficacy compared to single-drug therapies. The ability to target specific tissues, such as tumors or immune cells, further boosts their therapeutic potential, making them an ideal platform for precision medicine. However, despite these promising developments, several challenges remain. Regulatory hurdles continue to delay the widespread clinical use of nanoemulsion-based therapies, with inconsistent guidelines and safety concerns about long-term exposure to nanoparticles. Large-scale production also poses challenges, particularly in maintaining the reproducibility, stability, and uniformity of nanoemulsion formulations. These issues must be addressed to ensure the safe and effective commercialization of nanoemulsion-based drug delivery systems. Looking ahead, the future of nanoemulsion-based therapies is bright, with innovations in formulation techniques, targeting strategies, and personalized medicine continuing to drive progress. The integration of advanced technologies such as artificial intelligence and computational modeling will accelerate the design of more efficient and targeted nanoemulsion systems. As research progresses, it is likely that nanoemulsions will play a critical role in the development of next-generation therapies, particularly in areas like cancer treatment, immunomodulation, and vaccine delivery. With further refinement and overcoming existing challenges, nanoemulsion-based therapies are poised to revolutionize the landscape of modern medicine.

References

[1] Helmy SA, El-Morsi RM, Helmy SAM, El-Masry SM. Towards novel nano-based vaccine platforms for SARS-CoV-2 and its variants of concern: Advances, challenges and limitations. J Drug Deliv Sci Technol. Editions de Sante, 2022;76.

[2] Preeti SS, Malik R, Bhatia S, Al Harrasi A, Rani C, et al. Nanoemulsion: An emerging novel technology for improving the bioavailability of drugs. Scientifica. Hindawi Limited, 2023;2023.

[3] Thakur N, Garg G, Sharma PK, Kumar N. Nanoemulsions: A review on various pharmaceutical application. Glob J Pharmacol, 2012;6(3):222–225.

[4] DA R, PS L, Namdev G, Gole V, Rode A, Shaikh S. A overview on nanoemulsion. Asian J Res Pharm Sci [Internet], 2022 Aug 10 [cited 2025 May 12];12(3):239–244. Available from: https://ajpsonline.com/AbstractView.aspx?PID=2022-12-3-14

[5] Shakeel F, Ramadan W. Transdermal delivery of anticancer drug caffeine from water-in-oil nanoemulsions. Colloids Surf B Biointerfaces [Internet], 2010 Jan 1 [cited 2025 May 12];75(1):356–362. Available from: https://www.sciencedirect.com/science/article/abs/pii/S0927776509004202

[6] Yukuyama MN, Kato ETM, Lobenberg R, Bou-Chacra NA. Challenges and future prospects of nanoemulsion as a drug delivery system. Curr Pharm Des, 2016 Nov 3;23(3):495–508.

[7] Ganta S, Talekar M, Singh A, Coleman TP, Amiji MM. Nanoemulsions in translational research – Opportunities and challenges in targeted cancer therapy. AAPS PharmSciTech [Internet], 2014 [cited 2025 May 12];15(3):694–708. Available from: https://pubmed.ncbi.nlm.nih.gov/24510526/

[8] Kaur G, Panigrahi C, Agarwal S, Khuntia A, Sahoo M. Recent trends and advancements in nanoemulsions: Production methods, functional properties, applications in food sector, safety and toxicological effects. Food Phys, 2024 Sep;1:100024.

[9] Safaya M, Rotliwala YC. Nanoemulsions: A review on low energy formulation methods, characterization, applications and optimization technique. Mater Today Proc [Internet], 2020 Jan 1 [cited 2025 May 12];27:454–459. Available from: https://www.sciencedirect.com/science/article/abs/pii/S2214785319339537

[10] Kumar M, Bishnoi RS, Shukla AK, Jain CP. Techniques for formulation of nanoemulsion drug delivery system: A review. Prev Nutr Food Sci [Internet], 2019 [cited 2025 May 12];24(3):225. Available from: https://pmc.ncbi.nlm.nih.gov/articles/PMC6779084/

[11] Sánchez-López E, Guerra M, Dias-Ferreira J, Lopez-Machado A, Ettcheto M, Cano A, et al. Current applications of nanoemulsions in cancer therapeutics. Nanomaterials [Internet], 2019 May 31 [cited 2025 May 28];9:821. Available from: https://www.mdpi.com/2079-4991/9/6/821/htm

[12] Krishna G, Thrimothi D, Krishna G, Thrimothi D. Nano emulsions: A novel targeted delivery of cancer therapeutics. Nanoemul Des Appl [Working Title] [Internet], 2024 Mar 26 [cited 2025 May 28]; Available from: https://www.intechopen.com/chapters/1180267

[13] Ganta S, Talekar M, Singh A, Coleman TP, Amiji MM. Nanoemulsions in translational research – Opportunities and challenges in targeted cancer therapy. AAPS PharmSciTech [Internet], 2014 [cited 2025 May 29];15(3):694. Available from: https://pmc.ncbi.nlm.nih.gov/articles/PMC4037485/

[14] Kumar A, Singh AK, Chaudhary RP, Sharma A, Yadav JP, Pathak P, et al. Unraveling the multifaceted role of nanoemulsions as drug delivery system for the management of cancer. J Drug Deliv Sci Technol [Internet], 2024 Oct 1 [cited 2025 May 29];100:106056. Available from: https://www.sciencedirect.com/science/article/abs/pii/S1773224724007251

[15] Kim JE, Park YJ. Paclitaxel-loaded hyaluronan solid nanoemulsions for enhanced treatment efficacy in ovarian cancer. Int J Nanomed [Internet], 2017 [cited 2025 May 12];12:645–658. Available from: https://www.tandfonline.com/doi/pdf/10.2147/IJN.S124158

[16] Salata GC, Lopes LB. Phosphatidylcholine-based nanoemulsions for paclitaxel and a P-glycoprotein inhibitor delivery and breast cancer intraductal treatment. Pharmaceuticals [Internet], 2022 Sep 1 [cited 2025 May 12];15(9):1110. Available from: https://pmc.ncbi.nlm.nih.gov/articles/PMC9503599/

[17] Ribeiro EB, de Marchi PGF, Honorio-França AC, França EL, Soler MAG. Interferon-gamma carrying nanoemulsion with immunomodulatory and anti-tumor activities. J Biomed Mater Res A [Internet], 2020 Feb 1 [cited 2025 May 29];108(2):234–245. Available from: 10.1002/jbm.a.36808

[18] Gabrielle A, Dantas B, Limongi De Souza R, Rodrigues De Almeida A, Humberto F, Júnior X, et al. Development, characterization, and immunomodulatory evaluation of carvacrol-loaded nanoemulsion. Molecules [Internet], 2021 Jun 25 [cited 2025 May 29];26(13):3899. Available from: https://www.mdpi.com/1420-3049/26/13/3899/htm

[19] Orzechowska BU, Kukowska-Latallo JF, Coulter AD, Szabo Z, Gamian A, Myc A. Nanoemulsion-based mucosal adjuvant induces apoptosis in human epithelial cells. Vaccine [Internet], 2015 May 5 [cited 2025 May 12];33(19):2289–2296. Available from: https://www.sciencedirect.com/science/article/abs/pii/S0264410X1500287X

[20] Tayeb HH, Sainsbury F. Nanoemulsions in drug delivery: Formulation to medical application. Nanomedicine [Internet], 2018 Oct 1 [cited 2025 May 12];13(19):2507–2525. Available from: https://www.tandfonline.com/doi/abs/10.2217/nnm-2018-0088

[21] Lodaya RN, Brito LA, Wu TYH, Miller AT, Otten GR, Singh M, et al. Stable nanoemulsions for the delivery of small molecule immune potentiators. J Pharm Sci [Internet], 2018 Sep 1 [cited 2025 May 29];107(9):2310–2314. Available from: https://www.sciencedirect.com/science/article/abs/pii/S0022354918303162

[22] Zhang Z, Kuo JCT, Zhang C, Huang Y, Zhou Z, Lee RJ. A squalene-based nanoemulsion for therapeutic delivery of resiquimod. Pharmaceutics [Internet], 2021 Dec 1 [cited 2025 May 29];13(12):2060. Available from: https://www.mdpi.com/1999-4923/13/12/2060/htm

[23] Preeti SS, Malik R, Bhatia S, Al Harrasi A, Rani C, et al. Nanoemulsion: An emerging novel technology for improving the bioavailability of drugs. Scientifica (Cairo) [Internet], 2023 [cited 2025 May 12];2023:6640103. Available from: https://pmc.ncbi.nlm.nih.gov/articles/PMC10625491/

[24] Kapil Gawai HR. Preparation of nanoemulsion to enhance delivery of hydrophobic drugs. Int J Pharm Sci [Internet], 2025 [cited 2025 May 12];3:1970–1984. Available from: https://www.ijpsjournal.com/article/Preparation+Of+Nanoemulsion+to+Enhance+Delivery+of+Hydrophobic+Drugs

[25] Georgakopoulou VE, Papalexis P, Trakas N. Nanotechnology-based approaches for targeted drug delivery for the treatment of respiratory tract infections. J Biol Methods [Internet], 2024 [cited 2025 May 12];11(4):e99010032. Available from: https://pmc.ncbi.nlm.nih.gov/articles/PMC11744063/

[26] Nakamura Y, Mochida A, Choyke PL, Kobayashi H. Nano-drug delivery: Is the enhanced permeability and retention (EPR) effect sufficient for curing cancer? Bioconjug Chem [Internet], 2016 Oct 19 [cited 2025 May 12];27(10):2225. Available from: https://pmc.ncbi.nlm.nih.gov/articles/PMC7397928/

[27] Wilson RJ, Li Y, Yang G, Zhao CX. Nanoemulsions for drug delivery. Particuology [Internet], 2022 May 1 [cited 2025 May 12];64:85–97. Available from: https://www.sciencedirect.com/science/article/pii/S1674200121001176

[28] Chatterjee B, Gorain B, Mohananaidu K, Sengupta P, Mandal UK, Choudhury H. Targeted drug delivery to the brain via intranasal nanoemulsion: Available proof of concept and existing challenges. Int J Pharm [Internet], 2019 Jun 30 [cited 2025 May 12];565:258–268. Available from: https://www.sciencedirect.com/science/article/abs/pii/S0378517319303862

[29] Nirale P, Paul A, Yadav KS. Nanoemulsions for targeting the neurodegenerative diseases: Alzheimer's, Parkinson's and Prion's. Life Sci [Internet], 2020 Mar 15 [cited 2025 May 12];245. Available from: https://pubmed.ncbi.nlm.nih.gov/32017870/

[30] Tayeb HH, Sainsbury F. Nanoemulsions in drug delivery: Formulation to medical application. Nanomedicine [Internet], 2018 Oct 1 [cited 2025 May 12];13(19):2507–2525. Available from: https://www.tandfonline.com/doi/abs/10.2217/nnm-2018-0088

[31] Dudhe PS, Thakare R. Targeted drug delivery in cancer therapy. Int J Adv Res Sci, Comm Technol [Internet], 2024 Dec 4 [cited 2025 May 12];109–24. Available from: https://scispace.com/papers/targeted-drug-delivery-in-cancer-therapy-2036ede2n56k

[32]   Zeytunluoglu A, Arslan I. Current perspectives on nanoemulsions in targeted drug delivery. 2022
        Jan 1 [cited 2025 May 12];118–140. Available from: https://scispace.com/papers/current-perspectives
        -on-nanoemulsions-in-targeted-drug-3pd6nw6q

[33]   Lankalapalli S, Vinai Kumar Tenneti VS. Drug delivery through liposomes. Smart Drug Deliv
        [Internet], 2022 Jul 6 [cited 2025 May 12]; Available from: https://scispace.com/papers/drug-delivery
        -through-liposomes-bh3hq8fy

[34]   Ashutosh G, Verma S, Singh B, Yashwant JB. Liposomes: Current approaches for development and
        evaluation. J Drug Deliv [Internet], 2017 Dec 25 [cited 2025 May 12];7(04):269–275. Available from:
        https://scispace.com/papers/liposomes-current-approaches-for-development-and-evaluation
        -34srn60y4s

[35]   Wakaskar RR. Polymeric Micelles and their properties. J Nanomed Nanotechnol [Internet], 2017
        Mar 31 [cited 2025 May 12];2017(02):1–2. Available from: https://scispace.com/papers/polymeric-
        micelles-and-their-properties-4mln9bbo84

[36]   Farhoudi L, Maryam Hosseinikhah S, Vahdat-Lasemi F, Sukhorukov VN, Kesharwani P, Sahebkar
        A. Polymeric micelles paving the Way: Recent breakthroughs in camptothecin delivery for enhanced
        chemotherapy. Int J Pharm [Internet], 2024 May 1 [cited 2025 May 12];659:124292–124292. Available
        from: https://scispace.com/papers/polymeric-micelles-paving-the-way-recent-breakthroughs-in-
        daorrz3czc

[37]   Munir M, Zaman M, Waqar MA, Khan MA, Alvi MN. Solid lipid nanoparticles: A versatile approach
        for controlled release and targeted drug delivery. J Liposome Res [Internet], 2023 Oct 15 [cited
        2025 May 12];34(2):1–14. Available from: https://scispace.com/papers/solid-lipid-nanoparticles-a-
        versatile-approach-for-fw7w2655va

[38]   Yin Y, Zhang J, Zhou X. Solid lipid nanoparticles: A nano drug carrying system in treatment of
        nervous diseases. Highl Sci Eng Technol [Internet], 2022 Aug 23 [cited 2025 May 12];11:58–66.
        Available from: https://scispace.com/papers/solid-lipid-nanoparticles-a-nano-drug-carrying-system-
        in-3mbl9b33

[39]   Hadžiabdić J, Brekalo S, Rahić O, Tucak A, Sirbubalo M, Vranić E. Importance of stabilizers of
        nanocrystals of poorly soluble drugs. Macedonian Pharm Bull [Internet], 2020 Oct 29 [cited
        2025 May 12];66(03):145–146. Available from: https://scispace.com/papers/importance-of-stabilizers
        -of-nanocrystals-of-poorly-soluble-2j9kskpf8x

[40]   Zhou Y, Du J, Lulu W, Yancai W. Nanocrystals technology for improving bioavailability of poorly
        soluble drugs: A mini-review. J Nanosci Nanotechnol [Internet], 2017 Jan 1 [cited 2025 May 12];17
        (1):18–28. Available from: https://scispace.com/papers/nanocrystals-technology-for-improving-
        bioavailability-of-3rx42ejnwu

[41]   Kinge A, Kohale N, Yadav S, Bisen S, Sawarkar HSS. Microemulsions: A paradigm shift in drug
        delivery systems. Int J Pharm Res Appl [Internet], 2024 Jun 1 [cited 2025 May 12];09(06):610–618.
        Available from: https://scispace.com/papers/microemulsions-a-paradigm-shift-in-drug-delivery-
        systems-7gifywmx2c82

[42]   Mittal R, Md. Akhtar S, Bee R. A review: Microemulsion-based polymer matrix transdermal patch for
        the treatment of inflammation. J Pharm Negat Results [Internet], 2022 Oct 19 [cited
        2025 May 12];580–591. Available from: https://scispace.com/papers/a-review-microemulsion-based-
        polymer-matrix-transdermal-1b9y3nft

[43]   Annisa R, Mutiah R, Yuwono M, Hendradi E. Nanotechnology approach-self nanoemulsifying drug
        delivery system (snedds). Int J Appl Pharm [Internet], 2023 Jul 7 [cited 2025 May 12];15(4):12–19.
        Available from: https://scispace.com/papers/nanotechnology-approach-self-nanoemulsifying-drug-
        delivery-t1hp4qy9

[44]   Buya AB, Beloqui A, Memvanga PB, Préat V. Self-nano-emulsifying drug-delivery systems: From the
        development to the current applications and challenges in oral drug delivery. Pharmaceutics

[Internet], 2020 Dec 9 [cited 2025 May 12];12(12):1194. Available from: https://scispace.com/papers/self-nano-emulsifying-drug-delivery-systems-from-the-2nbmalg2hi

[45] Madaan K, Kumar S, Poonia N, Lather V, Pandita D. Dendrimers in drug delivery and targeting: Drug-dendrimer interactions and toxicity issues. J Pharm Bioallied Sci [Internet], 2014 Jul 1 [cited 2025 May 12];6(3):139–150. Available from: https://scispace.com/papers/dendrimers-in-drug-delivery-and-targeting-drug-dendrimer-bzyoi2f7fq

[46] Kheiriabad S, Dolatabadi JEN, Hamblin MR. Dendrimers for gene therapy. Dendrimer-Based Nanotherapeutics, 2021 Jan 1;285–309.

[47] Jiang S, Win KY, Liu S, Teng CP, Zheng Y, Han MY. Surface-functionalized nanoparticles for biosensing and imaging-guided therapeutics. Nanoscale [Internet], 2013 Mar 28 [cited 2025 May 12];5(8):3127–3148. Available from: https://scispace.com/papers/surface-functionalized-nanoparticles-for-biosensing-and-1s3eshsnnl

[48] Fraix A, Sortino S. Light-triggered unconventional therapies with engineered inorganic nanoparticles. Adv Inorg Chem [Internet], 2022 Jan 1 [cited 2025 May 12];80:171–203. Available from: https://scispace.com/papers/light-triggered-unconventional-therapies-with-engineered-3ssf3efn

[49] Aldosari H. Review of carbon nanotube toxicity and evaluation of possible implications to occupational and environmental health. Nano Hybrids Compos [Internet], 2023 Jul 31 [cited 2025 May 12];40:35–49. Available from: https://scispace.com/papers/review-of-carbon-nanotube-toxicity-and-evaluation-of-2c8ez98ynz

[50] Mendhe R, Dalal A. Carbon nanotubes an excellent targeted drug delivery system. RJPPD [Internet], 2024 Sep 2 [cited 2025 May 12];213–220. Available from: https://scispace.com/papers/carbon-nanotubes-an-excellent-targeted-drug-delivery-system-4untjrxroh02

[51] Kumar P, Chandra P, Verma N, Sharma A D, Mani M, et al. Gold nanoparticles: An emerging novel technology for targeted delivery system for site-specific diseases. Curr Drug Ther [Internet], 2024 Jul 18 [cited 2025 May 12];19. Available from: https://scispace.com/papers/gold-nanoparticles-an-emerging-novel-technology-for-targeted-1edxucpilz

[52] Fadel M, El-Kholy AI. Gold nanoparticles in photodynamic and photothermal therapy. Gold Nanoparticles Drug Deliv [Internet], 2024 Jan 1 [cited 2025 May 12];365–91. Available from: https://scispace.com/papers/gold-nanoparticles-in-photodynamic-and-photothermal-therapy-42gbi8jyi5

[53] Cotin G, Piant S, Mertz D, Felder-Flesch D, Begin-Colin S. Iron oxide nanoparticles for biomedical applications: Synthesis, functionalization, and application. Iron Oxide Nanoparticles Biomed Appl [Internet], 2018 Jan 1 [cited 2025 May 12];43–88. Available from: https://scispace.com/papers/iron-oxide-nanoparticles-for-biomedical-applications-3mzko3hold

[54] Bora M. Recent Bio-medical applications of iron oxide magnetic nanoparticles. J of ISAS [Internet], 2023 Apr 30 [cited 2025 May 12];1(4):56–72. Available from: https://scispace.com/papers/recent-bio-medical-applications-of-iron-oxide-magnetic-1gez24wr

[55] Dhumal GJ, Kulkarni AS, Bandal KK, Dambe RH, Thorat VR, Patil TT, et al. NANOSPONGES : A new approach for drug delivery system. Int J Sci Res Sci Technol [Internet], 2022 Dec 30 [cited 2025 May 12];9(6):170–178. Available from: https://ijsrst.com/IJSRST229617

[56] Meshram SI, Hatwar PR, Bakal RL, Rotake SB. An outlook towards nano-sponges: A unique drug delivery system and its application in drug delivery. GSC Biol Pharm Sci [Internet], 2024 Dec 11 [cited 2025 May 12];29(3):089–98. Available from: https://scispace.com/papers/an-outlook-towards-nano-sponges-a-unique-drug-delivery-4o3ju5ylkdh4

[57] Nadeem A, Saqib Z, Arif A, Abid L, Shahzadi H, Saghir A, et al. Hybrid hydrogels incorporating nanoparticles, their types, development, and applications: A comprehensive review of nanogels. Univers J Pharm Res [Internet], 2024 Nov 15 [cited 2025 May 12]; Available from: https://scispace.com/papers/hybrid-hydrogels-incorporating-nanoparticles-their-types-1c5i2pc0iki2

[58] Ren C, Wang T, Luo W, Pan X, Hu B, Li G, et al. Near-infrared-responsive nanofiber hydrogel with gradual drug release properties for wound healing. ACS Appl Nano Mater [Internet], 2024 Jun 21 [cited 2025 May 12];7(13):15517–15525. Available from: https://scispace.com/papers/near-infrared-responsive-nanofiber-hydrogel-with-gradual-427xkjfxap

[59] Shirsat Konkan Gyanpeeth Rahul NS, Mishra Konkan Gyanpeeth Rahul AC, Waghulde Konkan Gyanpeeth Rahul SO, Kale Konkan Gyanpeeth Rahul MK, Sheetal Roy Konkan Gyanpeeth Rahul CG, Roy SG, et al. A review on chitosan nanoparticles applications in drug delivery. J Pharmacogn Phytochem [Internet], 2018 Sep 1 [cited 2025 May 12];7(SP6):01–4. Available from: https://scispace.com/papers/a-review-on-chitosan-nanoparticles-applications-in-drug-1jjdr1n7ca

[60] Vichare R, Garner I, Paulson RJ, Tzekov R, Sahiner N, Panguluri SK, et al. Biofabrication of chitosan-based nanomedicines and its potential use for translational ophthalmic applications. Appl Sci [Internet], 2020 Jun 18 [cited 2025 May 12];10(12):4189. Available from: https://scispace.com/papers/biofabrication-of-chitosan-based-nanomedicines-and-its-3hctk0nxun

[61] Khaliq NU, Lee J, Kim J, Kim Y, Yu S, Kim J, et al. Mesoporous silica nanoparticles as a gene delivery platform for cancer therapy. Pharmaceutics [Internet], 2023 May 1 [cited 2025 May 12];15(5):1432–1432. Available from: https://scispace.com/papers/mesoporous-silica-nanoparticles-as-a-gene-delivery-platform-1fbacgak

[62] Mehmood A, Ghafar H, Yaqoob S, Gohar UF, Ahmad B. Mesoporous silica nanoparticles: A review. J Dev Drugs [Internet], 2017 Aug 16 [cited 2025 May 12];06(02):1–14. Available from: https://scispace.com/papers/mesoporous-silica-nanoparticles-a-review-4ber8ov28q

[63] Aljabali AAA, Rezigue M, Alsharedeh RH, Obeid MA, Mishra V, Serrano-Aroca Á, et al. Protein-based nanomaterials: A new tool for targeted drug delivery. Ther Deliv [Internet], 2022 Aug 4 [cited 2025 May 12];13(6):321–338. Available from: https://scispace.com/papers/protein-based-nanomaterials-a-new-tool-for-targeted-drug-13lfzfj8

[64] Kaltbeitzel J, Wich PR. Protein-based Nanoparticles: From Drug Delivery to Imaging, Nanocatalysis and Protein Therapy. In *Angewandte Chemie – International Edition*, vol. 62, John Wiley and Sons Inc, 2023.

[65] Zhang L, Huang Y, Lindstrom AR, Lin TY, Lam KS, Li Y. Peptide-based materials for cancer immunotherapy. Theranostics [Internet], 2019 Jan 1 [cited 2025 May 12];9(25):7807–7825. Available from: https://scispace.com/papers/peptide-based-materials-for-cancer-immunotherapy-2tnlnxq2vk

[66] Dai J, Ashrafizadeh M, Aref AR, Sethi G, Ertas YN. Peptide-functionalized, -assembled and -loaded nanoparticles in cancer therapy. Drug Discov Today [Internet], 2024 Apr 1 [cited 2025 May 12];29(7):103981–103981. Available from: https://scispace.com/papers/peptide-functionalized-assembled-and-loaded-nanoparticles-in-22rcem3kac

[67] Podder S. Fluorescent quantum dots, A technological marvel for optical bio-imaging: A perspective on associated in vivo toxicity. Appl Quantum Dots Biol Med Recent Adv [Internet], 2022 Jan 1 [cited 2025 May 12];143–163. Available from: https://scispace.com/papers/fluorescent-quantum-dots-a-technological-marvel-for-optical-3hacmmol

[68] Rasheed PA, Ankitha M, Pillai VK, Alwarappan S. Graphene quantum dots for biosensing and bioimaging. RSC Adv [Internet], 2024 May 15 [cited 2025 May 12];14(23):16001–16023. Available from: https://scispace.com/papers/graphene-quantum-dots-for-biosensing-and-bioimaging-2qn3t0r3h1

[69] Mobasher M, Ansari R, Castejon AM, Barar J, Omidi Y. Advanced nanoscale delivery systems for mRNA-based vaccines. Biochim Biophys Acta Gen Subj [Internet], 2024 Jan 1 [cited 2025 May 12];1868(3):130558–130558. Available from: https://scispace.com/papers/advanced-nanoscale-delivery-systems-for-mrna-based-vaccines-3pnh1zs9i4

[70] Tenchov R, Bird R, Curtze AE, Zhou Q. Lipid nanoparticles from liposomes to mRNA vaccine delivery, a landscape of research diversity and advancement. ACS Nano [Internet], 2021 Nov 23 [cited 2025 May 12];15(11):16982–17015. Available from: 10.1021/acsnano.1c04996

[71] A comprehensive review on the recent applications and development in nanocarriers. Asian J Pharm [Internet], 2024 Dec 1 [cited 2025 May 12];18(04). Available from: https://scispace.com/papers/a-comprehensive-review-on-the-recent-applications-and-525nkih0t9n4

[72] Kurtay G, Yılmaz M, Kuralay F, Demirel GB. Multifunctional therapeutic hybrid nanocarriers for targeted and triggered drug delivery: Recent trends and future prospects. Nanostructures Drug Deliv [Internet], 2017 Jan 1 [cited 2025 May 12];461–493. Available from: https://scispace.com/papers/multifunctional-therapeutic-hybrid-nanocarriers-for-targeted-3h9y7xrpi6

[73] Shanmugam K. Nanobioceramics for tissue engineering application. J Clin Nurs Res [Internet], 2024 Jul 18 [cited 2025 May 12];8(6):359–372. Available from: https://scispace.com/papers/nanobioceramics-for-tissue-engineering-application-3x8soz2bwf

[74] Kora AJ. Applications of inorganic metal oxide and metal phosphate-based nanoceramics in dentistry. Ind Appl Nanoceram [Internet], 2024 Jan 1 [cited 2025 May 12];63–77. Available from: https://scispace.com/papers/applications-of-inorganic-metal-oxide-and-metal-phosphate-34rkxf5rh7

[75] Tang P, Song F, Chen Y, Gao C, Ran X, Li Y, et al. Preparation and characterization of extracellular vesicles and their cutting-edge applications in regenerative medicine. Appl Mater Today [Internet], 2024 Apr 1 [cited 2025 May 14];37. Available from: https://scispace.com/papers/preparation-and-characterization-of-extracellular-vesicles-3jlbx7khhj

[76] Lu M, Shao W, Xing H, Huang Y. Extracellular vesicle-based nucleic acid delivery. Interdiscip Med [Internet], 2023 Jan 15 [cited 2025 May 14];1(2). Available from: https://scispace.com/papers/extracellular-vesicle-based-nucleic-acid-delivery-oqm10m89

[77] Zhang Q, Yan S, Yan X, Lv Y. Recent advances in metal-organic frameworks: Synthesis, application and toxicity. Sci Total Environ [Internet], 2023 Aug 1 [cited 2025 May 14];902:165944–165944. Available from: https://scispace.com/papers/recent-advances-in-metal-organic-frameworks-synthesis-5gb87r2htf

[78] Benny A, Kalathiparambil Rajendra Pai SD, Pinheiro D, Chundattu SJ. Metal organic frameworks in biomedicine: Innovations in drug delivery. Results Chem [Internet], 2024 Mar 1 [cited 2025 May 14];7:101414–101414. Available from: https://scispace.com/papers/metal-organic-frameworks-in-biomedicine-innovations-in-drug-1athsis1y5

[79] Mahendiratta S, Bansal S, Kumar S, Sarma P, Prakash A, Medhi B. Current practices in oncology drug delivery. Adv Drug Deliv Syst Manag Cancer [Internet], 2021 Jan 1 [cited 2025 May 14];17–26. Available from: https://scispace.com/papers/current-practices-in-oncology-drug-delivery-1sg3nxpmfd

[80] Wakaskar RR. Cancer therapy with drug delivery systems. J Pharmacogenomics Pharmacoproteomics [Internet], 2017 Jan 1 [cited 2025 May 14];8(1):1–2. Available from: https://scispace.com/papers/cancer-therapy-with-drug-delivery-systems-2y7pjdi72p

[81] Gulyakin ID, Oborotova NA, Pechennikov VM. Solubilization of hydrophobic antitumor drugs (Review). Pharm Chem J [Internet], 2014 Jul 20 [cited 2025 May 14];48(3):209–213. Available from: https://scispace.com/papers/solubilization-of-hydrophobic-antitumor-drugs-review-31btrwjnom

[82] Sparreboom A, Van Tellingen O, Nooijen WJ, Beijnen JH. Preclinical pharmacokinetics of paclitaxel and docetaxel. Anticancer Drugs [Internet], 1998 Jan 1 [cited 2025 May 14];9(1):1–17. Available from: https://scispace.com/papers/preclinical-pharmacokinetics-of-paclitaxel-and-docetaxel-1w6340xyri

[83] Was H, Borkowska A, Bagues A, Tu L, Liu JYH, Lu Z, et al. Mechanisms of chemotherapy-induced neurotoxicity. Front Pharmacol [Internet], 2022 Mar 28 [cited 2025 May 14];13. Available from: https://scispace.com/papers/mechanisms-of-chemotherapy-induced-neurotoxicity-3pucv5jy

[84] Stirland DL, Nichols JW, Denison TA, Bae YH. Targeted drug delivery for cancer therapy. Biomater Cancer Ther Diagn Prev Ther [Internet], 2013 Jan 1 [cited 2025 May 14];31–56. Available from: https://scispace.com/papers/targeted-drug-delivery-for-cancer-therapy-uipntgtbpc

[85] Dudhe PS, Thakare R. Targeted drug delivery in cancer therapy. Int J Adv Res Sci, Comm Technol [Internet], 2024 Dec 4 [cited 2025 May 14];109–124. Available from: https://scispace.com/papers/targeted-drug-delivery-in-cancer-therapy-2036ede2n56k

[86] Stylianopoulos T, Munn LL, Jain RK. Reengineering the tumor vasculature: Improving drug delivery and efficacy. Trends Cancer [Internet], 2018 Apr 1 [cited 2025 May 14];4(4):258–259. Available from: https://scispace.com/papers/reengineering-the-tumor-vasculature-improving-drug-delivery-2313karm0q

[87] Gout DY, Groen LS, van Egmond M. The present and future of immunocytokines for cancer treatment. Cell Mol Life Sci [Internet], 2022 Sep 6 [cited 2025 May 14];79(10):509. Available from: https://scispace.com/papers/the-present-and-future-of-immunocytokines-for-cancer-2s6yt6mt

[88] Bohmer M, Xue Y, Jankovic K, Dong Y. Advances in engineering and delivery strategies for cytokine immunotherapy. Expert Opin Drug Deliv, 2023 May 3 [cited 2025 May 14];20(5):579–595. Available from: https://scispace.com/papers/advances-in-engineering-and-delivery-strategies-for-cytokine-18iazkh1

[89] Scherließ R. Delivery of antigens used for vaccination: Recent advances and challenges. Ther Deliv [Internet], 2011 Oct 13 [cited 2025 May 14];2(10):1351–1368. Available from: https://scispace.com/papers/delivery-of-antigens-used-for-vaccination-recent-advances-304ploxtwj

[90] Han S, Lee P, Choi HJ. Non-invasive vaccines: Challenges in formulation and vaccine adjuvants. Pharmaceutics [Internet], 2023 Aug 1 [cited 2025 May 14];15(8). Available from: https://scispace.com/papers/non-invasive-vaccines-challenges-in-formulation-and-vaccine-25iy5rp7s2

[91] Mani G, Danasekaran R, Annadurai K. Controlled temperature chain: Reaching the unreached in resource-limited settings. Bangladesh J Med Sci [Internet], 2017 Jun 9 [cited 2025 May 14];16(3):477–479. Available from: https://scispace.com/papers/controlled-temperature-chain-reaching-the-unreached-in-4cxyvztwgd

[92] Seya T, Tatematsu M, Matsumoto M. Toward establishing an ideal adjuvant for non-inflammatory immune enhancement. Cells [Internet], 2022 Dec 1 [cited 2025 May 14];11(24):4006–4006. Available from: https://scispace.com/papers/toward-establishing-an-ideal-adjuvant-for-non-inflammatory-11ddmde5

[93] Tanuku S, Velisila D, Thatraju D, Vadaga AK. Nanoemulsion formulation strategies for enhanced drug delivery. J Pharm Insights Res [Internet], 2024 Aug 4 [cited 2025 May 14];2(4):125–138. Available from: https://scispace.com/papers/nanoemulsion-formulation-strategies-for-enhanced-drug-px5g0kvbn0a8

[94] Dhumal N, Yadav V, Borkar S. Nanoemulsion as novel drug delivery system: Development, characterization and application. Asian J Pharm Res Dev [Internet], 2022 Dec 14 [cited 2025 May 14];10(6):120–127. Available from: https://scispace.com/papers/nanoemulsion-as-novel-drug-delivery-system-development-2uk0ke43

[95] Shah MR, Imran M, Ullah S. Nanotechnological strategies involved in the targeted delivery of anticancer drugs. Nanocarriers Cancer Diagn Targeted Chemother [Internet], 2019 Jan 1 [cited 2025 May 14];23–41. Available from: https://scispace.com/papers/nanotechnological-strategies-involved-in-the-targeted-1kwza2pyx8

[96] Singh A, Rana V. Ligands for tumor targeting. Role Nanotechnol Cancer Ther [Internet], 2023 Aug 30 [cited 2025 May 14];89–139. Available from: https://scispace.com/papers/ligands-for-tumor-targeting-1bb1dkl84l

[97] Sokolov Yu V. Nanoemulsions as prospective drug delivery systems. Vìsnik farmacìï [Internet], 2014 Apr 6 [cited 2025 May 14];1(77):21–25. Available from: https://scispace.com/papers/nanoemulsions-as-prospective-drug-delivery-systems-5f6f7gqccg

[98] Kharkar PB, Talkar SS, Kadwadkar NA, Patravale VB. Nanosystems for oral delivery of immunomodulators. Nanostructures Oral Med [Internet], 2017 Apr 14 [cited 2025 May 14];295–334.

Available from: https://scispace.com/papers/nanosystems-for-oral-delivery-of-immunomodulators
-2c75vk393r

[99] Hamouda T, Simon J, Fattom A, Baker J. NanoBio™ nanoemulsion for mucosal vaccine delivery.
Novel Immune Potentiators and Delivery Technologies for Next Generation Vaccines [Internet],
2013 Nov 1 [cited 2025 May 14];269–286. Available from: https://scispace.com/papers/nanobio-
nanoemulsion-for-mucosal-vaccine-delivery-3lmjzdngna

[100] Li GC, Li HF, Jin Z, Feng R, Deng Y, Cheng H, et al. Cationic nanoemulsion-encapsulated retinoic acid
as an adjuvant to promote OVA-specific systemic and mucosal responses. J Vis Exp [Internet], 2024
Feb 23 [cited 2025 May 14];204(204). Available from: https://scispace.com/papers/cationic-
nanoemulsion-encapsulated-retinoic-acid-as-an-1mii2suncv

[101] Nanoemulsion vaccine adjuvant (2016) | Pan Hao | 1 Citations [Internet]. [cited 2025 May 14].
Available from: https://scispace.com/papers/nanoemulsion-vaccine-adjuvant-nhwtnmq7d5

[102] Water-in-oil type nanoemulsion vaccine preparation [Internet]. 2010 [cited 2025 May 14]. Available
from: https://scispace.com/papers/water-in-oil-type-nanoemulsion-vaccine-preparation
-2o4kgntem4

[103] Foyez T, Imran AB. Nanotechnology in vaccine development and constraints. NanoVacc Target Ther
[Internet], 2022 Jun 15 [cited 2025 May 14];1–20. Available from: https://scispace.com/papers/nano
technology-in-vaccine-development-and-constraints-1xfj8cwn

[104] Malik S, Kishore S, Kumar SA, Kumari A, Kumari M, Dhasmana A. Application of nanoemulsions in
the vaccination process. 2022 Jan 1 [cited 2025 May 14];494–516. Available from: https://scispace.
com/papers/application-of-nanoemulsions-in-the-vaccination-process-1jcqld67

[105] Borrajo ML, Lou G, Anthiya S, Lapuhs P, Moreira D, Tobío A, et al. Nanoemulsions and nanocapsules
as carriers for the development of intranasal mRNA vaccines (preprint). 2024 Apr 12 [cited
2025 May 14]; Available from: https://scispace.com/papers/nanoemulsions-and-nanocapsules-as-
carriers-for-the-590ied7vy3

[106] Dewangan N, Kothale D, Verma U, Jain D. Nanoemulsion as a nano carrier system in drug delivery:
A review. Int J Pharm Sci Nanotechnol [Internet], 2021 Mar 1 [cited 2025 May 15];14(2):5353–5363.
Available from: https://scispace.com/papers/nanoemulsion-as-a-nano-carrier-system-in-drug-
delivery-a-i88sxddspfd6

[107] Dhumal N, Yadav V, Borkar S. Nanoemulsion as novel drug delivery system: Development,
characterization and application. Asian J Pharm Res Dev [Internet], 2022 Dec 14 [cited
2025 May 15];10(6):120–127. Available from: https://scispace.com/papers/nanoemulsion-as-novel-
drug-delivery-system-development-2uk0ke43

[108] Chime SA, Kenechukwu FC, Attama AA. Nanoemulsions – Advances in formulation, characterization
and applications in drug delivery. Appl Nanotechnol Drug Deliv [Internet], 2014 Jul 25 [cited
2025 May 15]; Available from: https://scispace.com/papers/nanoemulsions-advances-in-formulation
-characterization-and-4hsmv6170g

[109] Eastoe J, Tabor RF. Surfactants and nanoscience. Colloidal foundations of nanoscience. Second Ed
[Internet], 2022 Jan 1 [cited 2025 May 15];153–82. Available from: https://scispace.com/papers/sur
factants-and-nanoscience-1rvh98tn

[110] Bahuguna A, Ramalingam S, Kim M. Formulation, characterization, and potential application of
nanoemulsions in food and medicine. Nanotechnol Life Sci [Internet], 2020 Jan 1 [cited
2025 May 15];39–61. Available from: https://scispace.com/papers/formulation-characterization-and-
potential-application-of-ctstegusd2

[111] Purnomo H, Nuraini N, Makmur T. The influence of alcohol type and concentration on the phase
behavior and interfacial tension in oil-surfactant cosurfactant-brine mixture system. SCOG
[Internet], 2022 Mar 30 [cited 2025 May 15];27(2):43–49. Available from: https://scispace.com/pa
pers/the-influence-of-alcohol-type-and-concentration-on-the-phase-3zu41rev03

[112] Algahtani MS, Ahmad MZ, Ahmad J. Investigation of factors influencing formation of nanoemulsion by spontaneous emulsification: Impact on droplet size, polydispersity index, and stability. Bioengineering [Internet], 2022 Aug 12 [cited 2025 May 15];9(8):384–384. Available from: https://scispace.com/papers/investigation-of-factors-influencing-formation-of-37ladb7i

[113] Gharibzahedi SMT, Jafari SM. Fabrication of nanoemulsions by ultrasonication. Nanoemulsions: Formul Appl Charact [Internet], 2018 Mar 5 [cited 2025 May 15];233–85. Available from: https://scispace.com/papers/fabrication-of-nanoemulsions-by-ultrasonication-263gmd6a9v

[114] Jasmina H, Džana O, Alisa E, Edina V, Ognjenka R. Preparation of nanoemulsions by high-energy and low energy emulsification methods. IFMBE Proc [Internet], 2017 Jan 1 [cited 2025 May 15];62:317–322. Available from: https://scispace.com/papers/preparation-of-nanoemulsions-by-high-energy-and-4msnn4a0rc

[115] Håkansson A. Fabrication of nanoemulsions by high-pressure valve homogenization. Nanoemulsions: Formul Appl Charact [Internet], 2018 Mar 5 [cited 2025 May 15];175–206. Available from: https://scispace.com/papers/fabrication-of-nanoemulsions-by-high-pressure-valve-25erfh9mql

[116] Azmi NAN, Elgharbawy AAM, Motlagh SR, Samsudin N, Salleh HM. Nanoemulsions: Factory for food, pharmaceutical and cosmetics. Processes [Internet], 2019 Sep 11 [cited 2025 May 15];7(9):617. Available from: https://scispace.com/papers/nanoemulsions-factory-for-food-pharmaceutical-and-cosmetics-2n0bfu91pl

[117] Gohtani S, Prasert W. Nano-emulsions; emulsification using low energy methods. Jpn J Food Eng [Internet], 2014 Sep 15 [cited 2025 May 15];15(3):119–130. Available from: https://scispace.com/papers/nano-emulsions-emulsification-using-low-energy-methods-j4upva08fy

[118] Tadros T, Izquierdo P, Esquena J, Solans C. Formation and stability of nano-emulsions. Adv Colloid Interface Sci, 2004 May 20;108–109:303–318.

[119] Singh Y, Meher JG, Raval K, Khan FA, Chaurasia M, Jain NK, et al. Nanoemulsion: Concepts, development and applications in drug delivery. J Control Release, 2017 Apr 28;252:28–49.

[120] Chaudhary A, Shivalika TB, Verma KK. Role of nanoemulsion as drugs delivery systems in opthalmology: A comprehensive review. Res J Pharm Technol [Internet], 2022 Jul 29 [cited 2025 May 15];15(7):3285–3294. Available from: https://scispace.com/papers/role-of-nanoemulsion-as-drugs-delivery-systems-in-1rz7k0xs

[121] Preeti SS, Malik R, Bhatia S, Al Harrasi A, Rani C, et al. Nanoemulsion: An emerging novel technology for improving the bioavailability of drugs. Scientifica (Cairo) [Internet], 2023 Oct 28 [cited 2025 May 15];2023. Available from: https://scispace.com/papers/nanoemulsion-an-emerging-novel-technology-for-improving-the-57ghpm13rx

[122] Zeytunluoglu A, Arslan I. Current perspectives on nanoemulsions in targeted drug delivery. 2022 Jan 1 [cited 2025 May 15];118–40. Available from: https://scispace.com/papers/current-perspectives-on-nanoemulsions-in-targeted-drug-3pd6nw6q

[123] Abdulelah FM, Taghi HS, Abdulbaqi HR, Abdulbaqi MR. Review on nanoemulsion: Preparation and evaluation. J Drug Deliv [Internet], 2020 Mar 25 [cited 2025 May 15];10(01):187–189. Available from: https://scispace.com/papers/review-on-nanoemulsion-preparation-and-evaluation-11hw1ae7v9

[124] Fernandes DA, Fernandes DD, Wang YJ, Li Y, Gradinaru CC, Rousseau D, et al. Phase change nanoemulsions for cancer therapy and imaging. 2024 Nov 27 [cited 2025 May 15]; Available from: https://scispace.com/papers/phase-change-nanoemulsions-for-cancer-therapy-and-imaging-75o7u2g1toru

[125] Li X, Tsibouklis J, Weng T, Zhang B, Yin G, Feng G, et al. Nano carriers for drug transport across the blood–brain barrier. J Drug Target [Internet], 2017 Jan 2 [cited 2025 May 15];25(1):17–28. Available from: https://scispace.com/papers/nano-carriers-for-drug-transport-across-the-blood-brain-vbppplf1s2

[126] Kralj S, Rojnik M, Romih R, Jagodič M, Kos J, Makovec D. Effect of surface charge on the cellular uptake of fluorescent magnetic nanoparticles. J Nanopart Res [Internet], 2012 Sep 7 [cited 2025 May 15];14(10):1151. Available from: https://scispace.com/papers/effect-of-surface-charge-on-the-cellular-uptake-of-44e79q7y9l

[127] Foglietta F, Bozza A, Ferraris C, Cangemi L, Bordano V, Serpe L, et al. Surface functionalised parenteral nanoemulsions for active and homotypic targeting to melanoma. Pharmaceutics [Internet], 2023 Apr 28 [cited 2025 May 15];15(5):1358–1358. Available from: https://scispace.com/papers/surface-functionalised-parenteral-nanoemulsions-for-active-38zcjyxb

[128] Praveen Kumar G, Divya A, Kumar GP. Nanoemulsion based targeting in cancer therapeutics. Med Chem (Los Angeles) [Internet], 2015 Jun 24 [cited 2025 May 15];5(6):1–13. Available from: https://scispace.com/papers/nanoemulsion-based-targeting-in-cancer-therapeutics-191h3n8sby

[129] Moosavian SA, Kesharwani P, Sahebkar A. Nanoemulsion-based curcumin delivery systems as cancer therapeutics. Curcumin-Based Nanomed Cancer Ther [Internet], 2024 Jan 1 [cited 2025 May 15];147–63. Available from: https://scispace.com/papers/nanoemulsion-based-curcumin-delivery-systems-as-cancer-4gn4wdsbcs

[130] Kanawade S, Jagdale V, Shinde A, Poojari D, Pawar H. Review on nanoemulsion technology: A formulation for enhanced therapeutic efficacy. Res J Pharm Dosage Forms Technol [Internet], 2024 Feb 22 [cited 2025 May 15];113–8. Available from: https://scispace.com/papers/review-on-nanoemulsion-technology-a-formulation-for-enhanced-gk1khxx0yt

[131] D'Angelo NA, Noronha MA, Câmara MCC, Kurnik IS, Feng C, Araujo VHS, et al. Doxorubicin nanoformulations on therapy against cancer: An overview from the last 10 years. Biomater Adv [Internet], 2022 Feb 1 [cited 2025 May 15];133:112623–112623. Available from: https://scispace.com/papers/doxorubicin-nanoformulations-on-therapy-against-cancer-an-2avxkum2

[132] Bagul M, Kakumanu S, Wilson T, Nicolosi R. In vitro evaluation of antiproliferative effects of self-assembling nanoemulsion of paclitaxel on various cancer cell lines. Nano Biomed Eng [Internet], 2010 Jun 26 [cited 2025 May 15];2(2):100–108. Available from: https://scispace.com/papers/in-vitro-evaluation-of-antiproliferative-effects-of-self-42kcguadgp

[133] Khan MA, Khan A, Khan AA, Simrah MM, Javed Naquvi K, et al. Anti-carcinogenic potential of Nano-curcumin -"Killing Cancerous Cells". Curr Bioact Compd [Internet], 2024 May 15 [cited 2025 May 15];20. Available from: https://scispace.com/papers/anti-carcinogenic-potential-of-nano-curcumin-killing-261cpwjkjq

[134] Binh NT, Huong NTM, Huong LTT, Thuy PT, Tinh NT, Melnikova G, et al. Curcumin nanoemulsion: Evaluation of stability and anti-cancer activity in vitro. J Nano Res [Internet], 2020 Nov 1 [cited 2025 May 15];64:21–37. Available from: https://scispace.com/papers/curcumin-nanoemulsion-evaluation-of-stability-and-anti-2rv2f0b9cp

[135] Lin S, Xie P, Luo M, Li Q, Li L, Zhang J, et al. Efficiency against multidrug resistance by co-delivery of doxorubicin and curcumin with a legumain-sensitive nanocarrier. Nano Res [Internet], 2018 Aug 2 [cited 2025 May 15];11(7):3619–3635. Available from: https://scispace.com/papers/efficiency-against-multidrug-resistance-by-co-delivery-of-3dx627q1sh

[136] Li J, Li Z, Li M, Zhang H, Xie Z. Synergistic effect and drug-resistance relief of paclitaxel and cisplatin caused by co-delivery using polymeric micelles. J Appl Polym Sci [Internet], 2015 Feb 10 [cited 2025 May 15];132(6). Available from: https://scispace.com/papers/synergistic-effect-and-drug-resistance-relief-of-paclitaxel-3d4kho4ulk

[137] Wang Q, Wu C, Li X, Yang D, Shi L. Cisplatin and paclitaxel co-delivery nanosystem for ovarian cancer chemotherapy. Regen Biomater [Internet], 2021 Apr 25 [cited 2025 May 15];8(3). Available from: https://scispace.com/papers/cisplatin-and-paclitaxel-co-delivery-nanosystem-for-ovarian-4190pclfqz

[138] Subhan MA, Yalamarty SSK, Filipczak N, Parveen F, Torchilin VP. Recent advances in tumor targeting via EPR effect for cancer treatment. J Pers Med [Internet], 2021 Jun 1 [cited 2025 May 29];11(6):571. Available from: https://pmc.ncbi.nlm.nih.gov/articles/PMC8234032/

[139] Tan A, Jeyaraj R, De Lacey SF. Nanotechnology in neurosurgical oncology. Nanotechnol Cancer, 2017;139–70.

[140] Dudhe PS, Thakare R. Targeted drug delivery in cancer therapy. Int J Adv Res Sci, Comm Technol [Internet], 2024 Dec 4 [cited 2025 May 15];109–24. Available from: https://scispace.com/papers/targeted-drug-delivery-in-cancer-therapy-2o36ede2n56k

[141] Prajapati A, Rangra S, Patil R, Desai N, Jyothi VGSS, Salave S, et al. Receptor-targeted nanomedicine for cancer therapy. Receptors [Internet], 2024 Jul 3 [cited 2025 May 15];3(3):323–361. Available from: https://scispace.com/papers/receptor-targeted-nanomedicine-for-cancer-therapy-5cvvhr44if

[142] Song B, Wu S, Li W, Chen D, Hu H. Folate modified long circulating nano-emulsion as a promising approach for improving the efficiency of chemotherapy drugs in cancer treatment. Pharm Res [Internet], 2020 Nov 13 [cited 2025 May 15];37(12):242–242. Available from: https://scispace.com/papers/folate-modified-long-circulating-nano-emulsion-as-a-8yx9axrk3d

[143] Lee YH, Ma YT. Synthesis, characterization, and biological verification of anti-HER2 indocyanine green–doxorubicin-loaded polyethyleneimine-coated perfluorocarbon double nanoemulsions for targeted photochemotherapy of breast cancer cells. J Nanobiotechnology [Internet], 2017 May 18 [cited 2025 May 15];15(1):41–41. Available from: https://scispace.com/papers/synthesis-characterization-and-biological-verification-of-2oz20z9hxe

[144] Cao X, Luo J, Gong T, Zhang ZR, Sun X, Fu Y. Coencapsulated doxorubicin and bromotetrandrine lipid nanoemulsions in reversing multidrug resistance in breast cancer in vitro and in vivo. Mol Pharm [Internet], 2015 Jan 5 [cited 2025 May 15];12(1):274–286. Available from: https://scispace.com/papers/coencapsulated-doxorubicin-and-bromotetrandrine-lipid-1bxitnnrtv

[145] Kim JE, Park YJ. Paclitaxel-loaded hyaluronan solid nanoemulsions for enhanced treatment efficacy in ovarian cancer. Int J Nanomed [Internet], 2017 Jan 17 [cited 2025 May 15];12:645–658. Available from: https://scispace.com/papers/paclitaxel-loaded-hyaluronan-solid-nanoemulsions-for-3gcts7v0m7

[146] Kumar A, Singh AK, Chaudhary RP, Sharma A, Yadav JP, Pathak P, et al. Unraveling the multifaceted role of nanoemulsions as drug delivery system for the management of cancer. J Drug Deliv Sci Technol [Internet], 2024 Oct 1 [cited 2025 Jun 2];100:106056. Available from: https://www.sciencedirect.com/science/article/abs/pii/S1773224724007251

[147] Hwang J, Park JY, Kang J, Oh N, Li C, Yoo CY, et al. Enhanced drug delivery with oil-in-water nanoemulsions: Stability and sustained release of doxorubicin. Macromol Rapid Commun [Internet], 2024 Jul 31 [cited 2025 May 15];45. Available from: https://scispace.com/papers/enhanced-drug-delivery-with-oil-in-water-nanoemulsions-7oeuennmfh3x

[148] Curcumin in Reducing Joint Pain in Breast Cancer Survivors With Aromatase Inhibitor-Induced Joint Disease | Division of Cancer Prevention [Internet]. [cited 2025 Jun 2]. Available from: https://prevention.cancer.gov/clinical-trials/clinical-trials-search/nct03865992

[149] Superficial Basal Cell Cancer's Photodynamic Therapy: Comparing Three Photosensitizers: HAL and BF-200 ALA Versus MAL [Internet]. [cited 2025 Jun 2]. Available from: https://ctv.veeva.com/study/superficial-basal-cell-cancers-photodynamic-therapy-comparing-three-photosensitizers-hal-and-bf2

[150] Study Details | Pilot Study of Curcumin for Women With Obesity and High Risk for Breast Cancer | ClinicalTrials.gov [Internet]. [cited 2025 Jun 2]. Available from: https://clinicaltrials.gov/study/NCT01975363

[151] Sharma Y, Arora M, Bala K. The potential of immunomodulators in shaping the future of healthcare. Deleted J [Internet], 2024 Sep 3 [cited 2025 May 15];1(1). Available from: https://scispace.com/papers/the-potential-of-immunomodulators-in-shaping-the-future-of-qz8v9gj2uche

[152] Palupi PD, Safwan M, Khan A, Amalina ND. Role of immunosuppressive and immunomodulatory agents in cancer. Immunosuppression Immunomodulation [Internet], 2022 Dec 11 [cited 2025 May 15]; Available from: https://scispace.com/papers/role-of-immunosuppressive-and-immunomodulatory-agents-in-1brulunz

[153] Liebman HA. Immunomodulatory Drugs and Monoclonal Antibodies. Antibody Therapy: Substitution – Immunomodulation – Monoclonal Immunotherapy [Internet]. 2018 Jan 1 [cited 2025 May 15];85–100. Available from: https://scispace.com/papers/immunomodulatory-drugs-and-monoclonal-antibodies-358fjpg223

[154] (Open Access) Immunomodulators: a pharmacological review (2012) | U.S. Patil | 94 Citations [Internet]. [cited 2025 May 15]. Available from: https://scispace.com/papers/immunomodulators-a-pharmacological-review-3mf0kwniwv

[155] Chime SA, Kenechukwu FC, Attama AA. Nanoemulsions – Advances in formulation, characterization and applications in drug delivery. Appl Nanotechnol Drug Deliv [Internet], 2014 Jul 25 [cited 2025 May 15]; Available from: https://scispace.com/papers/nanoemulsions-advances-in-formulation-characterization-and-4hsmv6170g

[156] Li G, Zhang Z, Liu H, Hu L. Nanoemulsion-based delivery approaches for nutraceuticals: Fabrication, application, characterization, biological fate, potential toxicity and future trends. Food Funct [Internet], 2021 Mar 15 [cited 2025 May 15];12(5):1933–1953. Available from: https://scispace.com/papers/nanoemulsion-based-delivery-approaches-for-nutraceuticals-rd7wdm78ax

[157] Liu Y, Zeng W, Na W, Wei X, Song K, Wang Y, et al. Stable IL-2 nano-assembly for improved anti-tumor effect. Adv Ther [Internet], 2023 Oct 19 [cited 2025 May 15];7(3). Available from: https://scispace.com/papers/stable-il-2-nano-assembly-for-improved-anti-tumor-effect-17y4xjbyib

[158] Izadiyan Z, Webster TJ, Kia P, Kalantari K, Misran M, Rasouli E, et al. Nanoemulsions based therapeutic strategies: Enhancing targeted drug delivery against breast cancer cells. Int J Nanomed [Internet], 2025 [cited 2025 May 29];20:6133–6162. Available from: https://www.tandfonline.com/doi/pdf/10.2147/IJN.S488545

[159] Selepe CT, Dhlamini KS, Tshweu L, Moralo M, Kwezi L, Ray SS, et al. Trastuzumab-based nanomedicines for breast cancer therapy: Recent advances and future opportunities. Nano Select [Internet], 2024 May 1 [cited 2025 May 29];5(5):2300191. Available from: https://onlinelibrary.wiley.com/doi/pdf/10.1002/nano.202300191

[160] Kharkar PB, Talkar SS, Kadwadkar NA, Patravale VB. Nanosystems for oral delivery of immunomodulators. Nanostructures Oral Med [Internet], 2017 Apr 14 [cited 2025 May 15];295–334. Available from: https://scispace.com/papers/nanosystems-for-oral-delivery-of-immunomodulators-2c75vk393r

[161] Yocum DE. Cyclosporine, fk-506, rapamycin, and other immunomodulators. Rheum Dis Clin North Am [Internet], 1996 Feb 1 [cited 2025 May 15];22(1):133–154. Available from: https://scispace.com/papers/cyclosporine-fk-506-rapamycin-and-other-immunomodulators-1erd2nfpsz

[162] Kharkar PB, Talkar SS, Kadwadkar NA, Patravale VB. Nanosystems for oral delivery of immunomodulators. Nanostructures Oral Med [Internet], 2017 Apr 14 [cited 2025 May 15];295–334. Available from: https://scispace.com/papers/nanosystems-for-oral-delivery-of-immunomodulators-2c75vk393r

[163] Gupta K, Dangi K. Immunomodulatory potential of nanomaterials: Interaction with the immune system. Indian J Biochem Biophys [Internet], 2022 Jan 1 [cited 2025 May 15];59(12):1159–1162. Available from: https://scispace.com/papers/immunomodulatory-potential-of-nanomaterials-interaction-with-1i11qr9n

[164] Tanuku S, Velisila D, Thatraju D, Vadaga AK. Nanoemulsion formulation strategies for enhanced drug delivery. J Pharm Insights Res [Internet], 2024 Aug 4 [cited 2025 May 15];2(4):125–138. Available from: https://scispace.com/papers/nanoemulsion-formulation-strategies-for-enhanced-drug-px5g0kvbn0a8

[165] Al-Hussaniy HA, Almajidi YQ, Oraibi AI, Alkarawi AH. Nanoemulsions as medicinal components in insoluble medicines. Pharmacia [Internet], 2023 Jul 27 [cited 2025 May 15];70(3):537–547. Available from: https://scispace.com/papers/nanoemulsions-as-medicinal-components-in-insoluble-medicines-3zyc3bocxx

[166] Zeytunluoglu A, Arslan I. Current perspectives on nanoemulsions in targeted drug delivery. 2022 Jan 1 [cited 2025 May 15];118–40. Available from: https://scispace.com/papers/current-perspectives-on-nanoemulsions-in-targeted-drug-3pd6nw6q

[167] Liu Y, Zeng W, Na W, Wei X, Song K, Wang Y, et al. Stable IL-2 nano-assembly for improved anti-tumor effect. Adv Ther [Internet], 2023 Oct 19 [cited 2025 May 15];7(3). Available from: https://scispace.com/papers/stable-il-2-nano-assembly-for-improved-anti-tumor-effect-17y4xjbyib

[168] Bhargava S, Satheesh Madhav NV. Cyclosporine A loaded nanoemulsions using bio-oil fractions of sesame oil. Indian J Pharm Educ Res [Internet], 2020 Mar 3 [cited 2025 May 15];54(2):357–366. Available from: https://scispace.com/papers/cyclosporine-a-loaded-nanoemulsions-using-bio-oil-fractions-5352sgj8ni

[169] MJ N, R N. Nanoemulsions in cancer therapeutics. J Nanomed Nanotechnol [Internet], 2016 Mar 31 [cited 2025 May 15];2016(02). Available from: https://scispace.com/papers/nanoemulsions-in-cancer-therapeutics-l3o7kcufd8

[170] Croitoru GA, Niculescu AG, Epistatu D, Mihaiescu DE, Antohi AM, Grumezescu AM, et al. Nanostructured drug delivery systems in immunotherapy: An updated overview of nanotechnology-based therapeutic innovations. Appl Sci [Internet], 2024 Oct 4 [cited 2025 May 15];14(19):8948–8948. Available from: https://scispace.com/papers/nanostructured-drug-delivery-systems-in-immunotherapy-an-50wr96jnl57d

[171] Panday MK, Agarwal MB, Mehrotra M, Mittal P, Goyal P, Agarwal S, et al. New developments and challenges in nanomaterial-based vaccine delivery. Adv Chem Mater Eng [Internet], 2024 Oct 22 [cited 2025 May 15];379–404. Available from: https://scispace.com/papers/new-developments-and-challenges-in-nanomaterial-based-39tf0a1ok3g5

[172] Tabarzad M, Mohit E, Ghorbani-Bidkorbeh F. Nanovaccines delivery approaches against infectious diseases. Emerg Nanomater Nano-based Drug Deliv Approaches Combat Antimicrob Resist [Internet], 2022 Jan 1 [cited 2025 May 15];425–84. Available from: https://scispace.com/papers/nano vaccines-delivery-approaches-against-infectious-diseases-1iuyiin4

[173] Shi X, Yang K, Song H, Teng Z, Zhang Y, Ding W, et al. Development and efficacy evaluation of a novel nano-emulsion adjuvant for a foot-and-mouth disease virus-like particles vaccine based on squalane. Nanomaterials [Internet], 2022 Nov 1 [cited 2025 May 29];12(22):3934. Available from: https://www.mdpi.com/2079-4991/12/22/3934/htm

[174] Koh J, Kim S, Lee SN, Kim SY, Kim JE, Lee KY, et al. Therapeutic efficacy of cancer vaccine adjuvanted with nanoemulsion loaded with TLR7/8 agonist in lung cancer model. Nanomedicine [Internet], 2021 Oct 1 [cited 2025 May 29];37. Available from: https://pubmed.ncbi.nlm.nih.gov/34174421/

[175] Malik S, Kishore S, Kumar SA, Kumari A, Kumari M, Dhasmana A. Application of nanoemulsions in the vaccination process. 2022 Jan 1 [cited 2025 May 15];494–516. Available from: https://scispace.com/papers/application-of-nanoemulsions-in-the-vaccination-process-1jcqld67

[176] Petkar KC, Patil SM, Chavhan SS, Kaneko K, Sawant KK, Kunda NK, et al. An overview of nanocarrier-based adjuvants for vaccine delivery. Pharmaceutics [Internet], 2021 Mar 27 [cited 2025 May 15];13(4):455. Available from: https://scispace.com/papers/an-overview-of-nanocarrier-based-adjuvants-for-vaccine-4gwnhpryd7

[177] Hendy DA, Amouzougan EA, Young IC, Bachelder EM, Ainslie KM. Nano/microparticle formulations for universal influenza vaccines. AAPS J [Internet], 2022 Jan 1 [cited 2025 May 15];24(1). Available from: https://scispace.com/papers/nano-microparticle-formulations-for-universal-influenza-1jvv1xms

[178] Zhao L, Zhu Z, Ma L, Li Y. O/W nanoemulsion as an adjuvant for an inactivated H3N2 influenza vaccine: Based on particle properties and mode of carrying. Int J Nanomed [Internet], 2020 Mar 25 [cited 2025 May 15];15:2071–2083. Available from: https://scispace.com/papers/o-w-nanoemulsion-as-an-adjuvant-for-an-inactivated-h3n2-3xndjueffk

[179] Grimaudo MA. Nanotechnology for the development of nanovaccines in cancer immunotherapy. Adv Exp Med Biol [Internet], 2021 Jan 1 [cited 2025 May 15];1295:303–315. Available from: https://scispace.com/papers/nanotechnology-for-the-development-of-nanovaccines-in-cancer-2k0a1tmszx

[180] Poudel K, Vithiananthan T, Kim JO, Tsao H. Recent progress in cancer vaccines and nanovaccines. Biomaterials [Internet], 2024 Sep 28 [cited 2025 May 15];314:122856–122856. Available from: https://scispace.com/papers/recent-progress-in-cancer-vaccines-and-nanovaccines-7bs4rlpgdp3c

[181] Subiza J, El-Qutob D, Fernandez-Caldas E. Immunogenic compositions comprising nanoemulsions. Recent Pat Inflamm Allergy Drug Discov [Internet], 2013 Sep 30 [cited 2025 May 15];9(1):4–15. Available from: https://scispace.com/papers/immunogenic-compositions-comprising-nanoemulsions-xxe0sptl1i

[182] Lopes Chaves L, Dourado D, Prunache IB, Manuelle Marques da Silva P, Tacyana dos Santos Lucena G, Cardoso de Souza Z, et al. Nanocarriers of antigen proteins for vaccine delivery. Int J Pharm [Internet], 2024 Apr 1 [cited 2025 May 15];659:124162–124162. Available from: https://scispace.com/papers/nanocarriers-of-antigen-proteins-for-vaccine-delivery-2crgqt1jmp

[183] Chime SA, Kenechukwu FC, Attama AA. Nanoemulsions – Advances in formulation, characterization and applications in drug delivery. Appl Nanotechnol Drug Deliv [Internet], 2014 Jul 25 [cited 2025 May 15]; Available from: https://scispace.com/papers/nanoemulsions-advances-in-formulation-characterization-and-4hsmv6170g

[184] Ruiz Ceja A, Varela Arzate A, Cornejo Cornejo L. Nanoemulsions as coadjuvants in intranasal vaccines. 2019 Jul 9 [cited 2025 May 15];6239. Available from: https://scispace.com/papers/nanoemulsions-as-coadjuvants-in-intranasal-vaccines-1f7c6hs94m

[185] Zou Y, Wu N, Miao C, Yue H, Wu J, Ma G. A novel multiple emulsion enhanced immunity via its biomimetic delivery approach. J Mater Chem B [Internet], 2020 Aug 26 [cited 2025 May 15];8(33):7365–7374. Available from: https://scispace.com/papers/a-novel-multiple-emulsion-enhanced-immunity-via-its-34awq14itk

[186] Lonappan D, Krishnakumar K, Dineshkumar B. Nanoemulsion in pharmaceuticals. Am J PharmTech Res [Internet], 2018 Apr 8 [cited 2025 May 15];8(2):1–14. Available from: https://scispace.com/papers/nanoemulsion-in-pharmaceuticals-g3on6j2i6u

[187] Desai JL, Patel PB, Patel AD, Dave RR, Swayamprakash P, Shah P. Controlled-release injectables. Novel Drug Deliv Syst: Part 2 [Internet], 2024 Dec 9 [cited 2025 May 15];199–229. Available from: https://scispace.com/papers/controlled-release-injectables-2mjno9asrxdz

[188] Arunachalam PS, Walls AC, Golden N, Atyeo C, Fischinger S, Li C, et al. Adjuvanting a subunit COVID-19 vaccine to induce protective immunity. Nature [Internet], 2021 Apr 19 [cited 2025 May 15];594 (7862):253–258. Available from: https://scispace.com/papers/adjuvanting-a-subunit-covid-19-vaccine-to-induce-protective-4tflbjgeb0

[189] Tran VA, Vo V, Dang VQ, Vo GNL, Don TN, Doan VD, et al. Nanomaterial for adjuvants vaccine: Practical applications and prospects. Indones J Chem [Internet], 2024 Feb 1 [cited 2025 May 15];24 (1):284–302. Available from: https://scispace.com/papers/nanomaterial-for-adjuvants-vaccine-practical-applications-468k4vsb70

[190] O'Hagan DT, Tsai T, Reed S. Emulsion-based adjuvants for improved influenza vaccines. Birkhauser Adv Infect Dis [Internet], 2011 Jan 1 [cited 2025 May 15];327–57. Available from: https://scispace.com/papers/emulsion-based-adjuvants-for-improved-influenza-vaccines-2pv1g78iui

[191] Brazzoli M, Magini D, Bonci A, Buccato S, Giovani C, Kratzer R, et al. Induction of broad-based immunity and protective efficacy by self-amplifying mRNA vaccines encoding influenza virus

hemagglutinin. J Virol [Internet], 2016 Jan 1 [cited 2025 May 15];90(1):332–344. Available from: https://scispace.com/papers/induction-of-broad-based-immunity-and-protective-efficacy-by -3dj3dfa3nl

[192] Desai N, Chavda V, Singh TRR, Thorat ND, Vora LK. Cancer nanovaccines: Nanomaterials and clinical perspectives. Small [Internet], 2024 May 1 [cited 2025 May 15];20(35). Available from: https://scispace.com/papers/cancer-nanovaccines-nanomaterials-and-clinical-perspectives-2es qaaje62

[193] Poudel K, Vithiananthan T, Kim JO, Tsao H. Recent progress in cancer vaccines and nanovaccines. Biomaterials [Internet], 2024 Sep 28 [cited 2025 May 15];314:122856–122856. Available from: https://scispace.com/papers/recent-progress-in-cancer-vaccines-and-nanovaccines-7bs4rlpgdp3c

[194] Fangueiro JF, Severino P, Souto SB, Souto EB. Nanomedicines for immunization and vaccines. Patenting Nanomed: Legal Aspects, Intellectual Property and Grant Opportunities [Internet], 2012 Oct 1 [cited 2025 May 15];9783642292651:435–450. Available from: https://scispace.com/papers/ nanomedicines-for-immunization-and-vaccines-2w1pqyfztg

[195] Priyanka AMAH, Chopra H, Sharma A, Mustafa SA, Choudhary OP, et al. Nanovaccines: A game changing approach in the fight against infectious diseases. Biomed Pharmacother [Internet], 2023 Sep 30 [cited 2025 May 15];167:115597–115597. Available from: https://scispace.com/papers/nanovac cines-a-game-changing-approach-in-the-fight-against-y33civ7lwq

[196] Borrajo ML, Lou G, Anthiya S, Lapuhs P, Álvarez DM, Tobío A, et al. Nanoemulsions and nanocapsules as carriers for the development of intranasal mRNA vaccines. Drug Deliv Transl Res [Internet], 2024 May 29 [cited 2025 May 15];14(8):2046–2061. Available from: https://scispace.com/ papers/nanoemulsions-and-nanocapsules-as-carriers-for-the-3b1mcjfdkl

[197] Makidon PE, Nigavekar SS, Bielinska AU, Mank N, Shetty AM, Suman J, et al. Characterization of stability and nasal delivery systems for immunization with nanoemulsion-based vaccines. J Aerosol Med Pulm Drug Deliv [Internet], 2010 Apr 12 [cited 2025 May 15];23(2):77–89. Available from: https://scispace.com/papers/characterization-of-stability-and-nasal-delivery-systems-for -45kp2u3osg

[198] Fan Y, Zhang Y, Yokoyama W, Yi J. Endocytosis of corn oil-caseinate emulsions In Vitro: Impacts of droplet sizes. Nanomaterials [Internet], 2017 Oct 26 [cited 2025 May 29];7(11):349. Available from: https://www.mdpi.com/2079-4991/7/11/349/htm

[199] Zhang B, Lei M, Huang W, Liu G, Jiang F, Peng D, et al. Improved storage properties and cellular uptake of casticin-loaded nanoemulsions stabilized by whey protein-lactose conjugate. Foods [Internet], 2021 Jul 1 [cited 2025 May 29];10(7):1640. Available from: https://pmc.ncbi.nlm.nih.gov/ articles/PMC8303442/

[200] Hansen ME, Ibrahim Y, Desai TA, Koval M. Nanostructure-mediated transport of therapeutics through epithelial barriers. Int J Mol Sci [Internet], 2024 Jun 28 [cited 2025 May 15];25(13):7098– 7098. Available from: https://scispace.com/papers/nanostructure-mediated-transport-of- therapeutics-through-43ego5ri2x

[201] Nanoparticle-based drug delivery. | Zihni Basar Bilgicer | 39 Citations [Internet]. 2012 [cited 2025 May 15]. Available from: https://scispace.com/papers/nanoparticle-based-drug-delivery -44zgocn7p1

[202] Rizwanullah M, Ahmad MZ, Ghoneim MM, Alshehri S, Imam SS, Md S, et al. Receptor-mediated targeted delivery of surface-modified nanomedicine in breast cancer: Recent update and challenges. Pharmaceutics [Internet], 2021 Nov 29 [cited 2025 May 15];13(12):2039. Available from: https://scispace.com/papers/receptor-mediated-targeted-delivery-of-surface-3miwnxpg3f

[203] Eloy JO, Petrilli R, Lee RJ. Targeting of drug nanocarriers. 2021 Jan 1 [cited 2025 May 15];107–26. Available from: https://scispace.com/papers/targeting-of-drug-nanocarriers-29ylv95c7p

[204] Das V, Chikkaputtaiah C, Pal M. Aptamer-conjugated functionalized nano-biomaterials for diagnostic and targeted drug delivery applications. Funct Polysac Biomed Appl [Internet], 2019 Jan 1

[cited 2025 May 15];469–94. Available from: https://scispace.com/papers/aptamer-conjugated-functionalized-nano-biomaterials-for-580r14abqi

[205] Javid H, Oryani MA, Rezagholinejad N, Esparham A, Tajaldini M, Karimi-Shahri M. RGD peptide in cancer targeting: Benefits, challenges, solutions, and possible integrin–RGD interactions. Cancer Med [Internet], 2024 Jan 1 [cited 2025 May 15];13(2). Available from: https://scispace.com/papers/rgd-peptide-in-cancer-targeting-benefits-challenges-21xte6qt8m

[206] Asati S, Pandey V, Soni V. RGD peptide as a targeting moiety for theranostic purpose: An update study. Int J Pept Res Ther [Internet], 2019 Mar 1 [cited 2025 May 15];25(1):49–65. Available from: https://scispace.com/papers/rgd-peptide-as-a-targeting-moiety-for-theranostic-purpose-an-mimebff8ui

[207] Prajapati A, Rangra S, Patil R, Desai N, Jyothi VGSS, Salave S, et al. Receptor-targeted nanomedicine for cancer therapy. Receptors [Internet], 2024 Jul 3 [cited 2025 May 15];3(3):323–361. Available from: https://scispace.com/papers/receptor-targeted-nanomedicine-for-cancer-therapy-5cvvhr44if

[208] Pandey R, Dhiman R, Bazad N, Mukerjee R, Vidic J, Leal E, et al. Enhancing breast cancer therapy: Nanocarrier-based targeted drug delivery. 2023 Dec 9 [cited 2025 May 15]; Available from: https://scispace.com/papers/enhancing-breast-cancer-therapy-nanocarrier-based-targeted-3fd6it8p2w

[209] McClements DJ, Rao J. Food-Grade nanoemulsions: Formulation, fabrication, properties, performance, Biological fate, and Potential Toxicity. Crit Rev Food Sci Nutr [Internet], 2011 Apr [cited 2025 May 29];51(4):285–330. Available from: https://pubmed.ncbi.nlm.nih.gov/21432697/

[210] Li G, Zhang Z, Liu H, Hu L. Nanoemulsion-based delivery approaches for nutraceuticals: Fabrication, application, characterization, biological fate, potential toxicity and future trends. Food Funct. R Soc Chem, 2021;12:1933–1953.

[211] Dib N, Lépori CMO, Correa NM, Silber JJ, Falcone RD, García-Río L. Biocompatible solvents and ionic liquid-based surfactants as sustainable components to formulate environmentally friendly organized systems. Polymers (Basel) [Internet], 2021 Apr 23 [cited 2025 May 15];13(9):1378. Available from: https://scispace.com/papers/biocompatible-solvents-and-ionic-liquid-based-surfactants-as-4ud847dk63

[212] Patel D, Patel B, Thakkar H. Lipid based nanocarriers: Promising drug delivery system for topical application. Eur J Lipid Sci Technol [Internet], 2021 Mar 12 [cited 2025 May 15];123(5):2000264. Available from: https://scispace.com/papers/lipid-based-nanocarriers-promising-drug-delivery-system-for-9koucsg28a

[213] Roy A, Nishchaya K, Rai VK. Nanoemulsion-based dosage forms for the transdermal drug delivery applications: A review of recent advances. Expert Opinion on Drug Delivery, 2022 Feb 23 [cited 2025 May 15];19(3):303–319. Available from: https://scispace.com/papers/nanoemulsion-based-dosage-forms-for-the-transdermal-drug-14kni8t5

[214] Leary JF. Quality assurance and regulatory issues of nanomedicine for the pharmaceutical industry. Fundam Nanomed [Internet], 2022 Apr 30 [cited 2025 May 15];279–309. Available from: https://scispace.com/papers/quality-assurance-and-regulatory-issues-of-nanomedicine-for-snl8kwim

[215] Sandbhor P, Palkar P, Bhat S, John G, Goda JS. Nanomedicine as a multimodal therapeutic paradigm against cancer: On the way forward in advancing precision therapy. Nanoscale [Internet], 2024 Mar 12 [cited 2025 May 15];16(13):6330–6364. Available from: https://scispace.com/papers/nanomedicine-as-a-multimodal-therapeutic-paradigm-against-3nskzgge34

[216] Liu Y, Sun X, Wei C, Guo S, Song C, Zhang J, et al. Targeted drug nanodelivery and immunotherapy for combating tumor resistance. Comb Chem High Throughput Screen [Internet], 2024 Apr 25 [cited 2025 May 15];27. Available from: https://scispace.com/papers/targeted-drug-nanodelivery-and-immunotherapy-for-combating-4flk7mtx94

[217] Jia L, Pang M, Fan M, Tan X, Wang Y, Huang M, et al. A pH-responsive Pickering Nanoemulsion for specified spatial delivery of Immune Checkpoint Inhibitor and Chemotherapy agent to Tumors. Theranostics [Internet], 2020 Aug 7 [cited 2025 May 15];10(22):9956–9969. Available from: https://scispace.com/papers/a-ph-responsive-pickering-nanoemulsion-for-specified-spatial -1x2vyj9jq9

[218] Qian X, Hu W, Yan J. Nano-Chemotherapy synergize with immune checkpoint inhibitor- A better option? Front Immunol [Internet], 2022 Aug 9 [cited 2025 May 15];13. Available from: https://scispace.com/papers/nano-chemotherapy-synergize-with-immune-checkpoint-inhibitor -30j6ev69

[219] Praveen Kumar G, Divya A, Kumar GP. Nanoemulsion based targeting in cancer therapeutics. Med Chem (Los Angeles) [Internet], 2015 Jun 24 [cited 2025 May 15];5(6):1–13. Available from: https://scispace.com/papers/nanoemulsion-based-targeting-in-cancer-therapeutics-191h3n8sby

[220] Su Z, Xiao Z, Wang Y, Huang J, An Y, Wang X, et al. Codelivery of Anti-PD-1 antibody and paclitaxel with matrix metalloproteinase and pH dual-sensitive micelles for enhanced tumor chemoimmunotherapy. Small [Internet], 2020 Jan 28 [cited 2025 May 15];16(7):1906812. Available from: https://scispace.com/papers/codelivery-of-anti-pd-1-antibody-and-paclitaxel-with-matrix -2a8vrqz3an

[221] Buonaguro L, Tagliamonte M. Peptide-based vaccine for cancer therapies. Front Immunol [Internet], 2023 Aug 16 [cited 2025 May 15];14. Available from: https://scispace.com/papers/peptide- based-vaccine-for-cancer-therapies-2hgytyivkk

[222] Fang X, Lan H, Jin K, Gong D, Qian J. Nanovaccines for cancer prevention and immunotherapy: An update review. Cancers (Basel) [Internet], 2022 Aug 1 [cited 2025 May 15];14(16):3842–3842. Available from: https://scispace.com/papers/nanovaccines-for-cancer-prevention-and- immunotherapy-an-1ou59zn8

[223] Singh V, Verma A, Chouhan APS, Saxena R, Koranga S, Husain T. Safety and toxicity of nanomaterials in medicine. AMDTC Book Series [Internet], 2023 Nov 3 [cited 2025 May 15];429–49. Available from: https://scispace.com/papers/safety-and-toxicity-of-nanomaterials-in-medicine-3ob lcc5keg

[224] Rather GA, Gul MZ, Riyaz M, Chakravorty A, Khan MH, Nanda A, et al. Toxicity and risk assessment of nanomaterials. Handbook of Res Nano-Strategies Combating Antimicrob Resist Cancer [Internet], 2021 Jan 1 [cited 2025 May 15];391–416. Available from: https://scispace.com/papers/toxicity-and- risk-assessment-of-nanomaterials-331e74cd2c

[225] Wani TA, Masoodi FA, Jafari SM, McClements DJ. Safety of Nanoemulsions and Their Regulatory Status. In *Nanoemulsions: Formulation, Applications, and Characterization*, Elsevier Inc., 2018, pp. 613–628.

[226] Modarres-Gheisari SMM, Gavagsaz-Ghoachani R, Malaki M, Safarpour P, Zandi M. Ultrasonic nano- emulsification – A review. Ultrason Sonochem [Internet], 2019 Apr 1 [cited 2025 May 28];52:88–105. Available from: https://www.sciencedirect.com/science/article/pii/S1350417718312690

[227] Peshkovsky AS, Peshkovsky SL, Bystryak S. Scalable high-power ultrasonic technology for the production of translucent nanoemulsions. Chem Eng Process: Process Intensif [Internet], 2013 Jul 1 [cited 2025 May 28];69:77–82. Available from: https://www.sciencedirect.com/science/article/abs/ pii/S0255270113000573

[228] Ashaolu TJ. Nanoemulsions for health, food, and cosmetics: A review. Environ Chem Lett [Internet], 2021 Mar 15 [cited 2025 May 15];19(4):1–15. Available from: https://scispace.com/papers/nanoemul sions-for-health-food-and-cosmetics-a-review-3j6km5wqfy

[229] Li Y, Weng Y, Hui Y, Wang J, Xu L, Yang Y, et al. Design of stimuli-responsive minimalist heptad surfactants for stable emulsions. Commun Mater [Internet], 2024 Dec 1 [cited 2025 May 28];5(1):1– 11. Available from: https://www.nature.com/articles/s43246-024-00670-6

[230] Wang S, Wang X, Luo Y, Liang Y. A comprehensive review of conventional and stimuli-responsive delivery systems for bioactive peptides: From food to biomedical applications. Adv Compos Hybrid Mater [Internet], 2024 Nov 30 [cited 2025 May 28];8(1):1–31. Available from: https://link.springer.com/article/10.1007/s42114-024-01053-8

[231] Mukherjee S, Holliday DL, Banjara N, Hettiarachchy N. Nanoemulsions for antitumor activity. Food, Med, Environ Appl Nanomater [Internet], 2022 Jan 1 [cited 2025 May 15];435–54. Available from: https://scispace.com/papers/nanoemulsions-for-antitumor-activity-31lhl5ea

[232] Vishwajeet Chatekar PSSSSBPP. Pharmacokinetic modeling of nanoemulsions in systematic drug delivery. Int J Pharm Sci [Internet], 2025 [cited 2025 May 28];3:742–747. Available from: https://www.ijpsjournal.com/article/Pharmacokinetic+Modeling+of+Nanoemulsions+in+Systematic+Drug+Delivery+

[233] Sharma H, Kaundal T, Prakash A, Medhi B. Artificial Intelligence: A catalyst for breakthroughs in nanotechnology and pharmaceutical research. Int J Pharm Sci Nanotechnol [Internet], 2024 Aug 15 [cited 2025 May 15];17(4):7439–7445. Available from: https://scispace.com/papers/artificial-intelligence-a-catalyst-for-breakthroughs-in-321cb41qy3q2

[234] Ribeiro LNM, Rodrigues da Silva GH, Couto VM, Castro SR, Breitkreitz MC, Martinez CS, et al. Functional hybrid nanoemulsions for sumatriptan intranasal delivery. Front Chem [Internet], 2020 Nov 12 [cited 2025 May 28];8:589503. Available from: https://pmc.ncbi.nlm.nih.gov/articles/PMC7689160/

[235] Kour P, Dar AA. Interfacially modified hybrid nanoemulsion-based alginate capsules: A novel food grade system with enhanced oxidative stability of linoleic acid with a potential antioxidant property. ACS Food Sci Technol [Internet], 2023 May 19 [cited 2025 May 28];3(5):838–849. Available from: 10.1021/acsfoodscitech.3c00032

[236] Villalobos-Castillejos F, Granillo-Guerrero VG, Leyva-Daniel DE, Alamilla-Beltrán L, Gutiérrez-López GF, Monroy-Villagrana A, et al. Fabrication of nanoemulsions by microfluidization. Nanoemulsions: Formul Appl Charact [Internet], 2018 Mar 5 [cited 2025 May 15];207–32. Available from: https://scispace.com/papers/fabrication-of-nanoemulsions-by-microfluidization-4t9hsjw9ma

[237] Gu T, Yeap EWQ, Somasundar A, Chen R, Hatton TA, Khan SA. Droplet microfluidics with a nanoemulsion continuous phase. Lab Chip [Internet], 2016 Jul 5 [cited 2025 May 15];16(14):2694–2700. Available from: https://scispace.com/papers/droplet-microfluidics-with-a-nanoemulsion-continuous-phase-2jfx994eb3

[238] Namratha Vinaya K, Mary John A, Mangsatabam M, Anna PM. Ultrasound-assisted synthesis and characterization of sesame oil based nanoemulsion. IOP Conf Ser Mater Sci Eng [Internet], 2021 Mar 1 [cited 2025 May 15];1114(1):012085. Available from: https://scispace.com/papers/ultrasound-assisted-synthesis-and-characterization-of-sesame-v5uf9sw9jc

[239] Thakur P, Sonawane S, Potoroko I, Sonawane SH. Recent advances in ultrasound-assisted synthesis of nano-emulsions and their industrial applications. Curr Pharm Biotechnol [Internet], 2021 Jan 1 [cited 2025 May 15];22(13):1748–1758. Available from: https://scispace.com/papers/recent-advances-in-ultrasound-assisted-synthesis-of-nano-3kjt6iodwl

[240] Caianiello C, D'avino M, Cavasso D, Paduano L, D'errico G. Bioinspired nanoemulsions stabilized by phosphoethanolamine and phosphoglycerol lipids. Nanomaterials [Internet], 2020 Jun 1 [cited 2025 May 28];10(6):1–14. Available from: https://pubmed.ncbi.nlm.nih.gov/32570696/

[241] Liu F, Lin S, Zhang Z, Hu J, Liu G, Tu Y, et al. PH-responsive nanoemulsions for controlled drug release. Biomacromolecules [Internet], 2014 Mar 10 [cited 2025 May 28];15(3):968–977. Available from: 10.1021/bm4018484

[242] Alayoubi A, Alqahtani S, Kaddoumi A, Nazzal S. Effect of PEG surface conformation on anticancer activity and blood circulation of nanoemulsions loaded with tocotrienol-rich fraction of palm oil. AAPS J [Internet], 2013 Aug 30 [cited 2025 May 15];15(4):1168–1179. Available from: https://scispace.com/papers/effect-of-peg-surface-conformation-on-anticancer-activity-227pvj571d

[243] Fernandes DA. Review on the applications of nanoemulsions in cancer theranostics. J Mater Res [Internet], 2022 Jun 10 [cited 2025 May 15];37(12):1953–1977. Available from: https://scispace.com/papers/review-on-the-applications-of-nanoemulsions-in-cancer-17b298gw

[244] Jacob S, Kather FS, Boddu SHS, Shah J, Nair AB. Innovations in nanoemulsion technology: Enhancing drug delivery for oral, parenteral, and ophthalmic applications. Pharmaceutics [Internet], 2024 Oct 17 [cited 2025 May 28];16(10):1333. Available from: https://www.mdpi.com/1999-4923/16/10/1333/htm

[245] Li H, Lu H, Zhang Y, Liu D, Chen J. Oil-in-water nanoemulsion with reversible charge prepared by the phase inversion composition method. J Mol Liq [Internet], 2021 Aug 15 [cited 2025 May 28];336:116174. Available from: https://www.sciencedirect.com/science/article/abs/pii/S0167732221009016

[246] Prakasha R, Vinay GM, Srilatha P, Pandey H. Nanoemulsions as carriers of bioactive compounds in functional foods: Preparation and application. Eur J Nutr Food Saf [Internet], 2025 Jan 9 [cited 2025 May 15];17(1):78–95. Available from: https://scispace.com/papers/nanoemulsions-as-carriers-of-bioactive-compounds-in-6o7bag6nhqt1

[247] Murugan K, Abd-Elsalam KA. Sustainable nanoemulsions for agri-food applications: Today and future trends. Bio-based Nanoemulsions Agri-Food Appl [Internet], 2022 Jan 1 [cited 2025 May 15];1–11. Available from: https://scispace.com/papers/sustainable-nanoemulsions-for-agri-food-applications-today-3kswl1cn

[248] Yan Z, Yu T, Wu X, Deng M, Wei P, Su N, et al. Nanoemulsion based lipid nanoparticles for effective demethylcantharidin delivery to cure liver cancer. Chem Biol Drug Des [Internet], 2024 Jul 1 [cited 2025 May 15];104(1). Available from: https://scispace.com/papers/nanoemulsion-based-lipid-nanoparticles-for-effective-39ixnd4i69

[249] Santosh P, Kate V, Desai S, Sandesh B, Deshmukh P, Deshmukh A. Composition of lipid based nanoemulsion for oral delivery of Orlistat. Middle East Res J Pharm Sci [Internet], 2023 Nov 8 [cited 2025 May 15];3(06):70–81. Available from: https://scispace.com/papers/composition-of-lipid-based-nanoemulsion-for-oral-delivery-of-3ozyma8mdb

Hend I Shahin and Lipika Chablani*

Chapter 7
Cutting-edge innovations: advances and applications of micro/nanoemulsions in gel formulations

Abstract: Micro- and nanoemulsions (ME/NE) in a gel-based formulation have been explored significantly for pharmaceutical and cosmetic applications. This chapter will summarize the role of ME/NE topical gel formulations, discuss the application of various excipients, describe the characterization tests used for formulation evaluation, list the dermatological/cosmetic applications of such formulations, and elaborate on future trends in this field. Overall, the studies highlight the significance of ME/NE in gel-based topical formulations and their role in formulation innovation for future applications.

Keywords: Microemulsions, nanoemulsions, emulgel, topical gels

7.1 Introduction

Micro/nanoemulsions (ME/NE) have revolutionized topical drug delivery, offering enhanced drug stability, skin permeation, and bioavailability compared to traditional formulations [1]. Microemulsions (MEs), typically with droplet sizes ranging from 1nm to 500 nm, and nanoemulsions (NEs), with droplet sizes <200 nm, are colloidal systems that are considered kinetically stable and consist of an oil phase, aqueous phase, surfactants, and cosurfactants. It is difficult to differentiate between NEs and MEs based solely on their particle size [2]. There is an overlapping size range between these two formulations. Additional characterizations that could be used to differentiate between them are their particle size distribution (PSD), particle structure, and thermodynamical stability [2]. MEs tend to show a single narrow peak for PSD, while NEs could have single or multiple peaks with various widths (narrow- or broadband). MEs and NEs are formed of colloidal particles that are usually spherical. However, MEs could be formed from nonspherical particles due to the lower interfacial tension of surfactant monolayer for MEs compared to NEs. Finally, MEs are thermodynamically stable, while NEs are thermodynamically unstable [2].

*__Corresponding author: Lipika Chablani,__ Department of Pharmaceutical Sciences, Wegmans School of Pharmacy, St. John Fisher University, 3690 East Ave, Rochester, NY 14618, USA, e-mail: lchablani@sjf.edu
__Hend I Shahin,__ Department of Pharmaceutical Sciences, Wegmans School of Pharmacy, St. John Fisher University, 3690 East Ave, Rochester, NY 14618, USA

https://doi.org/10.1515/9783111593654-008

MEs and NEs are divided into three types: a) the aqueous phase is dispersed in a continuous oil phase (W/O type), b) the oil phase is dispersed in a continuous aqueous phase (O/W type), and c) the aqueous and oil phases are inter-dispersed (bi-continuous type) [3]. These systems play a significant role in overcoming challenges associated with the delivery of lipophilic active ingredients, making them particularly attractive for topical and transdermal applications [4]. Due to their small particle size, the use of MEs and NEs is a practical approach for overcoming the skin barrier, without the need for penetration enhancers, through intra- and inter-cellular transport mechanisms offering increased therapeutic efficacy with fewer side effects [5]. In addition, MEs and NEs provide high-drug loading capacities, enhanced drug solubility, controlled and sustained drug release profiles, and ease of production [6]. This leads to more efficient drug delivery across the stratum corneum, the skin's outermost barrier. Additionally, NE can encapsulate hydrophilic and lipophilic substances, providing versatility in formulation design for various skin conditions [7].

However, the main drawback of using ME and NE for topical or transdermal drug delivery is their low viscosity, which leads to low spreadability and skin retention time [4]. Fortunately, adding gelling agent to ME and NE formulations can help resolve this limitation and provide a new formulation known as emulgel [8]. Micro/nano emulgels have gained attention in dermatological and cosmetic industries by providing additional functionality, enhancing spreadability and ease of application, and offering a nongreasy formulation while increasing skin retention time and improving the patient experience [8, 9]. Being hydrophilic, gels allow for better skin hydration, while the embedded ME/NE protects the drug from enzymatic degradation, facilitates the prolonged release of active ingredients, and ensures better skin absorption [4]. This combination of emulsions and gels provides a potent vehicle for delivering pharmaceuticals [10], cosmetics [11], and even wound-healing agents [12], making them a critical focus of research and innovation in topical formulations for the treatment of different skin disorders, including skin cancer [13]. Table 7.1 shows the differences between microemulgel and nanoemulgel.

Several studies have demonstrated the potential use of nanoemulgel and microemulgel for topical drug delivery. These formulations have shown enhanced skin permeations and thus improved drug bioavailability. In addition, nanoemulgel and microemulgel have been successfully used to deliver skin care active ingredients (such as antioxidants [14], antiaging [15], and skin whitening [16]) into the deep skin layer with improved long-term outcomes. Nanoemulgel and microemulgel have also been shown to enhance the stability of the active ingredients and protect them against enzymatic degradation [17].

Advantages of using ME/NE in topical gel formulation:

- Better formulation stability: Microemulgel and nanoemulgel are thermodynamically stable, enhancing API stability over time and providing an extended shelf-life.

–   Improved skin permeation: Due to their small droplet size, nanoemulgel enhances drug penetration by overcoming the stratum corneum's natural barrier.
–   Enhanced patient experience: Gel-based formulations with micro/nano emulgel offer superior aesthetics, ease of application, and improved patient compliance. Their lightweight, nongreasy textures make them ideal for cosmetics and dermatological applications.
–   Controlled release: When incorporated into gel systems, micro/nano emulgel enables sustained release of active ingredients, thus prolonging the formulation's therapeutic effect.

In addition to topical drug delivery, nanoemulgel can be used for transdermal drug delivery [18]. Transdermal drug delivery has many advantages (as shown in Figure 1), such as avoiding first-pass metabolism, controlled drug delivery, fewer side effects, ease of administration, and improved patient compliance [6]. Studies have shown that nanoemulgel has been used for transdermal drug delivery of anti-inflammatory agents [19], statins [18], antineoplastic [20, 21], NSAIDs [22], caffeine [23], as well as anti-cancer vaccines [24].

In this chapter, we will explore the latest innovations in the formulation of ME/NE specifically for topical and transdermal gel applications, with a focus on their challenges, manufacturing, advances, applications for various skin-related and systemic

Table 7.1: Primary differences between microemulgel and nanoemulgel.

Feature	Nanoemulgel	Microemulgel
Droplet size	Typically < 100 nm	100–500 nm
Skin penetration	Deeper penetration into skin layers, including the dermis	Limited to upper skin layers (more superficial)
Stability	Often more kinetically stable, less prone to coalescence	Thermodynamically stable but may need stabilizers to prevent separation
Preparation method	High-energy methods like ultrasonication and homogenization	Low-energy methods, such as spontaneous emulsification
Drug release	Typically, extended drug release	Immediate or modified drug release
Application focus	Topical and transdermal drug delivery	Topical drug delivery
Examples of use	Hormone delivery, NSAIDs, Anti-cancer	Antiaging, anti-fungal, and anti-microbial
Cost of production	Higher cost due to more complicated manufacturing process	Cost-effective

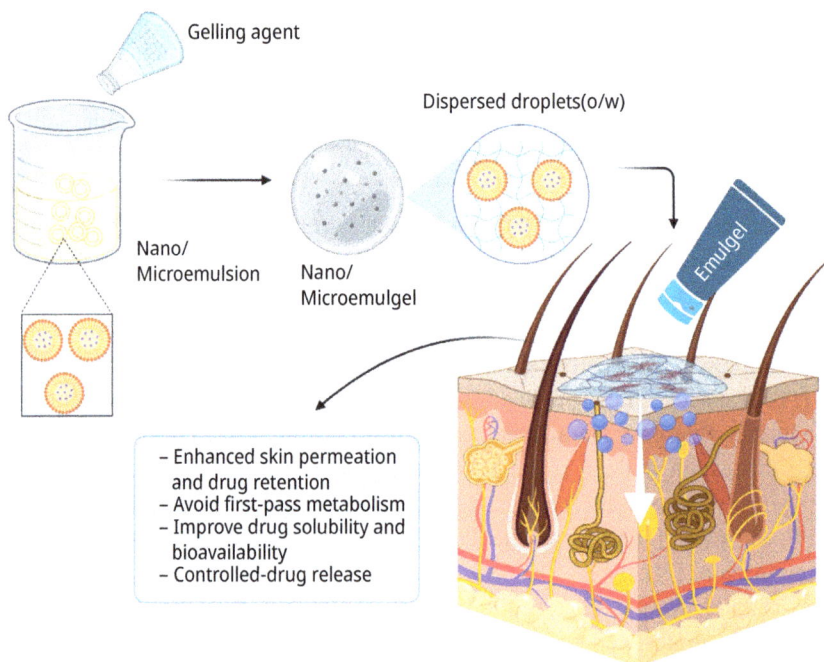

Figure 7.1: Diagram showing nano/microemulgel formulation and their advantages for topical and transdermal. Image generated using BioRender.

conditions, and their role in improving the efficacy of both pharmaceutical and cosmetic products.

7.2 Formulation advances in micro/nano emulgel

Recent advances in micro- and nano-emulgel formulations have helped develop several novel topical cosmetic and transdermal pharmaceutical products. The current goal is to formulate a product with enhanced skin penetration, better drug stability, higher efficacy, lower side effects, and improved patient satisfaction. This section introduces the components of emulgel with emphasis on the advances in formulation techniques, selection of each excipient and their outcomes, and stabilization strategies.

7.2.1 Components of micro/nano emulgel

Generally, micro- and nano-emulgels are formed of aqueous phase, oily phase, surfactants, and cosurfactants. In addition, the formulation of emulgel requires gelling agents and other additives such as antioxidants, preservatives, stabilizers, and humectants.

7.2.1.1 Aqueous phase

Purified water is usually used as an aqueous phase in preparing nano/microemulgel.

7.2.1.2 Oily phase

The selection of oil plays a crucial step in the preparation of emulgel. The type of emulsion (nanoemulgel or microemulgel), the intended application, and the physico-chemical properties of the active ingredient(s) influence the selection of the oil. The oil with the highest drug solubility and compatibility is preferred for lipophilic active ingredients. In addition, the type and quantity of the oil phase determine the viscosity, permeability, drug release, and stability of the final emulgel product [25]. Oils can be naturally derived (plant or animal), such as sesame oil or, jojoba oil, or can be synthetic [26]. Oils are broadly classified into long-chain triglycerides (LCT), medium-chain triglycerides (MCT), and short-chain triglycerides (SCT). LCTs comprise fatty acids with 13 or more carbon atoms [27]. Most of the natural oils used in the production of emulgel are composed of LCTs, which provide controlled drug release and high emollient properties [28]. Examples of LCTs are sunflower, jojoba, castor, and Emu oil. MCTs are composed of fatty acids with chain lengths of 6–12 carbon atoms [29]. Due to their lower molecular weight, MCTs are less dense than LCTs, offering deeper skin penetration and less greasy products. Examples of MCTs are caprylic and capric triglycerides [4]. SCTs are not used in the preparation of emulgel due to their lightweight and volatile properties. In addition, SCTs have lower skin penetration and emollient properties, making them unsuitable for topical application [30]. Other non-classified oils used in emulgel are isopropyl myristate and mineral oil, which are synthetic oils. Researchers are more interested in using natural oils due to their biocompatibility, biodegradability, and natural medicinal benefits that provide a pharmacological effect for treating various skin disorders. For instance, oleic acid is frequently used in the preparation of emulgel due to its antioxidant effect, biocompatibility, biodegradability, and enhanced absorption and permeation through the skin [4, 31]. Table 7.2 shows examples of oil used in emulgel formulations.

7.2.1.3 Surfactants and cosurfactants

Surfactants are surface active agents that are amphiphilic, which can be absorbed in the oil-water interface, forming a barrier that protects the emulsion from coalescence [32]. Using surfactants and cosurfactants is essential to stabilize emulgel formulation by reducing the interfacial tension between the hydrophobic and hydrophilic phases. Surfactants lead to the formation of stable emulsion by aligning at the interface with their hydrophobic tail in the oil and hydrophilic head in water. Surfactants can be classified into ionic (cationic, anionic, or zwitterionic) and nonionic surfactants based on their surface charge. The selection of the optimum surfactants depends on their hydrophilic-lipophilic balance (HLB) and emulsion type. Typically, surfactants with high HLB (>8), such as spans and tweens, are hydrophilic and prefer water as the external phase, thus yielding o/w emulsions, whereas surfactants with low HLB (<8) are lipophilic and result in the formation of w/o emulsions [33].

Since the formulation of nanoemulgel requires the formation of small-size droplets, surfactants with high HLB and sometimes two or more surfactants are needed to achieve optimal stability [32]. On the other hand, surfactants with lower HLB can be used to prepare microemulgel, where their stability depends not only on the particle size but also on thermodynamic factors [34]. A mixture of tween 20 and span 20 has been used to prepare different emulgels [35–37]. This mixture has resulted in emulgels with better stability than using either alone.

The toxicity profile of surfactants is an important factor as it may lead to skin irritation. Toxicity will depend on the nature of the surfactant and the concentration used. Ionic surfactants are less preferred for topical applications due to their toxicity and non-biocompatibility compared to nonionic surfactants [4]. Biosurfactants, derived from natural sources such as microbes or animals, are a safer alternative compared to chemical surfactants due to their biodegradability and reduced toxicity [32].

Cosurfactants work synergistically with surfactants to reduce interfacial tension and enhance emulsification. Cosurfactants increase the flexibility of the formed interfacial tension, leading to enhanced droplet formation. Cosurfactant can also be used to reduce the droplet size, which is essential in the formulation of nanoemulgel. Alcohol-based cosurfactants are widely used for their ability to partition between oil and water phases, thus improving their miscibility. Common cosurfactants are PEG-400, carbitol, ethyl alcohol, and transcutol HP [4].

The selection of surfactants and cosurfactants and their concentrations is a crucial step in emulgel formulation. Surfactants and cosurfactants affect the drug release from the formulation, viscosity, emulgel's spreadability, and emulgel absorption upon

Table 7.2: Example of oily phase in nanoemulgel and microemulgel formulations.

Oil phase	Source	Percent used	Advantages	Limitations	Emulgel type	References
Medium-chain triglycerides (MCTs) (derived from coconut or palm oil)	Synthetically processed	5–15%	Solubilize lipophilic drugs Enhance skin penetration Stable	Limited availability	Nanoemulgel	[27]
Isopropyl myristate (IPM)	Synthetic ester	2–10%	Skin penetration enhancer Nongreasy	Not for sensitive skin May cause skin irritation	Microemulgel	[39, 40]
Oleic acid (monounsaturated fatty acid)	Natural	1–10%	Potent penetration enhancer, good solubilizing agent, emollient properties	May irritate at high concentration	Nanoemulgel and microemulgel	[41]
Olive oil	Natural	5–15%	Rich in antioxidants. Suitable for dry skin	Unstable Can be greasy	Nanoemulgel and microemulgel	[39]
Sunflower oil	Natural	5–20%	Affordable Good skin compatibility High in linoleic acid	Less skin penetration properties compared to other oils	Nanoemulgel and microemulgel	[42, 43]
Caprylic/capric triglyceride	Synthetic oil from MCTs	3–8%	Lightweight Good penetration Stable	Too light for deep skin hydration needs	Nanoemulgel	[27, 39]
Mineral oil	Synthetic (derived from Petrolatum)	5–20%	Inexpensive Retain moisture	Limited skin penetration May clog pores	Microemulgel	[44]

(continued)

Table 7.2 (continued)

Oil phase	Source	Percent used	Advantages	Limitations	Emulgel type	References
Jojoba oil	Natural (derived from jojoba seeds)	3–10%	Similar to skin sebum Suitable for sensitive skin Noncomedogenic	Expensive Requires additional stabilizers	Nanoemulgel and microemulgel	[45, 46]
Sesame oil	Natural (similar in structure to long-chain triglycerides)	5–15%	Rich in vitamins and antioxidants High moisturizing	Potential for allergic reactions	Microemulgel	[47, 48]
Castor oil	Natural (long-chain fatty acid)	2–8%	Emollient	Greasy and viscous Slower penetration	Mainly used in microemulgel but can be used in nanoemulgel	[49, 50]
Emu oil	Natural (animal-derived) (long-chain fatty acid)	5–15%	Rich in essential fatty acids, including oleic acid, linoleic acid, and palmitic acid Anti-inflammatory Emollient Enhance skin permeability	Requires additional stabilizers and antioxidants to increase formulation stability and protect against oxidative degradation Could cause an allergic reaction Not vegan or cruelty-free (animal-derived) Has a strong natural odor that may not be appealing in cosmetic products Expensive	Nanoemulgel and microemulgel	[51, 52]

Table 7.3: Examples of common surfactants and cosurfactants in emulgel.

Surfactant/cosurfactant	Type	HLB value	Advantages	Limitations	Typical concentration (%)	References
Tween 80 (polysorbate 80)	Surfactant	15	High solubilizing power, compatible with many ingredients	Skin-irritant at high concentrations	10–18%	[38, 53]
Span 20 (sorbitan monolaurate)	Surfactant	8.6	Suitable for W/O emulsion	Not suitable alone for O/W nanoemulsions	0.5–3%	[54]
Lecithin	Surfactant	4–8	Natural, biodegradable, nonirritant	Limited emulsifying capacity when used alone	15–60%	[55]
Caprylocaproyl polyoxylglycerides (labrasol)	Surfactant	14	Enhances skin penetration	Expensive	1–5%	[56]
Cremphor RH40	Surfactant	14–16	Improves skin permeation, enhances stability, excellent solubilization enhancer	Can cause irritation Sensitive to extreme temperature changes	5–20%	[51]
Ethanol	Cosurfactant	–	Enhances skin permeation Solubilization enhancer	High concentrations can cause dryness	2–10%	[57]
Propylene glycol	Cosurfactant	–	Moisturizing, reduces surfactant requirement	Can irritate some skin types	5–20%	[58]
Transcutol P (diethylene glycol monoethyl ether)	Co-surfactant	–	Excellent penetration enhancer, enhances stability	High concentrations can irritate the skin	2–5%	[59]

application. In addition, surfactants and cosurfactants must be compatible with the gelling agent to prevent phase separation and coalescence.

In addition, the ratio of surfactant to oil impacts the droplet size of emulgel. As the ratio increases, an increase in the droplet size can be observed. This could happen due to excess free surfactant molecules that can form aggregates, leading to increased droplet size [38]. Table 7.3 summarizes the most used surfactants and cosurfactants in emulgel.

7.2.1.4 Gelling agent

Gelling agents are essential components in the formulation of emulgel, and they provide the required stability, structure, and texture for topical application. The unique properties of emulsion and gel combined in emulgel formulation allow the dispersion of hydrophilic and lipophilic ingredients within the gel matrix, providing better stability and absorption of the active ingredients. Incorporating the gelling agent into the emulsion system makes it thixotropic, that is, it transforms reversibly from a gel into liquid upon application of shear stress [32].

The role of gelling agent in emulgel includes:
1. Controls formulation viscosity: The addition of a gelling agent transforms the formulation from a liquid dosage form into a semisolid dosage form, which improves topical application, absorption, and spreadability on the skin. It also provides a nongreasy texture, which enhances compliance.
2. Enhances emulsion stability: The increased viscosity of the emulsion allows better dispersion of the emulsion droplets, which reduces coalescence, creaming, and phase separation. This stabilization potentiates the extended shelf life of emulgel.
3. Provides extended drug release: The active ingredient is entrapped within the formed three-dimensional hydrogel network formed by the gelling agent, which provides extended drug release.

Gelling agents are classified into natural, synthetic, and semisynthetic. Each gelling agent provides unique properties that make it suitable for a specific emulgel, whether nanoemulgel or microemulgel, depending on the required viscosity, stability, and intent of use [32]. Synthetic polymers, such as carbomers and Pluronic F127, are commonly used in emulgel formulation as they provide controlled viscosity and stability. However, they can be sensitive to pH and ionic changes. Semisynthetic gelling agents such as hydroxypropyl methylcellulose (HPMC) are biocompatible and easier to use in various formulations, including nano/microemulgel. Natural polysaccharides, such as xanthan gum and alginates, are prevalent for their natural origin and biodegradability; however, they may present challenges in achieving consistent viscosity and are susceptible to microbial degradation [60]. Table 7.4 presents examples of the commonly used gelling agents in emulgel formulation.

The concentration of gelling agents could vary between nanoemulgel and microemulgel – nanoemulgel could require a high concentration of gelling agents to achieve better stability. In addition, it has been found that the concentration of gelling agents inversely affects the extent of drug release. Emulgel, composed of HPMC as a gelling agent, showed a better drug release rate compared to carbopol-containing emulgels [36]. Furthermore, the combination of two gelling agents could enhance formulation stability. It was found that a combination of carbopol and HPMC results in better emulgel stability compared to individual gelling agents [61].

In summary, gelling agents are essential in controlling the stability and drug release of actives from emulgel formulations. Selecting the appropriate gelling agent and their concentration depends on identifying formulation type, stability, active ingredient compatibility, and desired release profile.

7.2.1.5 Other additives

7.2.1.5.1 Penetration enhancers

Although nanoemulgel and microemulgel formulations do not require the use of penetration enhancers due to their small droplet size that can penetrate the skin barrier, the use of penetration enhancers could be encouraged based on the desired drug delivery site (topical or transdermal) and the physicochemical properties of the active ingredients. For example, some active ingredients could still have limited skin permeation and require penetration enhancers due to their molecular size, solubility, or polarities. On the other hand, microemulgel could benefit from penetration enhancers due to their relatively large droplet size compared to nanoemulgel. Furthermore, penetration enhancers can provide additional control on the drug release rate and assist in transdermal drug delivery to maintain drug levels for an extended time, thereby reducing the need for frequent dosing.

Despite this, penetration enhancers are not required if the active ingredient is highly permeable or if the goal is to achieve topical drug delivery. Different types of penetration enhancers can be used in emulgel preparation, such as clove oil, lecithin, and menthol (5%) [68, 69]

7.2.1.5.2 Humectant

Humectants are added to retain moisture in emulgel formulations and reduce drying at the application site. Examples of commonly used humectants include glycerin and propylene glycol [70]. Algahtani et al. [71] explored the formulation of retinyl palmitate-loaded nanoemulgel using the low-energy emulsification method. The nanoemulgel was prepared utilizing an oil mixture (Capex ®m355 and capriole 90, 1:2 ratio), surfactant mixture (Kolliphor EL and Transcutol HP, 2:1 ratio), Carbopol 940 as a gelling agent, and glycerin as a humectant. Glycerin provided a smooth texture for the

Table 7.4: Gelling agents in emulgel.

Gelling agent	Type	Advantages	Limitations	Typical concentration (%)	References
Carbopol 934 Carbopol 940	Synthetic polymer	– Forms a highly viscous gel at a low concentration – Stable in an acidic environment – Provides nongreasy texture	– Sensitive to high ionic strength and alkaline pH	0.5–2%	[62, 63]
Pluronic F127	Synthetic copolymer	– Thermosensitive (forms gels at body temperature) – Provides controlled drug release	– Must be stored at low temperature – Expensive – Potential skin irritation in sensitive skin	1–3%	[64]
HPMC (hydroxypropyl methylcellulose)	Semisynthetic cellulose derivative	– Nonionic – Biocompatible – Sustained release	– Less viscous compared to carbomers	1–3%	[36, 65]
CMC (carboxymethylcellulose)	Semisynthetic cellulose derivative	– Biocompatible – Nontoxic	– Not stable in an acidic environment. – Requires additional stabilizers	0.5–2%	[65]
Xanthan gum	Natural polysaccharide	– Biodegradable – Stable in acidic and alkaline pH – Nontoxic	– High viscosity can affect spreadability – Incompatibility with some preservatives	0.1–2%	[66]
Sodium alginate	Natural polysaccharide	– Biocompatible, – Forms stable gels in calcium ions	– Sensitive to ionic conditions	1–5%	[65]
Guar gum	Natural polysaccharide	– Biodegradable	– Unstable in acidic conditions – Incompatibility with some preservatives	0.5–2%	[67]

formulation. Triethanolamine was used to adjust the formulation pH to 5.5 to reduce skin irritation. Nanoemulgel formulation showed average droplet size <50 nm with enhanced spreadability (1.32 cm^2/g vs. 1.27 cm^2/g), drug deposition in the skin (835.5 µg/cm^2 vs. 204.29 µg/cm^2), and cumulative drug permeated (417.30 µg vs. 219.33 µg) for nanoemulgel vs. gel.

7.2.1.5.3 pH adjustment

The pH value of the emulgel formulation must be compatible with the skin pH (4.1–5.8) to avoid any skin irritation upon application. The pH of emulgel preparations can be adjusted using triethanolamine [72].

7.2.1.5.4 Preservatives

Preservatives are commonly used to protect emulgel formulation against microbial degradation and enhance product stability [69]. Commonly used preservatives include methylparaben, propylparaben, benzalkonium chloride, benzoic acid, and benzyl alcohol [72]. It is essential to consider any possible incompatibility between preservatives and some gelling agents. For example, there has been reported incompatibility between methylcellulose and methylparaben, propylparaben, butylparaben, and cetylpyridinium chloride [72]. Furthermore, the concentration of the surfactant may affect the performance of the preservative and reduce its effectiveness [73].

7.2.2 Advances in formulation techniques for nanoemulgel and microemulgel preparation

Developing nanoemulgel and microemulgel formulations requires advanced techniques that facilitate the production of nano- and micro-sized droplets, ensuring stable and effective drug delivery. The preparation of emulgel includes three steps: (1) preparation of NE/ME, (2) preparation of gel, and (3) mixing NE/ME and gel to obtain emulgel. The formulation techniques can be classified into high- and low-energy methods. The high-energy methods include high-pressure homogenization, ultrasound, and microfluidization, while low-energy methods include self-emulsification and phase-inversion. Generally, nanoemulsions require more energy input than microemulsions to obtain small droplet sizes. The high energy is needed to produce a disruptive force necessary to mechanically break the oil phase and increase the surface area between the oil and aqueous phase to facilitate the formation of dispersed small and homogenous droplets [6]. The droplet size achieved depends on the specific technique, the processing conditions (like time and temperature), and the formulation composition.

Traditional NE/ME preparation starts with forming a pre-emulsion by mixing all the ingredients. At this stage, devices like high-speed mixers (e.g., ultra-tax) are used to create an initial dispersion, but their ability to reduce droplet size is limited, as much of the energy is lost in yielding heat and friction [74]. Therefore, additional equipment like high-pressure homogenizers (HPHs) are often needed to further break down the droplet size. HPHs push the pre-emulsion through narrow spaces at high pressures, causing it to accelerate to speeds of around 300 m/s. This rapid flow exposes the emulsion to intense cutting forces, impact, and cavitation, resulting in nanoscale droplets [75, 76].

Ultrasound techniques have become more popular because they use less energy, need fewer surfactants, and can create finer droplets than other methods. However, high-energy processes are unsuitable for thermolabile active ingredients, like peptides, proteins, or nucleic acids, potentially reducing their effectiveness. Low-energy methods offer a viable alternative for emulsion production of such active ingredients [76].

Low-energy techniques leverage the natural interactions among the NE/ME components and only require simple agitation to form the emulsion structure. This approach cuts production costs and preserves the potency and stability of sensitive drugs due to the mild processing conditions. Additionally, these methods use nontoxic, nonirritating components, making the emulsions safe for therapeutic use [77].

7.2.2.1 High-pressure homogenization (HPH)

The high-pressure homogenization method is widely used in developing both NEs and MEs. HPH is effective in reducing droplet particle size through the production of high-pressure forces. The process includes passing the oil and water mixture through a narrow inlet orifice at extremely high pressures (up to 5,000 psi). The high pressure generates intense shear forces and cavitation, breaking down the droplet size to nano- or microscale [65]. This directly correlates with the applied pressure, homogenization cycles, and droplet particle size. Floury et al. reported that the HPH method affects the viscosity of the emulsion and converts it from sheat-thinning to a Newtonian system. There is a significant drop in the viscosity as the homogenization pressure increases [78]. Although this method requires high energy consumption and production and is unsuitable for thermolabile active ingredients, HPH effectively produces an emulgel with uniform particle distribution. In addition, it is ideal for industrial applications as it is easy to scale up (Khan et al. [22] reported the formulation of nanoemulgel for the topical delivery of curcumin using the HPH technique. The formulated nanoemulgel showed optimum thermodynamic stability, enhanced skin penetration, and stability due to its small droplet size.

7.2.2.2 Ultrasonication

The ultrasonication method depends on exposure of coarse emulsion to ultrasonic waves, which results in cavitation forces, known as acoustic cavitation, that break the emulsion droplets into nanosized droplets [65]. Cavitation involves the creation and expansion of tiny bubbles that subsequently collapse, triggered by pressure variations from ultrasonic waves. When these bubbles collapse, they generate intense turbulence, which breaks the oil and aqueous phases into tiny, nano-sized droplets. This method uses ultrasonicators, and the particle size can be controlled by adjusting the energy input and sonication time [79]. Ultrasonication is a better alternative for HPH regarding cleaning, cost, and operation. It is more suitable for experimental formulations rather than manufacturing [80].

7.2.2.3 Microfluidization

The microfluidization technique depends on using a microfluidizer device consisting of microchannels. The emulsion is forced through the microchannels under high pressure (up to 20,00 psi), resulting in particle collision and reduction to nano-or micro-sized droplets through shearing and cavitation forces [79]. The microfluidization method results in the formulation of NE/ME with uniform particle size distribution compared to other methods, such as HPH. This technique is preferred for nanoemulgel development as it could achieve small droplet sizes with narrow particle size distribution. In addition, low surfactant concentration can be used to prepare nano/microemulsion formulations. Several studies have reported using the microfluidization technique to formulate nanoemulgel and microemulgel, showing promising results [10, 81–83].

7.2.2.4 Phase inversion emulsification method

In this technique, the change in the surfactant's spontaneous curvature leads to a phase shift during emulsification. Adjustments in factors like temperature or composition can alter the surfactant's curvature. There are two main phase inversion methods: transitional phase inversion (TPI), which includes phase inversion temperature (PIT) and phase inversion composition (PIC), and catastrophic phase inversion (CPI), which includes emulsion inversion point (EIP) [84].

TPI inversion happens when changes in temperature or composition shift the surfactant's natural curvature or affinity. On the other hand, CPI occurs when the dispersed phase is gradually added until its droplets connect to form a continuous or lamellar structure. A sudden system change characterizes catastrophic phase inversion due to shifting conditions. For CPI to occur, most surfactants must be present in the dispersed phase, leading to unstable emulsion, high coalescence rates, and rapid phase inversion. The surfactant's curvature or affinity changes in TPI, whereas these properties remain constant in CPI. Nonionic surfactants, such as polyethoxylated surfactants, are commonly used in all phase inversion methods. In PIT, the gradual increase in temperature of o/w emulsion until it exceeds phase inversion temperature leads to dehydration of the surfactant, making it more lipophilic and resulting in phase inversion (w/o emulsion). Upon rapid cooling, a stable NE/ME is formed [79].

In PIC, the system composition is altered rather than the system temperature. The PIC method is similar to the PIT method; however, in PIC, phase inversion is achieved by changing the system composition rather than the system temperature [85]. Similar to PIT, nonionic surfactant and oil are mixed, and then water is gradually added. At equilibrium, the spontaneous curvature of the surfactant will change to zero. Further addition of water will result in a change of the surfactant curvature from zero to high positive, which leads to phase inversion and the formation of nanosized droplets. Other examples of PIC include changing system pH and adding salt [85].

Catastrophic phase inversion includes an emulsion inversion point, whereas the inversion is induced by changing the dispersion phase fraction volume rather than changing the properties of the surfactant. The concept depends on initially creating an unstable emulsion by dissolving the surfactant in the dispersed phase. This is against Bancroft's rule, which states that the surfactant must be present in the continuous phase to obtain a stable emulsion. This results in high coalescence and rapid phase inversion once the water amount exceeds the critical water content, which inverts w/o into o/w nanoemulsion or microemulsion. The size of the formed droplet will depend on the water addition rate and agitation speed [79, 86].

The phase inversion method depends on a low-energy approach suitable for sensitive active ingredients. It is commonly used to prepare microemulgel and nanoemulgel; however, the resulting droplet size formulated by PI is usually larger than that of other methods, such as HPH or ultrasonication.

7.2.2.5 Spontaneous emulsification

The most advantageous low-energy approach for NE/ME formulations is spontaneous emulsification, wherein the components are typically combined with minimal mechanical input, relying on manual mixing or gentle low-energy mixers, such as mechanical stirring or vortexing. This process is effective only when components are selected and proportioned precisely; the oils, surfactants, and cosurfactants must be fully miscible, and sufficiently high concentrations of surfactants/cosurfactants are essential to enhance formulation stability and homogeneity and achieve minimal droplet sizes [87].

Spontaneous emulsification relies on the chemical energy released during dilution with the continuous phase, maintaining a constant temperature and avoiding any phase transition within the system. This method enables NE/ME formation at ambient conditions without specialized equipment. In this mechanism, an oil phase containing a hydrophilic surfactant and active ingredient is added to the aqueous phase, where oil droplets promptly form due to the movement of water-dispersible materials from the oil to the aqueous phase, facilitating spontaneous droplet formation [88].

Ahmad et al. [89] reported the formulation of a topical anticancer drug in the form of nanoemulgel for targeted drug delivery for skin cancer treatment. Using the spontaneous emulsification method, a nanoemulsion containing 5-fluorouracil was prepared using castor oil (oil, 40%), Transcutol HP (surfactant, 27%), and polyethylene glycol (PEG)-400 (cosurfactant, 13%) Optimized formulation was mixed with Carbopol® to prepare nanoemulgel. The prepared formulation showed homogenous particle size distribution with average droplet size ranging between 66.97 nm and 206.45 nm, good stability for 90 days at refrigerator temperature with no significant changes in physicochemical properties, and sustained drug release over 24 h.

7.3 Characterization of nanoemulgel and microemulgel formulations

Characterizing nanoemulgels and microemulgels is essential for evaluating their physicochemical properties, performance, safety, and stability, which are crucial for optimizing their efficacy as a drug delivery system. These characterizations involve various analytical, physicochemical, and functional assessments. Below is a detailed exploration of their properties and the corresponding evaluation techniques.

7.3.1 Visual inspection

Emulgel formulations are visually evaluated for their appearance, homogeneity, and color. This test is essential to record any instability and phase separation reflected in a change of appearance [68].

7.3.2 Particle size and size distribution

Measuring the droplet size and polydispersity index (PDI) of nanoemulgel and microemulgel systems is an essential characterization as they determine their stability, drug release kinetics, and penetration efficiency. Particle size and PDI are commonly measured using Zetasizer. A sample is dissolved in purified water with agitation until a homogeneous dispersion is formed. The sample is then injected into the photocell of Zetasizer. Nanoemulgels typically exhibit droplet sizes <200 nm, while microemulgels range from 200 nm to several microns. A low PDI (<0.3) indicates uniform particle distribution and stability. The smaller the droplet sizes, the better skin penetration and drug release, while a narrow size distribution minimizes the risk of phase separation and indicates better stability [90].

7.3.3 Zeta potential

Zeta potential is an indicator of the surface charge of droplets, which affects the physical stability of the system by preventing coalescence. As the zeta potential increases, the repulsion between the globules increases, resulting in better formulation stability. Zeta potential can also be measured using Zetasizer [91].

7.3.4 Viscosity and rheology

The viscosity and flow behavior of nanoemulgels and microemulgels are vital for ensuring proper application consistency and adherence to the skin or mucosal surfaces. Studying the rheological behavior of the emulgel is essential to understanding the influence of the selected components, such as oils, surfactants, and gelling agents, and their concentration on the viscoelastic properties. In addition, the viscosity and flow properties affect the formulation stability and drug release. The viscosity of the formulation will affect the skin contact time/residence time. The higher the viscosity of the emulgel within the optimum range of its spreadability, the better the contact time. If the emulgel shows shear-thinning behavior, it tends to create a thin layer on the skin surface, enhancing permeation and absorption. If plastic flow is observed, other parameters such as yield stress and flow index should be evaluated, which provide a better understanding of the formulation's structural integrity. The viscosity of the emulgel could be measured by Brookfield viscometer at different shear rates at room temperature [92].

7.3.5 Thermodynamic stability studies

Thermal and mechanical stresses are applied to evaluate the physical stability of emulgel under extreme conditions. This includes centrifugation tests, heating-cooling cycles, freeze-thaw cycles, and phase separation tests. Any physical change, including changes in clarity, coagulation, precipitation of drug, color change), drug content, and pH are recorded. Centrifugation test is performed by centrifuging the formulation at 3,000–5,000 rpm for 30 mins. The heating–cooling cycle test is conducted by storing the emulgel at 4° to 45 °C. For freeze–thaw cycles, test tubes filled with the nano/microemulgel are sealed and vertically stored for 16 h in a freezer at −21 °C and then for

8 h at room temperature (25 °C) [31]. Phase separation test is done by adding 0.1 N HCl, pH 7.4 buffer (phosphate), and water, respectively. Lack of phase separation, creaming, or coalescence indicates robust stability [93].

Visual inspection is carried out, and grading of the formulation is as follows:

A Grade: Rapid formation of clear nano/microemulgel

B Grade: Rapid formation of slightly bluish nano/microemulgel

C Grade: Slow formation of turbid nano/microemulgel

D Grade: Slow formation of dull, grayish turbid nano/microemulgel

E Grade: Formation of turbid nano/microemulgel presenting the oil globules on their surface

7.3.6 Drug content and encapsulation efficiency

Measuring drug loading and encapsulation efficiency is essential to ensure that the prepared formulation contains the desired therapeutic dose and is consistent among all formulations. A sample of the emulgel formulations is weighted and dissolved in a suitable solvent (based on API solubility), such as phosphate buffer of pH 6, ethanol, or methanol, by sonication. The solution is then diluted and filtered through 0.45-mm membrane filters. The absorbance is measured at the optimum wavelength using a UV-visible spectrophotometer [91]:

$$\text{Drug content} = (\text{Concentration} \times \text{Dilution factor} \times \text{Volume taken}) \times (\text{Conversion factor})$$

7.3.7 In vitro drug release

Extended drug release is a characteristic feature of nanoemulgel and microemulgel systems. The in vitro release testing of emulgel formulations is critical to elucidate the system's drug release kinetics and overall performance. This study is typically executed utilizing a dialysis bag or modified Franz diffusion cell composed of a donor and a receptor compartment. A dialysis membrane, preconditioned in a phosphate buffer of pH 7.4 and possessing a surface area of 2.5 cm^2, separates these compartments. The membrane is secured over the donor compartment, where the emulgel formulation (approximately 500 mg) is uniformly applied.

The receptor compartment is filled with freshly prepared phosphate buffer (pH 7.4) and supplemented with a magnetic stirring bead to ensure uniform mixing. The system is maintained at a physiological temperature of 37 ± 0.5 °C, with constant agitation at 50 rpm to simulate in vivo conditions. At predetermined time intervals, aliquots of 1 mL are withdrawn from the receptor compartment for analysis, ensuring the volume is immediately replenished with an equal quantity of fresh buffer to maintain sink conditions.

The withdrawn samples are evaluated using UV-spectrophotometric analysis at the predetermined wavelength, and the cumulative percentage of drug release is quantified over time. This test provides an approach for studying the release behavior and optimizing the formulation for enhanced therapeutic efficacy [91].

The release data is fitted into zero-order, first-order, Higuchi, or Korsmeyer–Peppas models to determine the release mechanism. Nanoemulgels often demonstrate extended release due to their small droplet size and gel matrix.

7.3.8 Skin permeation studies

Permeation efficiency is critical for topical and transdermal applications that control formulation performance and efficiency. Ex vivo permeation studies are indispensable for evaluating the topical and transdermal delivery efficiency of emulgel formulations. These experiments commonly use freshly excised full-thickness skin from albino rats, typically weighing 200–250 g and aged 10–12 weeks. The excised skin is meticulously inspected to ensure its surface is devoid of any irregularities or damage, followed by thorough washing with physiological saline buffer to remove residual impurities. The permeation study uses Keshary-Chien or Franz diffusion cells to simulate the skin's barrier function during drug permeation analysis. The excised skin is carefully mounted between the donor and receptor compartments, with the stratum corneum side facing the donor compartment. The receptor compartment is maintained at 37 ± 1 °C and continuously stirred to mimic physiological conditions. The donor compartment, representing the epidermal surface, is loaded with the emulgel formulation. An equivalent volume of freshly prepared buffer is replenished into the receptor compartment at predetermined intervals to maintain sink conditions. Simultaneously, 3-mL aliquots are withdrawn from the receptor compartment for analysis, ensuring consistency in the diffusion medium. The collected samples are analyzed to determine the cumulative amount of drug permeated through the skin over time. This data provides critical insights into the formulation's drug release and permeation kinetics, enabling optimization for enhanced therapeutic efficacy and targeted delivery. Permeation parameters such as flux (J), permeability coefficient (K_p), and lag time are calculated to evaluate skin transport efficiency [91].

To determine the percent drug retention in the skin (skin retention), the skin is removed from the Franz cells, washed thoroughly, and cut into small pieces. The small skin pieces are immersed in methanol overnight to extract any remaining drug from the skin layers. The methanol is filtered and analyzed for the drug contents [94].

7.3.9 Morphological analysis

Visualizing the structural properties of droplets within the gel matrix provides insights into emulsion uniformity and stability. The morphology of emulgel can vary based on the manufacturing techniques and the gelling agent used [95]. Porous and core-shell structures have been reported in emulgel formulations. Transmission electron microscopy (TEM) and scanning electron microscopy (SEM) are commonly used. TEM confirms the nano-scale size of droplets, while SEM evaluates the gel surface morphology. Spherical droplets and homogenous dispersion indicate a well-formulated system.

7.3.10 pH measurement

The formulation's pH must be compatible with the skin or application site to avoid irritation. A 1% aqueous dispersion of emulgel is prepared, and the pH is measured using a pH meter [96]. For topical applications, a pH close to the skin's natural pH (4.1–5.8) is preferred [97]. Changes in pH can also influence formulation stability.

7.3.11 Stability studies

Long-term stability ensures the formulation remains effective over its intended shelf life. Stability studies for optimized formulations are carried out according to the International Council for Harmonization (ICH) guidelines, which involve storing formulations in collapsible aluminum tubes at various temperatures and humidity for 3 months (5°, 25°/60% RH, 30°/65% RH, and 40°/75% RH). Parameters such as droplet size, PDI, viscosity, and drug content are monitored over one, two, and three months [70].

7.3.12 Spreadability

Spreadability is the ability of the emulgel to spread uniformly over the skin or mucosal surfaces when applied with minimal force. It is determined by the formulation's rheological properties and the type of gel base. Emulgel with high viscosity shows lower spreadability. Formulation with shear-thinning behavior exhibits better spreadability. In addition, the higher the oil content, the lesser spreadability is observed.

Furthermore, the type of gelling agent (e.g., carbomers and cellulose derivatives) and its cross-linking density significantly influence the ease of spreading. Spreadability is a crucial parameter in the characterization of nanoemulgel and microemulgel formulations, as it directly affects the ease of application, uniformity of drug distribution, and patient compliance. Optimal spreadability ensures that the formulation can cover the target area evenly without requiring excessive effort or leaving residues. Spreadability can be assessed using the sliding plate method. A pre-weighed sample of the emulgel is placed between two glass plates. A fixed weight is applied to the upper plate, and the formulation can spread for 5 min. The top plate is pulled out using an 80-g force. The time required by the top slide to cover a distance of 7.5 cm is recorded. A shorter time and larger spread diameter indicate better spreadability. The diameter of the spread area is measured:

$$\textbf{Spreadability } (S): \frac{ML}{T}$$

where M is the weight in grams attached to the top surface, T is the time in minutes required for complete separation, and L is the length of the glass slides.

Optimization strategies, such as optimizing gelling concentration, selecting a surfactant, and adding glycerin, could enhance spreadability.

7.3.13 Skin irritation test

The biosafety and nontoxicity of nano- and microemulgels are critical for their successful application in topical and transdermal applications. These systems must be safe for human skin and underlying tissues without causing irritation, allergic reactions, or cytotoxicity. Skin irritation tests are usually conducted to evaluate their safety profile.

The test is conducted on small animals such as albino rats or guinea pigs. Hairs from the dorsal surface of rats are shaved, and a pre-weighed amount of emulgel is applied. The formulation is applied for 3 min, one hour, and four hours on the skin and covered with cotton gauze in compliance with Guidelines Test No. 404 (OECD). Animals are observed for 14 days following patch removal, and any signs of erythema or edema are recorded. If two or more rats out of eight showed any signs of skin irritation, the formulation is deemed ineffective [70].

7.3.14 Extrudability

Extrudability is essential to evaluate the ease of application of the emulgel formulation from its package. The formulation viscosity and rheological properties under applied pressure also control the extrudability of the emulgel. This test is essential to ensure product consistency and user convenience.

The extrudability test involves filling an aluminum collapsible tube with the emulgel formulation, ensuring the material reaches the crimped end of the tube. The tube is then sealed with its cap, and a predetermined weight (measured in grams) is applied externally to exert pressure on the tube. After removing the cap, the force required to extrude the formulation is recorded by measuring the length of the emul-gel ribbon extruded within a specific time frame, typically 0.5 cm in 10 s. The recorded extrusion force, expressed in grams, is a quantitative measure of the formulation's extrudability [91].

Extrudability is calculated as follows:

Extrudability = weight applied to emulgel extrusion from tube (gm) /area (cm^2)

7.4 Applications of micro/nanoemulsions in topical and transdermal gels

Nanoemulgels and microemulgels are advanced delivery systems that have gained in-sight into the pharmaceutical industry due to their unique properties. They combine the benefits of emulsions, gels, and nanotechnology. They are versatile platforms that enhance the bioavailability, stability, topical application, and extended release of ther-apeutic agents. Table 7.5 summarizes recent advances for emulgel in the pharmaceuti-cal industry and cosmetic applications. Additionally, Table 7.6 provides emulgel-based marketed formulations.

7.4.1 Topical and dermatological applications

Nanoemulgels and microemulgels are extensively used to deliver drugs targeting skin conditions. Their ability to incorporate high lipophilic drugs and provide nongreasy, easy-to-spread gel formulations make them ideal treatment choices for skin conditions such as psoriasis, eczema, and acne. Nanoemulgels enhance drug permeation through the stratum corneum, enabling better dermal and transdermal drug deposition. Emul-gel can be used to deliver antimicrobial agents such as clindamycin for the treatment of acne [98], anti-inflammatory drugs such as diclofenac for arthritis and localized pain relief [99], or even delivery of 5-fluorouracil for non-melanoma skin cancers [100].

7.4.2 Transdermal drug delivery

Emulgel can also be employed for transdermal drug delivery, enabling systemic ab-sorption of drugs through the skin. Transdermal drug delivery provides many advan-tages, such as avoiding first-pass metabolism, decreasing systemic side effects, provid-ing extended drug release, and allowing smaller drug doses. The small droplet size enhances the permeation of drugs like analgesics, hormones, and cardiovascular drugs. For example, emulgel can be used in hormonal therapies such as transdermal delivery of estradiol or testosterone for hormone replacement therapy. In addition, emulgel can be used for transdermal drug delivery of analgesics such fentanyl or ibuprofen for pain management [101, 102].

7.4.3 Cosmeceutical applications

Both nanoemulgels and microemulgels have been increasingly incorporated into cosmetics to deliver active ingredients such as vitamins, peptides, and herbal extracts. They enhance skin hydration, elasticity, and the penetration of antiaging and skin-lightening agents. For example, emulgels have been used to deliver vitamins C and E to combat oxidative stress and whitening agents [103].

Kim et al. reported enhanced solubility, skin deposition, and permeation of a microemulsion-based hydrogel formulation containing 20(*S*)-protopanaxadiol [59]. 20(*S*)-Protopanaxadiol (PPD) can be used as antiaging, but due to its hydrophobic properties and high molecular weight, it is challenging for topical drug delivery. The microemulsion-based hydrogel was composed of Capmul oil, a mixture of Labrasol and Tween 20 as a surfactant mixture, and Transcutol as cosurfactant, carbapol 941 as a gelling agent; the developed formulation demonstrated enhanced penetration of PPD into the skin compared to standard formulations. Characterization showed appropriate pH, viscosity, and droplet size, ensuring compatibility with the skin. In vitro and in vivo deposition showed improved PPD deposition in both the epidermis and dermis in animal studies. No significant irritation or adverse effects were observed, indicating the formulation's safety. Incorporating PPD into emulgel offered improved stability, controlled release, and increased bioavailability, making it suitable for treating skin-related conditions.

7.4.4 Wound healing and tissue regeneration

Nanoemulgels with encapsulated bioactive compounds or growth factors promote wound healing and tissue regeneration. Their biocompatible and hydrating properties

Table 7.5: Examples of recent advances in emulgel applications.

Drug	Application	Problem with the drug	Study outcome	References
Calcipotriol	Vit D3 analogue Treatment of psoriasis	Hydrophobic nature and low penetration	– Emulgel can be used for topical delivery of Calcipotriol. – Formulation showed enhanced physicochemical properties. – Emulgel had good drug release, skin permeation, and less skin irritation.	[105, 106]
Amphotericin B	Anti-fungal and antiprotozoal Treatment of cutaneous leishmaniasis	Could cause skin irritation	Emulgel formulation was stable, with good skin penetration, and nontoxic.	[107]
Metronidazole	Antiprotozoal	Poor water solubility	– Formulating metronidazole as a microemulgel using Campul 908 P as an oily phase, Acconon MC8-2, and propylene glycol as surfactant enhanced drug solubility. – Formulations were stable and showed improved permeation of the drug from the emulgel compared to conventional gel.	[108]
Mefenamic acid	NSAIDs	Hydrophobic No topical formulation available	– Formulated emulgel showed optimum physicochemical properties, stability, analgesic, and anti-inflammatory effects compared to market diclofenac topical gel.	[35]
Chrysin	Anticancer treatment of skin melanoma	Hydrophobic Systemic side effects Poor skin penetration	– Nanoemulgel was formulated using Caproyl 90 oil, Tween 80 surfactants, and Transcutol cosolvent. – The formulation showed a mean droplet size of 156.9 nm and a polydispersity index of 0.26 with a good hardness of 487 g and adhesiveness of 500 g. – Significant improvement in ex vivo percutaneous absorption was recorded. – In vitro cytotoxicity showed a significantly enhanced therapeutic effect using a lower dose. Biocompatibility testing on the L929 cell line showed that the formulation was safe for topical application.	[13]

(continued)

Table 7.5 (continued)

Drug	Application	Problem with the drug	Study outcome	References
Piroxicam	NSAIDs	Hydrophobic Systemic drug delivery	– Emulgel-containing proxicam can be used for transdermal drug delivery. Emulgel comprises oleic acid as oil, tween 80 and ethanol as a surfactant–cosurfactant mixture, and carbopol as a gelling agent. – Nanoemulgel showed a higher cumulative amount of drug permeated with less lag time than the marketed formulation.	[31]
Isotretinoin and erythromycin	For the treatment of acne	Side effects associated with the oral route	– Emulgel was prepared using isopropyl myristate as an oil phase and Tween 80 as a surfactant. – The formulated emulgel enhanced skin permeation, retention, and antimicrobial activity compared to market products.	[94]
Atorvastatin	For wound healing	Systemic side effects	– Nanoemulgel was prepared using liquid paraffin (oil), Tween 80 (surfactant), and carboxy methyl cellulose (gelling agent). – The formulated emulgel showed enhanced skin permeation, skin retention, in vitro drug release, spreadability, and wound healing efficiency compared to conventional gel. – The formulation of Atorvastatin in the form of emulgel showed high drug stability.	[109]
Chlorphenesin	Anti-fungal	It is present in the form of powder.	– Emulgel was formulated using liquid paraffin, Tween 20, and HPMC. – The formulation with low oil concentration and a high emulsifying agent concentration showed optimum drug release profile and antifungal activity.	[36]

| Curcumin | Anti-inflammatory Treatment of rheumatoid arthritis | Low solubility and skin permeation | – | Nanoemulgel was prepared using emu oil (oil), Cremphor RH 40 (surfactant), Labrafil M2125CS (cosurfactant), and carbopol as gelling agent. The anti-inflammatory efficacy was evaluated in carrageenan-induced paw edema and FCA-induced arthritic rat models. -Formulations containing curcumin in combination with emu oil showed significant improvement in anti-inflammatory activity compared to pure curcumin. | [51] |
| Ketocanazole | Antifungal | Poor solubility | – | The optimum emulgel formulation was composed of 2% ketoconazole, 20% oil mix (Oleic acid: coconut oil ratio 2:1), 34.06% Surfactant mix (Tween 80:PG ratio of 2:1), and 43.94% water. The gel base contained 1% HPMC K4M and 2% sodium alginate in the ratio of 1:1. -The microemulgel showed a slow drug release of 19.82% compared to 48,47% from market cream. The microemulgel was stable for two months. In vivo study showed no skin irritation on animal skin. | [110] |

Table 7.6: List of marketed formulations.

Brand	Active ingredient	Uses	Manufacturer
Voltaren	Diclofenac	Analgesic and anti-inflammatory	Novartis
Diclomax	Diclofenac sodium	Analgesic and anti-inflammatory	Torrent Pharmaceuticals Ltd
Avindo gel	Azithromycin	Antibacterial	Adcock Ingram Healthcare Pvt. Ltd.
Isofen emulgel	Ibuprofen	Analgesic and anti-inflammatory	Beit Jala Pharm.
Benzolait emulgel	Benzoyl peroxide	For acne	RoyaDermal
Miconaz H emulgel	Miconazole nitrate + hydrocortisone	Antifungal Antibacterial Anti-inflammatory	Medical Union Pharmaceuticals
Dermafeet emulgel	Urea	For dry skin with hyperkeratosis	Herbitas

make them ideal for chronic wounds and burns. Silver nanoparticles have been incorporated into emulgel for antimicrobial activity in wound care [104].

Emulgel can be used in wound healing, such as those caused by *S aureus*. A nanoemulgel containing simvastatin was developed using a high-shear homogenization method. The nanomulgel showed enhanced antibacterial efficiency in vivo compared to the drug solution and high stability for 72 days [38].

7.5 Technological challenges and innovations in topical emulgels

Emulgel has emerged as a novel drug delivery platform, combining the properties of emulsions and gels to create versatile formulations for dermatological and transdermal applications. However, the development of emulgels involves overcoming a series of technological challenges and exploring new techniques to enhance drug efficacy and user acceptability.

7.5.1 Formulation challenges

7.5.1.1 Stability concerns

One of the primary challenges in emulgel formulations is ensuring physical and chemical stability. Some problems associated with emulsion formulation include phase separation, creaming, coalescence, and Ostwald ripening. When an emulsion is integrated into a gel matrix to form an emulgel, these challenges are combined with maintaining formulation homogeneity and consistency over time. Stabilizers like hydrophilic polymers as carbomers can maintain gel integrity and overcome emulsion instability. Additionally, smaller and more stable emulsion droplets can be prepared using ultrasonic emulsification or homogenization techniques. Furthermore, including liposomes or solid lipid nanoparticles within the gel matrix could enhance the emulsion stability by isolating the emulsion phase [111].

It is important to take into consideration that the formulation and stability of emulgel are also influenced by different parameters, including the HLB value, solubility profiles of the oil and aqueous phases, emulsifier concentration, production temperature, mechanical agitation, and incorporation of specific additives. These considerations become even more critical during the scale-up process, transitioning the formulation from laboratory scale to commercial production. Key factors for successful scale-up include the meticulous selection of raw materials, solvents, production methodologies, and cost-effective approaches, ensuring the final product aligns with patient acceptability and regulatory standards.

Additionally, rigorous stability testing in compliance with International Council for Harmonization (ICH) Q1 guidelines is mandatory. Such assessments confirm the active pharmaceutical ingredient (API) retains its potency and efficacy under different storage conditions. Compatibility studies between excipients are crucial to maintaining formulation integrity and consistent performance throughout its shelf life [112].

7.5.1.2 High-energy methods unsuitable for thermolabile drugs

Many drugs, such as vitamins or proteins, are thermolabile and cannot withstand elevated temperatures associated with high-energy emulsification methods. Thus, low-energy methods like spontaneous emulsification or phase inversion are adopted to minimize thermal exposure. These techniques rely on carefully selecting surfactants and cosurfactants to facilitate emulsification under mild conditions. In addition, the cost of specialized emulsification equipment may hinder the scalability of emulgel production.

7.5.1.3 Skin irritation

Although emulgel can cause less skin irritation than conventional topical formulations, using some emulsifiers and penetration enhancers can cause skin irritation at high concentrations. Thus, proper selection of the emulgel components and testing for any possible skin sensitivity are essential in the development of emulgel. Biocompatible penetration enhancers like phospholipids or natural oils are encouraged to minimize skin irritation. In addition, natural emulsifiers like lecithin can be used to reduce adverse reactions [91].

7.5.1.4 Sensory properties

Topical formulations must balance efficacy with patient compliance, which is affected by smooth application and spreadability. An emulgel must feel nongreasy, absorb quickly, and leave minimal residue while delivering active ingredients effectively. Achieving a balance between oil content for solubilization and a gel structure that provides a light, pleasant texture is sometimes challenging. Incorporating humectants like glycerin enhances skin hydration without stickiness [113].

7.5.1.5 Long-term effect

The long-term effect of topical emulgel in treating chronic conditions is unknown. This can be a significant concern during the development of emulgel. Thus, long-term safety studies must be considered for clinical trials.

7.5.1.6 Risk assessment

The commercial risk assessment of emulgel formulations is critical to ensure product quality, safety, and adherence to regulatory standards. A thorough evaluation of formulation and process variables is essential, as these factors significantly impact the formulation's critical quality attributes (CQAs) and can potentially lead to product failure. Among the key considerations are the risks of skin irritation, inadequate viscosity, poor stability, and suboptimal drug release profiles. These factors highlight the importance of systematic risk analysis during product development.

Figure 7.2 represents the factors influencing emulgel formulation processes, providing a visual framework for understanding potential risks. Complementary qualitative and quantitative risk assessment methodologies, such as relative risk-based matrix assessments and failure mode effects analysis (FMEA), enable prioritizing critical risk factors in the development process. These analyses lead to the identification of

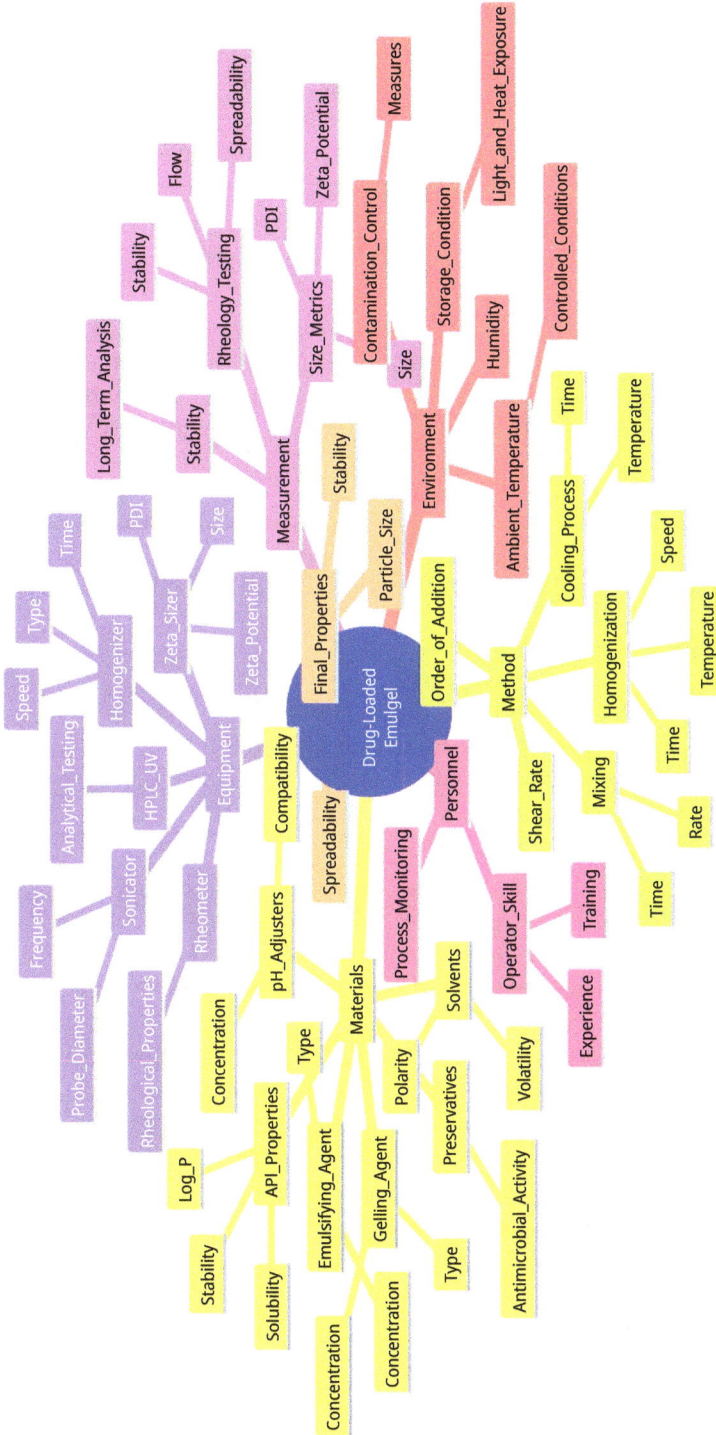

Figure 7.2: Mind map representation of the key factors influencing the formulation and characterization of drug-loaded emulgels, including materials, methods, equipment, measurement parameters, personnel considerations, environmental factors, and final properties. Image generated using AI software.

critical material attributes (CMAs) and critical process parameters (CPPs), which are subsequently integrated into a design of experiments (DoE) framework [112].

Furthermore, optimization designs like Factorial, Central Composite, and Box-Behnken are selected based on the complexity and number of identified critical parameters and attributes. By applying quality by design (QbD) principles, an optimized emulgel product is developed, and a design space is established. This design space ensures the robustness of the manufacturing process and provides a safety margin where troubleshooting can be applied without compromising product quality, facilitating regulatory approval.

The growing volume of patents filed in emulgel technology highlights its promise as a versatile and innovative drug delivery platform [70, 111, 112]. This surge emphasizes the commercial viability of emulgels and reflects the significant advances in the science behind the formulation, which is useful in both the pharmaceutical and cosmeceutical industries.

7.6 Future trends in micro/nano emulsion-based topical gels

The field of topical drug delivery is witnessing remarkable advancements with the development of novel micro- and nanoemulgel, providing a platform for the next generation of formulations. These trends are evident in technological innovation, personalized medicine approaches, and environmental sustainability efforts.

7.6.1 Smart formulations

One of the most promising innovations in emulsion-based topical gels is the development of smart formulations. These systems are designed to actively respond to environmental stimuli, such as pH, temperature, or oxidative stress, and achieve more targeted and extended drug delivery [114]. Stimuli-responsive gels are under active research, which releases active ingredients in response to conditions like inflammation reflected in pH changes or elevated local temperatures. For example, pH-sensitive nanoemulgels for delivering anti-inflammatory drugs can be used on a locally inflamed dermal site with an acidic pH. Such smart formulations target drug delivery, enhancing drug efficacy and reducing systemic drug exposure [113, 115].

7.6.2 Personalization of skincare

Personalized medicine extends into dermatology, including developing gels tailored to individual skin types and conditions. Innovations in diagnostics and skin analysis can guide the formulation of specific emulgels using micro- or nanoemulsions optimized for unique therapeutic needs, such as acne-prone, sensitive, or aged skin. Personalized nanoemulsion-based gels containing active molecules like peptides or hyaluronic acid are being explored to provide antiaging benefits based on individual skin hydration levels [116].

Advances in nanotechnology and delivery systems, like nanoemulsions, allow targeted and efficient cosmetic formulations, enhancing product stability and reducing active ingredient dosages. These innovations merge cosmetics and pharmaceutical industries, particularly in antiaging products, and require strict safety evaluations due to their structural impacts on the skin. While nanoemulsions promise sustainability and efficacy, regulatory gaps, especially for cosmeceuticals, need more studies to ensure long-term safety for consumers and the environment [117].

7.6.3 Sustainability and green materials

The increasing global need for sustainability encourages the development of eco-friendly emulsion-based gels. Biodegradable ingredients such as plant-derived surfactants and natural polymers such as alginate, xanthan gum, and chitosan are being incorporated to reduce the environmental impact of formulations [118]. Furthermore, green production methods such as solvent-free or low-energy manufacturing processes, such as phase inversion temperature techniques, are employed to minimize resource consumption and waste.

Furthermore, modern cosmetic consumers demand safe, effective, and environmentally friendly products. Concerns about synthetic cosmetic ingredients' long-term ecological and health impact have pushed manufacturers toward eco-friendly practices, like using natural alternatives and conducting studies on nanoparticle safety. Agencies like the EPA and European regulators emphasize sustainable development and proactive research into ingredients' effects on health and the environment.

7.6.4 Advanced therapeutics

Beyond conventional topical applications, micro- and nanoemulgels are being investigated to deliver advanced therapeutic agents. Recent studies highlight the potential of these formulations to deliver genetic materials, peptides, and proteins, enabling the treatment of genetic disorders like epidermolysis bullosa or psoriasis. RNA-loaded nanoemulgels could offer noninvasive options for treating genetic skin abnormalities by ensuring stability and effective skin penetration [119].

7.6.5 Hybrid systems

Combining micro- and nanoemulsion systems with other delivery technologies is a promising area of research. Multifunctional systems such as hybrid gels incorporating liposomes, dendrimers, or nanocapsules can simultaneously deliver multiple active ingredients, catering to complex conditions such as diabetic ulcers or chronic dermatitis [120]. These systems often show better stability under various environmental conditions and provide synergistic effects for skin repair and regeneration.

7.6.6 Enhanced toxicological profiling

While nanoemulgels hold great potential, concerns about nanoparticle toxicity and long-term effects on skin health persist. Comprehensive toxicological studies are needed to assess the safety of novel surfactants, stabilizers, and drug molecules in emulgels. Innovations are accompanied by efforts to establish standardized safety protocols for micro- and nanoemulsion-based topical products.

7.6.7 Advances in characterization techniques

To facilitate the scale-up of emulgel formulations, developing ex vivo and in vivo correlation models for topical emulgel systems is essential. This model will ease the path for testing and approval of novel emulgel formulations. In addition, in vitro–in vivo correlation (IVIVC) is cost-effective and provides insight into the medication's effectiveness in vivo. Furthermore, prediction-based software like GastroPlus can predict in vivo performance and study the pharmacokinetics and pharmacodynamics of new formulations [112].

In conclusion, the future of micro/nanoemulgel depends on developing smart drug delivery, enhanced sustainability, and employing personalized medicine. With continuous advancements in formulation science and production technologies, these systems are expected to reshape the field of topical and transdermal therapeutics.

References

[1]   Yang S, Wang F, Han H, Santos HA, Zhang Y, Zhang H, et al. Fabricated technology of biomedical micro-nano hydrogel. Biomed Technol, 2023;2:31–48.

[2]   McClements DJ. Nanoemulsions versus microemulsions: Terminology, differences, and similarities. Soft Matter, 2012;8(6):1719–1729.

[3]   Das S, Lee SH, Chow PS, Macbeath C. Microemulsion composed of combination of skin beneficial oils as vehicle: Development of resveratrol-loaded microemulsion based formulations for skin care applications. Colloids Surf B Biointerfaces, 2020;194:111161.

[4]   Donthi MR, Munnangi SR, Krishna KV, Saha RN, Singhvi G, Dubey SK. Nanoemulgel: A novel nano carrier as a tool for topical drug delivery. Pharmaceutics, 2023;15(1):164.

[5]   Eqbal A, Ansari VA, Hafeez A, Ahsan F, Imran M, Tanweer S. Recent applications of nanoemulsion based drug delivery system: A review. Res J Pharm Technol, 2021;14(5):2852–2858.

[6]   Souto EB, Cano A, Martins-Gomes C, Coutinho TE, Zielińska A, Silva AM. Microemulsions and nanoemulsions in skin drug delivery. Bioengineering, 2022;9(4):158.

[7]   Yadav KS, Soni G, Choudhary D, Khanduri A, Bhandari A, Joshi G. Microemulsions for enhancing drug delivery of hydrophilic drugs: Exploring various routes of administration. Med Drug Discov, 2023:100162.

[8]   Kushwah P, Sharma PK, Koka SS, Gupta A, Sharma R, Darwhekar G. Microemulgel: A novel approach for topical drug delivery. J Appl Pharm Res, 2021;9(3):14–20.

[9]   Aithal GC, Narayan R, Nayak UY. Nanoemulgel: A promising phase in drug delivery. Curr Pharm Des, 2020;26(2):279–291.

[10]   Ma Q, Zhang J, Lu B, Lin H, Sarkar R, Wu T, et al. Nanoemulgel for improved topical delivery of desonide: Formulation design and characterization. AAPS PharmSciTech, 2021;22(5):163.

[11]   Scomoroscenco C, Teodorescu M, Raducan A, Stan M, Voicu SN, Trica B, et al. Novel gel microemulsion as topical drug delivery system for curcumin in dermatocosmetics. Pharmaceutics, 2021;13(4):505.

[12]   Algahtani MS, Ahmad MZ, Shaikh IA, Abdel-Wahab BA, Nourein IH, Ahmad J. Thymoquinone loaded topical nanoemulgel for wound healing: Formulation design and in-vivo evaluation. Molecules, 2021;26(13):3863.

[13]   Nagaraja S, Basavarajappa GM, Attimarad M, Pund S. Topical nanoemulgel for the treatment of skin cancer: Proof-of-technology. Pharmaceutics, 2021;13(6):902.

[14]   Dos Santos RS, Bassi da Silva J, Vecchi CF, da Silva Souza Campanholi K, Rosseto HC, de Oliveira MC, et al. Formulation and performance evaluation of emulgel platform for combined skin delivery of curcumin and propolis. Pharm Dev Technol, 2023;28(6):559–570.

[15]   Yousuf M, Khan HMS, Rasool F, Khan KUR, Usman F, Ghalloo BA, et al. Chemical profiling, formulation development, in vitro evaluation and molecular docking of Piper nigrum seeds extract loaded emulgel for anti-aging. Molecules, 2022;27(18):5990.

[16]   Kothapalli L, Ozarkar R, Modak P, Deshkar S, Thomas A. Preparation and evaluation of nanoemulgel with seed oils for skin care. Curr Nanomed (Formerly: Recent Patents on Nanomedicine), 2024;14 (1):73–83.

[17]   Yousefpoor Y, Baharifar H, Esnaashari SS, Gheybi F, Mehrabi M, Osanloo M, et al. A review of topical micro-and nanoemulsions for common skin diseases. Nanomed J Accepted, 2024.

[18]   Sithole MN, Marais S, Maree SM, Du Plessis LH, Du Plessis J, Gerber M. Development and characterization of nano-emulsions and nano-emulgels for transdermal delivery of statins. Expert Opin Drug Delivery, 2021;18(6):789–801.

[19]   Shakeel F, Baboota S, Ahuja A, Ali J, Aqil M, Shafiq S. Nanoemulsions as vehicles for transdermal delivery of aceclofenac. AAPS PharmSciTech, 2007;8:191–199.

[20]   Sahu P, Kashaw SK, Kashaw V, Shabaaz J, Dahiya R. Synthesis and ex vivo evaluation of PLGA chitosan surface modulated double walled transdermal Pluronic nanogel for the controlled delivery of Temozolomide. Int J Biol Macromol, 2021;187:742–754.

[21] Yang C, Daoping Z, Xiaoping X, Jing L, Chenglong Z. Magnesium oil enriched transdermal nanogel of methotrexate for improved arthritic joint mobility, repair, and reduced inflammation. J Microencapsul, 2020;37(1):77–90.

[22] Khan BA, Haq SU, Khan MK, Khan MS, Menaa F. Fabrication of lornoxicam loaded into curcumin reinforced nanoemulgel: In-vitro characterization and in-vivo anti-inflammatory potential. J Drug Deliv Sci Technol, 2024;100:106090.

[23] Samah NHA, Heard CM. Enhanced in vitro transdermal delivery of caffeine using a temperature-and pH-sensitive nanogel, poly (NIPAM-co-AAc). Int J Pharm, 2013;453(2):630–640.

[24] Toyoda M, Hama S, Ikeda Y, Nagasaki Y, Kogure K. Anti-cancer vaccination by transdermal delivery of antigen peptide-loaded nanogels via iontophoresis. Int J Pharm, 2015;483(1–2):110–114.

[25] Bashir M, Ahmad J, Asif M, Khan S-U-D, Irfan M, Y Ibrahim A, et al. Nanoemulgel, an innovative carrier for diflunisal topical delivery with profound anti-inflammatory effect: In vitro and in vivo evaluation. Int J Nanomed, 2021:1457–1472.

[26] Azeez AR, Alkotaji M. Nanoemulgel as a recent drug delivery system. Mil Med Sci Lett/Vojenské Zdravotnické Listy, 2022;91(2).

[27] Anand K, Ray S, Rahman M, Shaharyar A, Bhowmik R, Bera R, et al. Nano-emulgel: Emerging as a smarter topical lipidic emulsion-based nanocarrier for skin healthcare applications. Recent patents on anti-infective drug discovery. 2019;14(1):16–35.

[28] Efendy Goon D, Sheikh Abdul Kadir SH, Latip NA, Rahim SA, Mazlan M. Palm oil in lipid-based formulations and drug delivery systems. Biomolecules, 2019;9(2):64.

[29] Jadhav HB, Annapure US. Triglycerides of medium-chain fatty acids: A concise review. J Food Sci Technol, 2023;60(8):2143–2152.

[30] Ahmad MZ, Ahmad J, Alasmary MY, Akhter S, Aslam M, Pathak K, et al. Nanoemulgel as an approach to improve the biopharmaceutical performance of lipophilic drugs: Contemporary research and application. J Drug Deliv Sci Technol, 2022;72:103420.

[31] Dhawan B, Aggarwal G, Harikumar S. Enhanced transdermal permeability of piroxicam through novel nanoemulgel formulation. Int J Pharm Investig, 2014;4(2):65.

[32] Alexander A, Khichariya A, Gupta S, Patel RJ, Giri TK, Tripathi DK. Recent expansions in an emergent novel drug delivery technology: Emulgel. J Contr Release, 2013;171(2):122–132.

[33] Dhawas V, Dhabarde D, Patil S. Emulgel: A comprehensive review for novel topical drug delivery system. Int J Recent Sci Res, 2020;11(4):38134–38138.

[34] López-Cervantes M, Escobar-Chávez JJ, Casas-Alancaster N, Quintanar-Guerrero D, Ganem-Quintanar A. Development and characterization of a transdermal patch and an emulgel containing kanamycin intended to be used in the treatment of mycetoma caused by Actinomadura madurae. Drug Dev Ind Pharm, 2009;35(12):1511–1521.

[35] Khullar R, Kumar D, Seth N, Saini S. Formulation and evaluation of mefenamic acid emulgel for topical delivery. Saudi Pharm J, 2012;20(1):63–67.

[36] Mohamed MI. Optimization of chlorphenesin emulgel formulation. AAPS J, 2004;6:81–87.

[37] Pranali S, Charushila S, Sayali C, Namrata M. Design and characterisation of emulgel of an antifungal drug. J Pharm Sci Res, 2019;11(6):2357–2361.

[38] Amoozegar H, Ghaffari A, Keramati M, Ahmadi S, Dizaji S, Moayer F, et al. A novel formulation of simvastatin nanoemulsion gel for infected wound therapy: In vitro and in vivo assessment. J Drug Deliv Sci Technol, 2022;72:103369.

[39] Ahmad J, Gautam A, Komath S, Bano M, Garg A, Jain K. Topical nano-emulgel for skin disorders: Formulation approach and characterization. Recent patents on anti-infective drug discovery. 2019;14(1):36–48.

[40] de Souza Ferreira SB, Bruschi ML. Investigation of the physicochemical stability of emulgels composed of poloxamer 407 and different oil phases using the Quality by Design approach. J Mol Liq, 2021;332:115856.

[41] Raju K, Sneha G, Khatoon R, Ashwini M, Shirisha G, Ajay B, et al. Formulation and evaluation of ornidazole topical emulgel. WJPPS, 2019;8(07):1179–1197.

[42] Kumar S, Patil A, Nile NP, Varma AK, Sahu D, Vishwakarma DK, et al. Formulation and characterization of Miconazole nanoemulgel for topical delivery by using natural oils. 2023.

[43] Kulawik-Pióro A, Goździcka W, Kruk J, Piotrowska A. Plant-origin additives from Boswellia species in emulgel formulation for radiotherapy skin care. Appl Sci, 2024;14(19):8648.

[44] Mazurkeviciute A, Matulyte I, Ivaskiene M, Zilius M. Modeling, the optimization of the composition of emulgels with Ciclopirox Olamine, and quality assessment. Polymers, 2024;16(13):1816.

[45] Elsewedy HS, Younis NS, Shehata TM, Mohamed ME, Soliman WE. Enhancement of anti-inflammatory activity of optimized niosomal colchicine loaded into jojoba oil-based emulgel using response surface methodology. Gels, 2021;8(1):16.

[46] Mohamed NK, Metwally AA, Fared SM, Farid A, Taha M. Formulation and characterization of Tea tree and Jojoba oils nano-emulgel for in-vivo wound healing assessment. Colloids Surf B Biointerfaces, 2024:114312.

[47] Yahaya ZS, Otaru LO, Mohammed BB, Adeleye OA, Abdullahi I. Evaluation of diclofenac emulgel prepared with sesame oil as a lipophilic carrier. Trop J Pharm Res, 2023;22(10):2021–2027.

[48] Khajavi M, Raoufi Z, Badr AA, Abdollahi S. A dual-layer nano dressing with enhanced antimicrobial and healing properties via sesame nanoemulgel with efficient levofloxacin delivery for wound management. J Drug Deliv Sci Technol, 2024;101:106211.

[49] Behera B, Biswal D, Uvanesh K, Srivastava A, Bhattacharya MK, Paramanik K, et al. Modulating the properties of sunflower oil based novel emulgels using castor oil fatty acid ester: Prospects for topical antimicrobial drug delivery. Colloids Surf B Biointerfaces, 2015;128:155–164.

[50] Okafo SE, Avbunudiogba JA, Ochonogor EA, Iwetan BB, Anie CO. Evaluation of the physicochemical, antimicrobial and in vivo wound healing properties of castor oil-loaded nanogels. Trop J Nat Prod Res, 2024;8(5).

[51] Jeengar MK, Rompicharla SVK, Shrivastava S, Chella N, Shastri NR, Naidu V, et al. Emu oil based nano-emulgel for topical delivery of curcumin. Int J Pharm, 2016;506(1–2):222–236.

[52] Akram M, Naqvi SBS, Khan A. Design and development of insulin emulgel formulation for transdermal drug delivery and its evaluation. Pak J Pharm Sci, 2013;26(2):323–332.

[53] Jyothi VGS, Veerabomma H, Kumar R, Khatri DK, Singh SB, Madan J. Meloxicam emulgel potently suppressed cartilage degradation in knee osteoarthritis: Optimization, formulation, industrial scalability and pharmacodynamic analysis. Colloids Surf B Biointerfaces, 2023;228:113399.

[54] Ghorbanzadeh M, Farhadian N, Golmohammadzadeh S, Karimi M, Ebrahimi M. Formulation, clinical and histopathological assessment of microemulsion based hydrogel for UV protection of skin. Colloids Surf B Biointerfaces, 2019;179:393–404.

[55] Talaat SM, Elnaggar YS, Abdalla OY. Lecithin microemulsion lipogels versus conventional gels for skin targeting of terconazole: In vitro, ex vivo, and in vivo investigation. AAPS PharmSciTech, 2019;20:1–20.

[56] Shabery AM, Widodo RT, Chik Z. Formulation and in vivo pain assessment of a novel niosomal lidocaine and prilocaine in an emulsion gel (Emulgel) of semisolid palm oil base for topical drug delivery. Gels, 2023;9(2):96.

[57] Raut S, Uplanchiwar V, Bhadoria S, Gahane A, Jain SK, Patil S. Comparative evaluation of zidovudine loaded hydrogels and emulgels. Res J Pharm Technol, 2012;5(1):41–45.

[58] Jagdale SC, Kothekar PV. Development of emulgel delivery of mupirocin for treatment of skin infection. Recent patents on anti-infective drug discovery. 2020;15(2):137–156.

[59] Kim K-T, Kim M-H, Park J-H, Lee J-Y, Cho H-J, Yoon I-S, et al. Microemulsion-based hydrogels for enhancing epidermal/dermal deposition of topically administered 20 (S)-protopanaxadiol: In vitro and in vivo evaluation studies. J Ginseng Res, 2018;42(4):512–523.

[60] Chaudhari KD, Ande SN, Bakal RL, Malode GP. Gelling agents: Can they formulate a perfect emulgel? 2022.

[61] Shahin M, Abdel Hady S, Hammad M, Mortada N. Novel jojoba oil-based emulsion gel formulations for clotrimazole delivery. AAPS PharmSciTech, 2011;12:239–247.

[62] Berdey I, Voyt O. Rheological properties of emulgel formulations based on different gelling agent. Pharm Innov, 2016;5(4, Part B):76.

[63] Daood NM, Jassim ZE, Gareeb MM, Zeki H. Studying the effect of different gelling agent on the preparation and characterization of metronidazole as topical emulgel. Asian J Pharm Clin Res, 2019;12:571–577.

[64] Shokri J, Azarmi S, Fasihi Z, Hallaj-Nezhadi S, Nokhodchi A, Javadzadeh Y. Effects of various penetration enhancers on percutaneous absorption of piroxicam from emulgels. Res Pharm Sci, 2012;7(4):225.

[65] Estabragh MAR, Bami MS, Dehghannoudeh G, Noudeh YD, Moghimipour E. Cellulose derivatives and natural gums as gelling agents for preparation of emulgel-based dosage forms: A brief review. Int J Biol Macromol, 2023;241:124538.

[66] Salih ZT, Gawhari F, Rajab NA. Preparation, release, rheology and stability of piroxicam emulgel. Int J Appl Pharm, 2018;10(1):26–29.

[67] Khule PK. Formulation and evaluation of itraconazole emulgel for various fungal infections. Asian J Pharm, 2019;13(01).

[68] Haneefa KM, Easo S, Hafsa P, Mohanta GP, Nayar C. Emulgel: An advanced review. J Pharm Sci Res, 2013;5(12):254.

[69] Pandey P, Minocha N, Vashist N, Shah R, Saini S, Makhija M, et al. Emulgel: An emerging approach towards effective topical drug delivery. Drug Deliv Lett, 2022;12(4):227–242.

[70] Bhadouria VS, Verma S, Mishra R, Kapoor B. Beyond creams and gels: The emergence of emulgels in pharmaceutical science. Curr Drug Ther, 2024.

[71] Algahtani MS, Ahmad MZ, Ahmad J. Nanoemulgel for improved topical delivery of retinyl palmitate: Formulation design and stability evaluation. Nanomaterials, 2020;10(5):848.

[72] Milutinov J, Krstonošić V, Ćirin D, Pavlović N. Emulgels: Promising carrier systems for food ingredients and drugs. Polymers, 2023;15(10):2302.

[73] Puschmann J, Herbig ME, Müller-Goymann CC. Correlation of antimicrobial effects of phenoxyethanol with its free concentration in the water phase of o/w-emulsion gels. Eur J Pharm Biopharm, 2018;131:152–161.

[74] Rao J, McClements DJ. Formation of flavor oil microemulsions, nanoemulsions and emulsions: Influence of composition and preparation method. J Agric Food Chem, 2011;59(9):5026–5035.

[75] Thiagarajan P. Nanoemulsions for drug delivery through different routes. Res Biotechnol, 2011;2(3).

[76] Anton N, Benoit J-P, Saulnier P. Design and production of nanoparticles formulated from nano-emulsion templates – A review. J Contr Release, 2008;128(3):185–199.

[77] Schalbart P, Kawaji M, Fumoto K. Formation of tetradecane nanoemulsion by low-energy emulsification methods. Int J Refrig, 2010;33(8):1612–1624.

[78] Floury J, Desrumaux A, Lardières J. Effect of high-pressure homogenization on droplet size distributions and rheological properties of model oil-in-water emulsions. Innovative Food Sci Emerg Technol, 2000;1(2):127–134.

[79] Kumar M, Bishnoi RS, Shukla AK, Jain CP. Techniques for formulation of nanoemulsion drug delivery system: A review. Preventive Nutr Food Sci, 2019;24(3):225.

[80] Landfester K, Eisenblätter J, Rothe R. Preparation of polymerizable miniemulsions by ultrasonication. JCT Res, 2004;1:65–68.

[81] Arun KJ, Khan A, Mahato N, Devi J, Jawahar SS. Sulindac-loaded topical nanoemulgel formulation and optimization. Bull Env Pharmacol Life Sci, 2023;12:237–245.

[82]   Kim GW, Yun S, Jang J, Lee JB, Kim SY. Enhanced stability, formulations, and rheological properties of nanoemulsions produced with microfludization for eco-friendly process. J Colloid Interface Sci, 2023;646:311–319.

[83]   Devadiga S, Sermasekaran A, Singh AD, Agrawal S, Sharma S, Choudhary D. Quality by design driven systematic development of nanoemulgel of clobetasol-17-propionate for effective treatment of psoriasis. J Drug Deliv Sci Technol, 2024;93:105422.

[84]   Perazzo A, Preziosi V, Guido S. Phase inversion emulsification: Current understanding and applications. Adv Colloid Interface Sci, 2015;222:581–599.

[85]   Sokolov YV. Nanoemulsion formation by low-energy methods: A review. News of Pharm, 2014;3 (79):16–19.

[86]   Armanet L, Hunkeler D. Phase inversion of polyacrylamide-based inverse-emulsions: Influence of inverting-surfactant type and concentration. J Appl Polym Sci, 2007;103(6):3567–3584.

[87]   Sghier K, Mur M, Veiga F, Paiva-Santos AC, Pires PC. Novel therapeutic hybrid systems using hydrogels and nanotechnology: A focus on nanoemulgels for the treatment of skin diseases. Gels, 2024;10(1):45.

[88]   Sadeq ZA. Review on nanoemulsion: Preparation and evaluation. Int J Drug Deliv Technol, 2020;10 (1):187–189.

[89]   Ahmad N, Ahmad R, Buheazaha TM, AlHomoud HS, Al-Nasif HA, Sarafroz M. A comparative ex vivo permeation evaluation of a novel 5-Fluorocuracil nanoemulsion-gel by topically applied in the different excised rat, goat, and cow skin. Saudi J Biol Sci, 2020;27(4):1024–1040.

[90]   Mazurkevičiūtė A, Matulytė I, Ivaškienė M, Žilius M. Assessment of physical, mechanical, biopharmaceutical properties of emulgels and bigel containing ciclopirox olamine. Polymers, 2022;14(14):2783.

[91]   Jain A, Kumar P, Verma A, Mohanta BC, Ashique S, Pal R, et al. Emulgel: A cutting edge approach for topical drug delivery system. Curr Drug Res Rev, 2024.

[92]   Khalil YI, Khasraghi AH, Mohammed EJ. Preparation and evaluation of physical and, rheological properties of clotrimazole emulgel. Iraqi J Pharm Sci, 2011;20(2):19–27.

[93]   Ahmad N, Khalid MS, Khan MF, Ullah Z. Beneficial effects of topical 6-gingerol loaded nanoemulsion gel for wound and inflammation management with their comparative dermatokinetic. J Drug Deliv Sci Technol, 2023;80:104094.

[94]   Alam A, Mustafa G, Agrawal GP, Hashmi S, Khan RA, Alkhayl FFA, et al. A microemulsion-based gel of isotretinoin and erythromycin estolate for the management of acne. J Drug Deliv Sci Technol, 2022;71:103277.

[95]   Siafaka PI, Özcan Bülbül E, Okur ME, Karantas ID, Üstündağ Okur N. The application of nanogels as efficient drug delivery platforms for dermal/transdermal delivery. Gels, 2023;9(9):753.

[96]   Sawant AA, Mohite S. Formulation and evaluation of itraconazole emulgel for topical drug delivery. Asian J Pharm Technol, 2015;5(2):91–96.

[97]   Prausnitz MR, Elias PM, Franz TJ, Schmuth M, Tsai J-C, Menon GK, et al. Skin barrier and transdermal drug delivery. Dermatology, 2012;3(18):2065–2073.

[98]   Apriani EF, Shiyan S, Hardestyariki D, Starlista V, Sari AL. Optimization of Hydroxypropyl Methylcellulose (HPMC) and Carbopol 940 in Clindamycin HCl ethosomal gel as anti-acne. Res J Pharm Technol, 2024;17(2):603–611.

[99]   Bhanu PV, Shanmugam V, Lakshmi P. Development and optimization of novel diclofenac emulgel for topical drug delivery. Int J Comp Pharm, 2011;2(9):1–4.

[100] Pachauri A, Chitme H, Visht S, Chidrawar V, Mohammed N, Abdel-Wahab BA, et al. Permeability-enhanced liposomal emulgel formulation of 5-fluorouracil for the treatment of skin cancer. Gels, 2023;9(3):209.

[101] Sultana SS, Parveen P, Rekha MS, Deepthi K, Sowjanya C, Devi AS. Emulgel-a novel surrogate approach for transdermal drug delivery system. Ind Am J Pharm Res, 2014;4:5250–5265.

[102] Chaithra N, Kulkarni PK, Siddartha H, Joshi H. A comprehensive review on formulation and evaluation of vitamin D3 emulgel for enhanced transdermal delivery. Int J Res Pharm Allied Sci, 2024;3(1):46–63.

[103] Stolić Jovanović A, Martinović M, Žugić A, Nešić I, Tosti T, Blagojević S, et al. Derivatives of L-Ascorbic acid in emulgel: Development and comprehensive evaluation of the topical delivery system. Pharmaceutics, 2023;15(3):813.

[104] Shakeel M, Kiani MH, Sarwar HS, Akhtar S, Rauf A, Ibrahim IM, et al. Emulgel-loaded mannosylated thiolated chitosan-coated silver nanoparticles for the treatment of cutaneous leishmaniasis. Int J Biol Macromol, 2023;227:1293–1304.

[105] Varma VNSK, Maheshwari P, Navya M, Reddy SC, Shivakumar H, Gowda D. Calcipotriol delivery into the skin as emulgel for effective permeation. Saudi Pharm J, 2014;22(6):591–599.

[106] Kaur A, Katiyar SS, Kushwah V, Jain S. Nanoemulsion loaded gel for topical co-delivery of clobitasol propionate and calcipotriol in psoriasis. Nanomed Nanotechnol Biol Med, 2017;13(4):1473–1482.

[107] Pinheiro IM, Carvalho IPS, Neto JAT, Lopes GLN, de Sousa Coêlho E, Sobrinho-Júnior EPC, et al. Amphotericin B-loaded emulgel: Effect of chemical enhancers on the release profile and antileishmanial activity in vitro. AAPS PharmSciTech, 2019;20:1–8.

[108] Rao M, Sukre G, Aghav S, Kumar M. Optimization of metronidazole emulgel. J Pharm, 2013;2013 (1):501082.

[109] Morsy MA, Abdel-Latif RG, Nair AB, Venugopala KN, Ahmed AF, Elsewedy HS, et al. Preparation and evaluation of atorvastatin-loaded nanoemulgel on wound-healing efficacy. Pharmaceutics, 2019;11 (11):609.

[110] Reddy EPK, Raghavamma S, Nadendla RR. Formulation and Evaluation of Ketoconazole Microemulgel with Mixture of Penetration Enhancers.

[111] Malavi S, Kumbhar P, Manjappa A, Chopade S, Patil O, Kataria U, et al. Topical emulgel: Basic considerations in development and advanced research. Indian J Pharm Sci, 2022;84(5).

[112] Kumbhar PR, Desai H, Desai VM, Priya S, Rana V, Singhvi G. Versatility of emulgel in topical drug delivery transforming its expedition from bench to bedside. Expert Opin Drug Delivery, 2025 (just-accepted).

[113] Sarella PNK, Godavari L. The expanding scope of emulgels: Formulation, evaluation and medical uses. Int J Curr Sci Res Rev, 2023;6(5):3030–3041.

[114] Campanholi KDSS, da Silva JB, Batistela VR, Gonçalves RS, Dos Santos RS, Balbinot RB, et al. Design and optimization of stimuli-responsive emulsion-filled gel for topical delivery of copaiba oil-resin. J Pharm Sci, 2022;111(2):287–292.

[115] Neumann M, di Marco G, Iudin D, Viola M, van Nostrum CF, van Ravensteijn BG, et al. Stimuli-responsive hydrogels: The dynamic smart biomaterials of tomorrow. Macromolecules, 2023;56 (21):8377–8392.

[116] Ayunin Q, Miatmoko A, Soeratri W, Erawati T, Susanto J, Legowo D. Improving the anti-ageing activity of coenzyme Q10 through protransfersome-loaded emulgel. Sci Rep, 2022;12(1):906.

[117] Che Marzuki NH, Wahab RA, Abdul Hamid M. An overview of nanoemulsion: Concepts of development and cosmeceutical applications. Biotechnol Biotechnol Equip, 2019;33(1):779–797.

[118] Gul R, Ahmed N, Ullah N, Khan MI, Elaissari A, Rehman A. Biodegradable ingredient-based emulgel loaded with ketoprofen nanoparticles. AAPS PharmSciTech, 2018;19:1869–1881.

[119] Zheng D, Giljohann DA, Chen DL, Massich MD, Wang X-Q, Iordanov H, et al. Topical delivery of siRNA-based spherical nucleic acid nanoparticle conjugates for gene regulation. Proc Natl Acad Sci, 2012;109(30):11975–11980.

[120] Pandey S, Shamim A, Shaif M, Kushwaha P. Development and evaluation of Resveratrol-loaded liposomes in hydrogel-based wound dressing for diabetic foot ulcer. Naunyn-Schmiedeberg's Arch Pharmacol, 2023;396(8):1811–1825.

Swarnali Das Paul, Disha Kesharwani*, Monika Bhairam,
Gunjan Kalyani, and Sandhya Mishra

Chapter 8
Clinical frontiers and regulatory status of micro/nanoemulsions

Abstract: Micro- and nanoemulsions represent a significant advancement in the field of drug delivery systems, offering enhanced bioavailability, stability, and targeted delivery of therapeutic agents. This chapter explores the clinical frontiers and regulatory status of micro/nanoemulsions, focusing on the evolving landscape of their development, approval, and market access across different global regions. The first section delves into the regulatory considerations, highlighting the regulatory pathways established by key agencies such as the US FDA, EMA, and CDSCO, as well as the guidelines for quality, safety, and efficacy assessments.

In the comparative analysis section, we explore efforts toward harmonizing regulatory standards, with a specific emphasis on the International Council for Harmonization (ICH) guidelines and their relevance to nanotechnology. Additionally, insights into emerging regulatory frameworks for nanoformulations are provided, shedding light on the future regulatory environment.

The chapter further discusses the latest innovations in formulation technologies, the role of micro/nanoemulsions in precision and personalized medicine, and the potential challenges facing their clinical and regulatory acceptance in the future. Finally, perspectives on the future of micro/nanoemulsions in healthcare are presented, followed by a conclusion that encapsulates the key themes of regulatory, clinical, and technological advancements in this rapidly evolving field. This chapter aims to provide a comprehensive overview for researchers, clinicians, and regulatory bodies engaged in the development and approval of micro/nanoemulsion-based products.

Keywords: Micro/nanoemulsions, innovative drug delivery, USFDA, EMA, regulatory and clinical applications

---

*Corresponding author: **Disha Kesharwani**, Columbia Institute of Pharmacy, Raipur, Chhattisgarh, India, e-mail: disha@columbiaiop.ac.in
Swarnali Das Paul, Shri Shankracharya Professional University, Bhilai, Chhattisgarh, India
Monika Bhairam, Gunjan Kalyani, Sandhya Mishra, Columbia Institute of Pharmacy, Raipur, Chhattisgarh, India

https://doi.org/10.1515/9783111593654-009

8.1 Introduction

Over the past few years, micro/nanoemulsions have emerged as a groundbreaking technology in the realm of drug delivery, significantly enhancing the bioavailability, stability, and therapeutic efficacy of a wide range of hydrophilic and lipophilic drugs. These submicron-sized emulsions are widely applicable, not only in pharmaceuticals but also in the food and cosmetic industries, owing to their unique attributes, such as enhanced solubility and controlled drug release [1–8].

Despite their promise, micro/nanoemulsions face significant hurdles in clinical translation. Key challenges include stability and scalability issues, along with difficulties in achieving consistent drug loading and release profiles. Additionally, patient safety and tolerability remain crucial concerns, as the biological interactions of nano-sized particles are complex and often unpredictable. The intricacies of formulation and scale-up processes also pose considerable obstacles, making it necessary to address these barriers through innovative approaches and robust quality control measures [9].

From a regulatory standpoint, navigating the path to clinical application is complex, given that global regulatory agencies like the US FDA, EMA, and Indian authorities have differing frameworks for assessing the safety and efficacy of nanoformulations. Harmonization efforts among international regulatory bodies aim to bridge these differences, but gaps remain, prolonging development timelines and complicating the approval of novel nanoemulsion-based therapeutics. Looking ahead, emerging trends in micro/nanoemulsion technology are set to reshape the landscape of personalized medicine [10, 11]. This chapter provides a comparative analysis of these regulatory pathways, shedding light on the challenges the companies face when working with inconsistent or evolving guidelines. Also, the ongoing efforts to establish harmonized standards for quality, safety, and efficacy across regions are discussed. Through the studies of approved products and an analysis of future challenges, the chapter underscores the potential of micro/nanoemulsions in modern healthcare. The conclusion emphasizes the need for regulatory alignment and continued innovation to unlock their full potential in clinical settings [12, 13].

8.2 Regulatory considerations and global perspectives

If we consider broader perspectives and uses of micro/nanoemulsions, the regulatory rules for these nano cargos are still in their infancy and are being formulated at many stages by various regulatory organizations such as the USFDA and EMA. It is anticipated that as the field of micro/nanoemulsion advances, manufacturers will have access to clear, concise recommendations.

8.2.1 Regulatory pathways in different regions: US FDA, EMA, India, etc

The translation of nanomedicines from laboratory-scale development to commercials-cale manufacturing poses numerous challenges, encompassing concerns related to reproducibility, scalability, pharmacodynamics, pharmacokinetics, process control, biocompatibility, and nanotoxicity, which must be meticulously addressed to ensure the successful transition of these innovative therapeutics [14, 15, 16]. The development of nanomedicines must also navigate complex requirements from multiple stakeholders: patients (clinical and therapeutic use), industry (manufacturing processes), and regulatory authorities (approval and compliance). Given the regulatory frameworks set by the Food and Drug Administration (FDA) and the European Medicines Agency (EMA), it is essential to examine the latest advancements and regulatory considerations surrounding micro/nanoemulsion drug delivery systems [17, 18]:

In European Union, the major approach of European Medicine Agency (EMA) towards nanomedicines involves assuring uniformity and cooperation throughout the EU by creating a platform that combines the necessary knowledge for the assessment of the regulatory and scientific elements.

The EMA framework specifies:

– Any nanomedicine should undergo a comprehensive evaluation using established risk-benefit analysis methodologies, considering not only the technology itself but also its potential environmental impact, to ensure a thorough assessment of its overall safety and efficacy, and to inform the development of effective risk management and mitigation strategies.
– Sophisticated multidisciplinary knowledge is needed; a group comprising specialists from academia and regulators was established in 2009 and was further strengthened in 2011, combining knowledge in quality, safety, and kinetics to facilitate evaluation and develop guidelines.
– Strong EU collaboration with other scientific networks (QNANO, ETP Nano), committees (SCENIHR, EFSA), and the European Commission.
– International collaboration: EMA is an expert group on international regulators, which includes the US FDA, Japan MHLW, Health Canada, and TGA Australia.

The European Union (EU) has a highly developed framework for evaluating the risk-benefit of medications, which has successfully included new technologies from the past, such as positron emission tomography (PET) and novel diagnostic modalities, as well as some nanoscale items. Since targeted sub technologies are emerging and scientific knowledge is being obtained, specific guidelines on quality, toxicity, clinical development, and monitoring elements are needed in this field [17, 19]. Despite significant efforts and advancements, the EU's regulatory environment for nanomedicines and nanocarriers is still fragmented and stratified. A harmonization approach, with specific recommendations for the advancement and assessment of initial phases of clinical trials (CTs), would be beneficial [20, 21].

Under US FDA, the growing interest in products utilizing nanotechnology resulted in the release of a significant guideline in 2014 that helps determine if a product regulated by the FDA uses nanotechnology, even though CTs were not specifically addressed. Currently, there is no dedicated regulatory framework for nanomedicines. The FDA assesses nanotechnology-based products using a product-specific evaluation process, relying on sponsor consultations to address potential regulatory challenges. The agency assumes that existing regulations are adequate to identify potential toxicity risks. Like other regulatory bodies, the FDA adopts a case-by-case approach when evaluating nanotechnology. However, in response to the expanding field of nanomedicine, the FDA has published foundational guidelines for pharmaceutical products, including biologics that incorporate nanomaterials. Under this direction, the structural characterization of medicinal products using nanoparticles and the quality features of nanomaterials were addressed; and finally, CTs were included in the scope. A proportionate approach to describing and characterizing the Investigational Medicinal Product, based on its stage of development, is recognized, as long as it ensures safety during CTs. This is a significant step in establishing a broad framework for determining the critical quality attributes (CQAs) of the drug product, while also recognizing that the function and possible influence of the nanomaterials on the product's performance should be taken into account when determining the CQAs of the nanomaterials. A regulatory guideline specifically addressing the development, evaluation, and safety of liposome-based medicinal products was recently published [22].

In India, the rules for evaluating nanopharmaceuticals provide a detailed framework for CT applications, outlining data requirements for physicochemical characterization, CQAs, contaminants, stability, and in vitro/in vivo release and degradation kinetics. These case-specific guidelines align with global standards, enhancing regulatory clarity for nanopharmaceutical assessment [23].

In Canada, since 2010, problems with nanomedicines have been recognized by *Health Canada*. The current legislative and regulatory frameworks are used for the risk-benefit assessment and approval of nanomedicines; however, nanomaterials are not specifically mentioned in acts or regulations, and there are no specific guidelines for the submission of nanotechnology products in CTs. Health Canada has established a functional definition for nanoparticles, which allows for the collection of specific information about nanomaterials. This information is used to better understand these materials, assess their potential benefits and risks in regulated products and substances, and implement risk management strategies. This applies to products and substances, including those used in computed tomography applications, which may contain or be composed of nanomaterials [24].

8.2.2 Guidelines for quality, safety, and efficacy assessment

Nanomedicine research is a multifaceted and intricate field with numerous unresolved legal and regulatory issues. The product quality evaluation (physicochemical properties, quality control, and manufacturing process) and the product safety assessment (pharmacokinetics, biodegradation, accumulation, and nanotoxicity) are generally the most important factors to take into account when it comes to nanoformulations. The quality-by-design (QbD) paradigm is a potential strategy for the development and marketing authorization of nanomedicines. To facilitate a seamless clinical translation of the innovative nanomedicines, the regulatory bodies both strongly advise and mandate the submission of QbD-based data [25, 26].

8.2.3 Navigating intellectual property and patent challenges

Maintaining a well-planned intellectual property (IP) portfolio is economically important, serving both offensive purposes (protecting innovations and securing market advantage) and defensive purposes (preventing infringement claims and ensuring freedom to operate). In terms of IP rights related to patents, the field of nanotechnology is new and highly competitive globally, with developing nations participating in this competition. The evolving technology presents both opportunities and difficulties for the patent system. Nanotechnology ideas are cross-sectional and interdisciplinary, and their industrial uses and challenges in meeting patent requirements make them problematic to patent outside of technical domains. As a result, it is impossible to overstate the worth and strategic significance of IP rights protection. Many wide, potentially overlapping patents are being given in this fast-developing field of technology, which has enormous promise and potential for technological, social, and economic benefits.

Additionally, certain nonexclusive licenses are being offered [27]. Current laws impose specific patentability criteria, particularly regarding novelty and inventive step, for nano-sized applications and their utility. Jurisdiction-specific studies and key global precedents, especially under India's Patent Act, 1970 (notably, restrictive section 3), address these criteria and highlight challenges in patenting biological innovations [28, 29].

8.3 Comparative analysis of regulatory guidelines

As the micro- and nanoemulsions are continuously growing due to their advanced benefits in treatment of diseases, regulatory requirements are the prerequisite in their development. However, the growth of nanotechnology-based products is exponential as compared to the regulatory framework. Because of their unique physical and chemical characteristics, nanoemulsions provide a wide range of functions that encourage their usage to satisfy the demands of several applications. The Global Summit on Regulatory Science (GSRS) provides the platform for a discussion on various matters related to regulatory issues, innovative technologies, and collaborative initiatives between various national and international member countries for the safety estimation and development of various nanotechnological-based products, including nanoemulsions [30].

Regulatory science research gaps and/or priorities were recognized at this summit's closing session, which concentrated on what needs to be done in these areas. The harmonization strategy, through collaboration, was also one of major objectives of this summit. It aimed to help international partners collaborate effectively by creating networks, communities of practice, platforms, and processes. Different countries have different regulatory authorities for controlling matters related to nanomedicine. As discussed in the earlier section, there are specific regulatory frameworks for nanomedicine in various countries, including India, EU, and Canada, but some of the major countries such as the United States work on a case-by-case approach. There is no specific framework for the assessment of safety of nanomedicine.

Also, the specific frameworks, work on the principle of risk/benefit analysis. Nanomaterials, including micro/nanoemulsions, are not included in the guiding framework of Health Canada. However, for better understanding the concept and evaluating the risks and benefits associated with it, a definition of nanomaterials is included in its system. The Canadian regulatory strategy for nanomaterials complies with the Council of the Organization for Economic Co-operation and Development's (OECD) advice for the evaluation and testing of produced nanomaterials for safety. With a steady rise in medical goods that use nanomaterials, they provide, for instance, new opportunities in healthcare diagnostics and therapies, and the chance to meet unmet medical requirements.

8.3.1 Harmonization efforts: ICH guidelines and nanotechnology standards

There is great potential for better disease prevention, detection, and treatment with nanomedicine. However, competing worries over the possible safety hazards of nanotechnologies, in general, and nanomedical products, in particular, pose a danger to halt or at least postpone the launch and assess the economic viability of numerous applications of nanomedicine. Nanotechnology offers a chance to adopt a different approach, aiming for international consensus before enacting domestic laws. Since nanotechnology is still relatively new and there are no specific regulatory frameworks dedicated to it, there is an opportunity to establish international regulations without being limited by preexisting national laws [31]. Most industrialized countries are simultaneously advancing in the field of nanomedicine, and they are at a similar stage in both technological development and regulatory assessment [32–34].

Governments worldwide are struggling to balance the potential benefits and risks of nanotechnology in medicine and other industries. It is becoming more and more obvious that the successful application of nanotechnology will depend on sensible, efficient, and predictable regulatory frameworks. So, international guiding standards are crucial for all the countries. As an effort, under the auspices of the International Council for Harmonization of Technical Requirements for Pharmaceuticals for Human Use (ICH), the International Pharmaceutical Regulators Program (IPRP) seeks to identify and address new challenges in pharmaceutical regulation for nanomedicines that are of common concern to all the nations of the world [35]. Government and agency representatives from nations in the Americas, Asia, Europe, and Oceania are part of the particular IPRP nanomedicines working group. Its goals include non-confidential information exchange, harmonization of regulations, cooperation amongst foreign regulators in organizing training, and outreach to nanomedicine inventor groups and other stakeholders [36].

More uniform international nanotechnology legislation could lead to several significant benefits, including enhanced safety standards, streamlined regulatory approvals, and improved global collaboration in research and innovation. Several advantages would result from having uniform national product requirements and regulations through some kind of harmonization system, as many nanotechnology products will be offered in worldwide trade. A unified set of regulations would help makers of these devices (nanomedicine-related products such as diagnostic tools, drug delivery systems, imaging agents, and other medical technologies that incorporate nanomaterials) by streamlining product testing, regulatory filings, and design and labelling specifications.

Multinational corporations in the nanotechnology sector would be able to implement uniform environmental, health, and safety programs worldwide, including compliance, product stewardship, worker training, and reporting components, through internationally consistent environmental and occupational safety and health requirements. This would further advance the objectives of efficiency and effectiveness. Additionally, globally standardized standards would promote global trade and avoid the trade barriers and disputes that have impeded the advancement of other technology.

Internationally harmonized standards would ensure that all citizens receive equal protection from the potential risks of nanotechnology products. They would also prevent a "race to the bottom" or the creation of "risk havens," where some governments might lower safety standards to attract business investment and jobs. Additionally, when national governments address the same risks and regulatory challenges independently, it can lead to inefficiencies and duplicated efforts. Establishing international standards could help avoid these issues by providing a unified framework. A single international forum would allow for greater efficiency through economies of scale and shared expertise.

Nonetheless, attempts to harmonize international regulation would not be free from hazards and expenses. The most important element is probably that international discussions need a lot of time and resources since delegates from countries with very different political, scientific, and economic backgrounds find it difficult to agree. In the interim, dangerous items or uses may remain unregulated since attempts to obtain international accords might take years or even decades. The informational benefits of various jurisdictions experimenting with various regulatory approaches – which can teach other countries about good or bad approaches – may also be lost as a result of international harmonization. Lastly, in terms of perceived hazards, risk-benefit analysis, and ethical viewpoints on technology that could support disparate regulation strategies, worldwide harmonization runs the risk of stifling actual national variations. The most formal method of international harmonization would be an international treaty or convention on nanotechnology, in general, or nanomedicine, in particular. As of now, there are no international treaties that specifically address nanomedicine, and it doesn't appear likely that any will be signed soon. Furthermore, there would probably be significant barriers to negotiations of such a convention, including the continuous dispute between countries regarding the proper application of the precautionary principle in international accords [37].

8.3.2 Insights into emerging regulatory frameworks for nanoformulations

According to Abbott et al., a framework convention, or "FC," would probably be the most sensible option if a formal international agreement on nanotechnology were to be considered at some point in the future. A framework convention first establishes an infrastructure and procedure before progressively adding significant requirements in the form of distinct protocols [38]. The **FC protocol** provides a structured approach for countries to gradually address scientific and technological uncertainties related to nanotechnology. It enables the adoption of precautionary measures that can be adjusted as research and understanding progress. In the absence of a forthcoming international treaty on nanotechnology, the **Organisation for Economic Co-operation and Development (OECD)** has emerged as a key driver of global efforts to achieve regulatory harmonization, fostering cooperation and coordination among its member countries to address the complex challenges posed by the development and application of nanotechnology. Within the OECD, the **Working Party on Manufactured Nanomaterials (WPMN)** and **Working Party on Nanotechnology** (WPN) are leading initiatives to standardize regulations, establish safety guidelines, and ensure thorough risk assessments for nanomaterials used in various applications.

The ISO's Technical Committee 229 (TC 229) promotes global nanotechnology standards, setting terminology, definitions, and guidelines for managing environmental, health, and safety (EHS) risks in nanomaterial production, handling, and disposal. Similarly, ASTM International and other standard organizations provide risk management standards. In addition, nongovernmental initiatives, like the Responsible Nanocode – a code developed by the Royal Society, Insight Investment – Nanotechnology

Industries Association, and Nanotechnology Knowledge Transfer Network, aim to standardize safe practices in nanotechnology research, production, and disposal globally [39].

Private collaborations, such as the Environmental Defense Fund and Dupont's Nano Risk Framework, guide businesses on systematically managing environmental, health, and safety risks throughout a nanomaterial life cycle, from sourcing to disposal. All nanoproducts are covered by the global harmonization efforts mentioned above. However, two international organizations, namely International Conference on Harmonization (ICH), focusing on harmonizing national regulations for pharmaceuticals at the international level and Global Harmonization Task Force (GHTF), aiming to harmonize regulations for medical devices worldwide that specialize in the field of medical products present, are possibly the most promising opportunities for the international harmonization of nanomedicine. Globally, harmonized guidance documents for nanomedical products could be enacted by the International Conference on Harmonization and the Global Harmonization Task Force, which aim to harmonize national regulations for pharmaceuticals and medical devices, respectively, at the international level [32].

8.4 Emerging trends and future directions

The field of drug delivery is undergoing rapid evolution, driven by advancements in technology and an increasing understanding of disease mechanisms [40]. Micro/nanoemulsions continue to play a crucial role in pharmaceutical development. The synergistic convergence of artificial intelligence and machine learning in drug delivery is transforming the field, yielding unprecedented efficiencies in formulation processes, predictive insights into drug behavior, and tailored therapies. By leveraging the analysis of vast datasets, researchers can pinpoint optimal parameters, bolster formulation stability, and usher in a new era of personalized medicine. This shift toward personalized medicine promotes the development of tailored drug delivery systems, particularly micro/nanoemulsions engineered to encapsulate patient-specific therapies for more effective and targeted treatments that minimize side effects [41]. Meanwhile, the burgeoning field of nanomedicine is yielding innovative solutions, including ligand-targeted nanoparticles that facilitate precise, tissue-specific delivery. This targeted approach not only amplifies therapeutic efficacy but also mitigates toxicity, paving the way for safer, more effective treatments. In response to environmental concerns, the pharmaceutical industry is adopting sustainable practices, such as green synthesis methods for micro/nanoemulsions and the use of biodegradable materials, aligning with global sustainability goals. Research is also expanding into novel delivery routes, including transdermal, intranasal, and oral administration, to improve drug absorption and bioavailability [42].

Additionally, the increasing use of nanotechnology necessitates updated regulatory frameworks to establish clear guidelines for evaluating the safety, efficacy, and quality of nanoformulations while addressing IP challenges. Lastly, the emergence of smart drug delivery systems with integrated biosensors for real-time monitoring of drug levels and physiological responses is paving the way for dynamic therapy adjustments, ultimately improving patient outcomes [43].

Various summits have been organized to address the challenges in clinical and regulatory pathways for nanotechnology [40]. These summits focused on topics such as standards for medical devices and drugs, safety assessments of nanomaterials, and emerging pollutants like nanoplastics and nanomaterials. They served as platforms for regulators, policymakers, and scientists to discuss innovative technologies, regulatory methods, and strategies for global collaboration. The 2019 GSRS attracted around 200 participants from over 30 countries, including the United States, Canada, and the EU. Major discussions on clinical and regulatory challenges are illustrated in Figure 8.1.

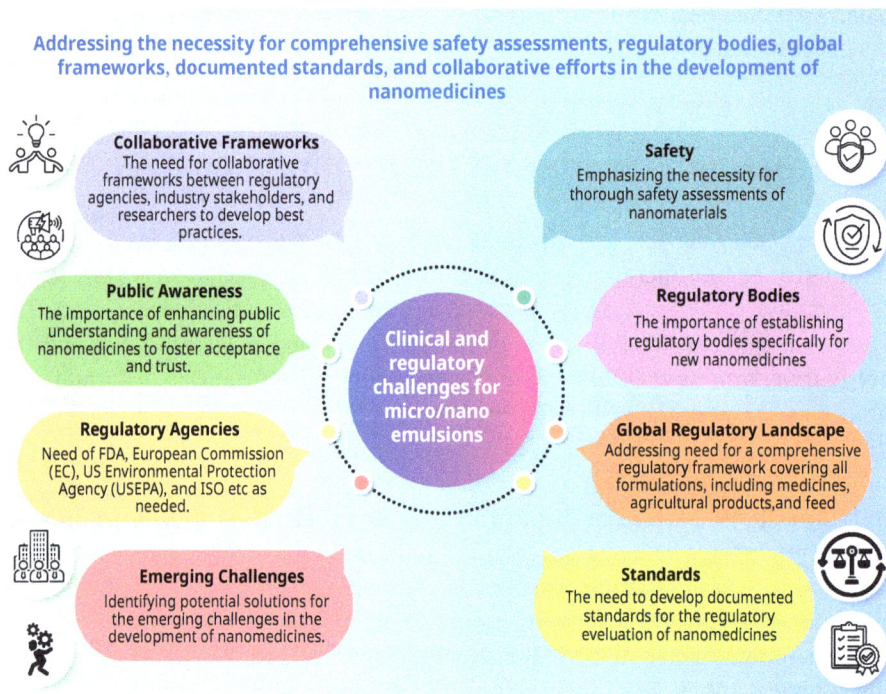

Figure 8.1: Addressing the need for comprehensive safety assessments, regulatory bodies, global frameworks, documented standards, and collaborative efforts in the development of nanomedicines, alongside the clinical and regulatory challenges for micro/nanoemulsions [44].

8.4.1 Innovations in formulation technology

Recent advancements in formulation technology have significantly broadened the potential applications of micro/nanoemulsions, making them a vital tool in modern drug delivery systems. One of the most impactful innovations is the use of microfluidics, which enables precise control over the mixing of components at the microscale [45]. This technique facilitates the generation of uniform droplets with controlled sizes, enhancing the stability and bioavailability of the emulsion products. Microfluidic devices also allow for continuous processing, leading to scalable production methods that can meet industrial demands without compromising product quality [46].

Supercritical fluid technology is another emerging method that offers several advantages over traditional solvent-based approaches [47]. By utilizing supercritical carbon dioxide as a solvent, this technology enables the formation of micro/nanoemulsions without the need for organic solvents, reducing toxicity and environmental impact. This method also allows for the creation of highly uniform and stable formulations, which can be particularly beneficial for delivering poorly soluble drugs [48].

The rise of green synthesis methods reflects a growing commitment to sustainability in pharmaceutical development. These methods prioritize the use of renewable resources and environmentally friendly solvents, minimizing waste and reducing the ecological footprint of drug production [49]. Innovations in this area have led to the development of biodegradable surfactants and excipients, which not only enhance the sustainability of formulations but also improve patient safety by reducing the risk of toxicity [50, 51].

Moreover, researchers are continually exploring new excipients and surfactants to optimize the performance of micro/nanoemulsions. The selection of surfactants can significantly affect the stability, droplet size, and release profile of the formulation. Novel surfactants derived from natural sources or designed through synthetic approaches offer improved compatibility with active pharmaceutical ingredients, enhancing the overall delivery of both hydrophilic and lipophilic drugs [52].

These advances not only enhance formulation stability but also increase the versatility of micro/nanoemulsions in various therapeutic applications. From targeted drug delivery in oncology to enhanced bioavailability for poorly soluble drugs, the innovations in formulation technology are opening new avenues for improving patient outcomes and expanding the capabilities of drug delivery systems. The ongoing research and development in this field hold great promise for the future of pharmaceuticals, as the formulation of micro/nanoemulsions becomes increasingly sophisticated and aligned with patient needs [53].

8.4.2 Micro/nanoemulsions in precision and personalized medicine

The integration of micro/nanoemulsions into precision and personalized medicine represents a pivotal shift in how therapies are designed and administered, particularly for complex diseases such as cancer and neurological disorders. This innovative approach leverages the unique properties of micro/nanoemulsions, enabling the encapsulation and targeted delivery of therapeutic agents with exceptional precision [54].

One of the defining characteristics of micro/nanoemulsions is their ability to form stable, homogenous mixtures that can encapsulate both hydrophilic and lipophilic drugs. This versatility allows for the effective delivery of a wide range of therapeutics, including poorly soluble compounds that are often challenging to administer. Furthermore, by optimizing the size and surface properties of these emulsions, researchers can enhance the targeting of specific tissues or cells, thus improving the therapeutic index of treatments. For instance, in cancer therapy, micro/nanoemulsions can be engineered to deliver chemotherapeutic agents directly to tumor cells, minimizing damage to surrounding healthy tissue and reducing systemic side effects [55].

The personalized aspect of this approach hinges on the ability to tailor formulations based on a patient's unique genetic makeup, environmental exposures, and lifestyle factors. By incorporating biomarkers or targeting ligands into the formulation, micro/nanoemulsions can be designed to respond to individual patient profiles [56]. For example, in the case of a cancer patient, the formulation can be customized to enhance uptake by cancer cells expressing specific receptors, thereby improving efficacy and reducing the required dosage. This precision not only optimizes therapeutic outcomes but also minimizes adverse effects, ultimately leading to improved patient quality of life [57].

The trend towards personalized medicine is further supported by advancements in drug design and formulation technologies. As researchers gain a deeper understanding of the molecular mechanisms underlying diseases, they can develop patient-specific treatments that align with the unique pathophysiology of individual patients. This includes leveraging data from genomics, proteomics, and metabolomics to identify the most effective therapeutic targets and optimize drug delivery methods [40]. Moreover, the use of micro/nanoemulsions in combination with real-time monitoring technologies offers the potential for dynamic adjustments to treatment plans based on patient responses. For instance, wearable biosensors can track drug levels and physiological parameters, allowing healthcare providers to tailor dosages or modify treatment strategies on-the-fly, further enhancing the personalization of care [49].

The integration of micro/nanoemulsions into precision and personalized medicine signifies a transformative approach to drug delivery, enabling highly targeted, effective, and individualized therapies. By harnessing the unique properties of these emulsions, researchers and clinicians can develop innovative treatments that not only enhance efficacy and safety but also align with the evolving landscape of personalized healthcare. As this field continues to evolve, it holds immense promise for addressing the complexities of modern diseases and improving patient outcomes [58].

8.4.3 Future challenges in clinical and regulatory pathways

Despite the significant advancements and potential of micro/nanoemulsions in drug delivery, several challenges hinder their clinical translation and regulatory approval. Addressing these challenges is crucial for the successful implementation of these innovative systems in real-world medical applications [59, 60].

8.4.4 Formulation scalability

One of the primary technical hurdles is the scalability of micro/nanoemulsion formulations. While laboratory-scale methods can produce small batches of highly controlled formulations, scaling up to meet commercial production demands poses challenges in maintaining product consistency and quality. Variability in production processes can lead to differences in droplet size, distribution, and stability, which can significantly affect drug delivery performance. Establishing robust manufacturing processes that ensure reproducibility and compliance with good manufacturing practices (GMP) is essential for large-scale production [61].

8.4.5 Long-term stability

The long-term stability of micro/nanoemulsions is another critical challenge. These systems are inherently thermodynamically unstable, and factors such as temperature, light exposure, and storage conditions can impact their stability over time [51]. Maintaining a stable formulation that preserves the integrity of the active pharmaceutical ingredient and delivers consistent performance is vital for clinical efficacy. Developing advanced formulation strategies, including the use of stabilizers and novel packaging techniques, is necessary to enhance stability and shelf life [58].

8.4.6 Variability in drug release profiles

Another challenge is the variability in drug release profiles observed with micro/nanoemulsions. Achieving consistent and predictable drug release is crucial for therapeutic effectiveness, yet many formulations exhibit variability due to factors such as droplet size, surfactant composition, and environmental conditions [46]. This inconsistency can complicate the translation of preclinical findings to clinical outcomes. Research into advanced characterization techniques and in vitro-in vivo correlation (IVIVC) models is essential for understanding and controlling release mechanisms [59].

8.4.7 Intellectual property concerns

IP concerns also pose challenges for the development and commercialization of micro/nanoemulsions. The rapidly evolving nature of nanotechnology often leads to uncertainties regarding patentability and freedom to operate [62]. Researchers and companies must navigate complex IP landscapes to protect their innovations, while ensuring compliance with existing patents. Establishing clear IP frameworks and collaborative approaches will be critical in promoting innovation and ensuring fair competition in the market [63].

8.5 Future perspectives of nanoemulsions

Recently, nanoemulsions have received attention as promising carriers for targeted cosmetic delivery and optimum dispersion of active substances in specific skin layers. In comparison to liposomes, nanoemulsions exhibit superior efficacy in delivering lipophilic substances, attributed to their lipophilic core, which facilitates the efficient penetration of active ingredients into the dermal layer and enhances their cutaneous concentration. Furthermore, nanoemulsions offer a distinct advantage over macroemulsions, as they are less prone to sedimentation, creaming, coalescence, and flocculation, rendering them an attractive formulation strategy for cosmetic applications. In the fields of hair care and cosmetics, nanoemulsions have been proven to be useful for scalp therapy and silicon deposition onto hair fibers.

Among the unique characteristics that are widely recognized in the current period is the parenteral administration of nanoemulsions to meet nutritional needs, controlled medication delivery, and targeted drug delivery. They provide an effective oral medication delivery system because of their small droplet size, which greatly enhances absorption from the gastrointestinal tract [64].

8.6 Conclusion

Microemulsions and nanoemulsions, with their deformable nano-sized droplets, present remarkable potential as novel carriers in the rapidly advancing field of nanotechnology. These systems offer enhanced bioavailability, a broader range of optical and flow properties, and the ability to transition from opaque to transparent appearances, making them valuable across diverse industries such as biotechnology, medicine, and cosmetics. Their ability to be produced with minimal emulsifiers further boosts their commercial appeal. Despite these promising advantages, micro- and nanoemulsions face several challenges, particularly related to stability and regulatory hurdles. The unique characteristics of nanoformulations, including their small size, reactivity, and high surface area, introduce potential risks. These formulations can penetrate cells, float in the air, and, in some cases, cause harmful adverse effects. Consequently, stringent regulatory oversight is essential to ensure their safe development and use. While nanoformulations offer significant benefits, their clinical and commercial success hinges on the establishment of robust regulatory frameworks.

Microemulsions and nanoemulsions have emerged as powerful tools in drug delivery systems, offering improved bioavailability, stability, and targeted therapeutic delivery. Their versatility across a range of applications, from biotechnology to medicine and cosmetics, underscores their significant potential. However, despite their promising advantages, these systems face substantial challenges, particularly in terms of regulatory approval, safety, and stability. The regulatory landscape for micro- and nanoemulsions is complex and evolving, with different regions and regulatory bodies such as the US FDA, EMA, and CDSCO establishing diverse frameworks for their approval and monitoring.

Currently, regulatory approaches for micro- and nanoemulsions are still in the early stages, with different countries adopting varying systems based on their own guidelines. There is a growing need for a unified, efficient, and globally harmonized regulatory framework to address the complexities of nanotechnology. International regulatory bodies are working toward developing common standards that can facilitate the safe and effective use of nanoemulsions worldwide. Achieving this harmonization will not only enhance the global accessibility of these innovative systems but also ensure their safe application in clinical settings. For micro- and nanoemulsions to realize their full potential in healthcare and improve patient outcomes, collaboration among researchers, regulatory bodies, and industry stakeholders is essential. By addressing the challenges of stability and regulatory complexities, these advanced formulations can be successfully integrated into clinical practices, offering a new avenue for the treatment of various diseases and improving therapeutic outcomes.

References

[1] Royer M, Prado M, ME G-P. Study of nutraceutical, nutricosmetics and cosmeceutical potentials of polyphenolic bark extracts from Canadian forest species. PharmaNutrition, 2013;1, 158–167.

[2] Katz LM, Dewan K, Bronaugh RL. Nanotechnology in Cosmetics. In *Food and Chemical Toxicology*, vol. 85, Elsevier, 2015, pp. 127–137.

[3] Singh R, Lillard JW. Nanoparticle-based targeted drug delivery. ExpMolPathol, 2009;86, 215–223.

[4] Tadros T, Izquierdo P, Esquena J, et al. Formation and stability of nano-emulsions. Adv Colloid Interface Sci, 2004;108–109, 303–318.

[5] Ali J, Pramod K, Ansari SH. Near-infrared spectroscopy for nondestructive evaluation of tablets. Sys Rev Pharm, 2010;1(1), 17–23. doi: 10.4103/0975-8453.59508

[6] Mason TG, Wilking JN, Meleson K, et al. Nanoemulsions: Formation, structure, and physical properties. J Phys: Condens Matter, 2006;18, R635–R666.

[7] Singh Y, Meher JG, Raval K, et al. Nanoemulsion: Concepts, development and applications in drug delivery. J Control Release, 2017;252, 28–49.

[8] Sonnevilleaubrun O. Nanoemulsions: A new vehicle for skincare products. Adv Colloid Interface Sci, 2004;108–109, 145–149.

[9] Wani TA, Masoodi FA, Jafari SM, McClements DJ. Safety of Nanoemulsions and Their Regulatory Status. In *Nanoemulsions*, Academic Press, 2018, pp. 613–628.

[10] Bouchemal K, Briancon S, Perrier E, et al. Nano-emulsion formulation using spontaneous emulsification: Solvent, oil and surfactant optimization. Int J Pharm, 2004;280, 241–251.

[11] Jacob S, Kather FS, Boddu SH, Shah J, Nair AB. Innovations in nanoemulsion technology: Enhancing drug delivery for oral, parenteral, and ophthalmic applications. Pharmaceutics, 2024 Oct 17;16 (10), 1333.

[12] Shafiq S, Shakeel F, Talegaonkar S, Ahmad FJ, Khar RK, Ali M. Development and bioavailability assessment of ramipril nanoemulsion formulation. Eur J Pharm Biopharm, 2007 May 1;66(2), 227–243.

[13] Simonnet JT, Sonneville O, Legret S inventors; LOreal SA, assignee. Nanoemulsion based on phosphoric acid fatty acid esters and its uses in the cosmetics, dermatological, pharmaceutical, and/ or ophthalmological fields. United States patent US 6,274,150. 2001 Aug 14.

[14] Amenta V, Aschberger K, Arena M, Bouwmeester H, Moniz FB, Brandhoff P, Gottardo S, Marvin HJ, Mech A, Pesudo LQ, Rauscher H. Regulatory aspects of nanotechnology in the agri/feed/food sector in EU and non-EU countries. Regul Toxicol Pharmacol, 2015 Oct 1;73(1), 463–476.

[15] Baker JR Jr, Hamouda T, Shih A, Myc A inventors. Methods of preventing and treating microbial infections. United States patent US 6,506,803. 2003 Jan 14.

[16] Kesharwani D, Paul SD, Paliwal R, Satapathy T. Exploring potential of diacerein nanogel for topical application in arthritis: Formulation development, QbD based optimization and pre-clinical evaluation. Colloids Surf B Biointerfaces, 2023 Mar 1; 223, 113160.

[17] Csóka I, Ismail R, Jójárt-Laczkovich O, Pallagi E. Regulatory considerations, challenges and risk-based approach in nanomedicine development. Curr Med Chem, 2021 Nov 1;28(36), 7461–7476.

[18] Paliwal R, Kumar P, Chaurasiya A, Kenwat R, Katke S, Paliwal SR. Development of nanomedicines and nano-similars: Recent advances in regulatory landscape. Curr Pharma Des, 2022 Jan 1;28(2), 165–177.

[19] De Angelis I, Barone F, Di Felice G, Grigioni M. Regulatory Perspectives on Medical Nanotechnologies. In *Nanomaterials for Theranostics and Tissue Engineering*, Elsevier, 2020, pp. 273–291.

[20] Bremer-Hoffmann S, Amenta V, Rossi F. Nanomedicines in the European translational process. Eur J Nanomed, 2015 Jun 1;7(3), 191–202.

[21] Pita R, Ehmann F, Papaluca M. Nanomedicines in the EU – Regulatory overview. AAPS J, 2016 Nov; 18(6), 1576–1582.

[22] Çetintaş HC, Tonbul H, Şahin A, Çapan Y. Regulatory guidelines of the US food and drug administration and the european medicines agency for actively targeted nanomedicines. Drug Deliv with Targeted Nanoparticles, 2021 Nov;29(1), 725–741.

[23] Malik R, Patil S. Nanotechnology: Regulatory outlook on nanomaterials and nanomedicines in United States, Europe and India. Appl Clin Res Clin Trials Regul Aff, 2020 Dec 1;7(3), 225–236.

[24] Storbeck SD, Semalulu S, Omara F, Vu D Regulatory control of nano-based therapies in Canada. In 14th IEEE International Conference on Nanotechnology 2014 Aug 18 (pp. 739–741). IEEE.

[25] Rana S, Yadav KK, Jha M. Environmental, Legal, Regulatory, Health, and Safety Issues of Nanoemulsions. In *Industrial Applications of Nanoemulsion*, Elsevier, 2024, pp. 219–247.

[26] Sunaina KK, Jha M. Challenges for commercialization of nanoemulsions. Industrial Applications of Nanoemulsion, 2023 Nov;7(1), 179–198. https://doi.org/10.1016/B978-0-323-90047-8.00009-1

[27] Afre RA. Nanotechnology Industry: Scenario of Intellectual Property Rights. In *Intellectual Property Issues in Nanotechnology*, CRC Press, 2020, pp. 345–358.

[28] Sen A. Intellectual Property Management in Nano-Biology Research. In *Biological Applications of Nanoparticles*, Springer Nature, 2023, pp. 233–261.

[29] Jain A, Hallihosur S, Rangan L. Dynamics of nanotechnology patenting: An Indian scenario. Technol Soc, 2011 Feb 1;33(1–2), 137–144.

[30] Marchant GE, Sylvester DJ, Abbott KW, Danforth TL. International harmonization of regulation of nanomedicine. Studies in Ethics, Law, and Technology, 2010 Jan 26;3(3), 1–14.

[31] Moradi M. Global developments in nano-enabled drug delivery markets. Nanotech L Bus, 2005;2, 139.

[32] Gaspar R. Regulatory issues surrounding nanomedicines: Setting the scene for the next generation of nanopharmaceuticals. Nanomedicine, 2007 Apr 1;2(2), 143–147.

[33] Marchant GE, Sylvester DJ. Transnational models for regulation of nanotechnology. J L Med & Ethics, 2006 Dec;34(4), 714–725.

[34] Schummer J, Pariotti E. Regulating nanotechnologies: Risk management models and nanomedicine. Nanoethics, 2008 Apr; 2, 39–42.

[35] International Pharmaceutical Regulators Programme, 2020. http://www.iprp.global/home.

[36] Johnston M The IPRP Nanomedicines Working Group: Current and Potential Future Activities.

[37] Applegate JS. The taming of the precautionary principle. Wm & Mary Envtl L & Pol'y Rev, 2002;27, 13.

[38] Abbott KW, Marchant GE, Sylvester DJ. A framework convention for nanotechnology?. Envtl L Rep, 2006;36, 10931–10942.

[39] ASTM International. 2007. ASTM E2535 – 07, Standard Guide for Handling Unbound Engineered Nanoscale Particles in Occupational Settings, available at http://www.astm.org/Standards/E2535.htm.

[40] Allan J, Belz S, Hoeveler A, Hugas M, Okuda H, Patri A, Rauscher H, Silva P, Slikker W, Sokull-Kluettgen B, Tong W. Regulatory landscape of nanotechnology and nanoplastics from a global perspective. Regul Toxicol Pharmacol, 2021 Jun 1; 122, 104885.

[41] Alexy P, Anklam E, Emans T, Furfari A, Galgani F, Hanke G, Koelmans A, Pant R, Saveyn H, SokullKluettgen B. Managing the analytical challenges related to micro- and nanoplastics in the environment and food: Filling the knowledge gaps. Food Addit Contam: Part A, 2020 Jan 2;37(1), 1–0.

[42] Amenta V, Aschberger K, Arena M, Bouwmeester H, Moniz FB, Brandhoff P, Gottardo S, Marvin HJ, Mech A, Pesudo LQ, Rauscher H. Regulatory aspects of nanotechnology in the agri/feed/food sector in EU and non-EU countries. Regul Toxicol Pharmacol, 2015 Oct 1;73(1), 463–476.

[43] Clifford CA, Stinz M, Hodoroaba VD, Unger WE, Fujimoto T. International Standards in Nanotechnologies. In *Characterization of Nanoparticles*, Elsevier, 2020, pp. 511–525.

[44] Akthar N. Challenges of developing commercial nanomedicines. Private Communication at Global Summit on Regulatory Science, 2019;1(1), 1–25.

[45] Basu B, Dutta A, Ash D, Prajapati B. Nanoemulsions in skin cancer therapy: A promising frontier. Curr Pharm Biotechnol, 2024;1. https://doi.org/10.2174/0113892010302313240610111842

[46] Md S, Mujtaba MA, Lim WM, Alhakamy NA, Zeeshan F, Hosny KM. Nanoemulsions to Preserve/process Bioactive and Nutritional Food Compounds: Contemporary Research and Applications. In *Nanoemulsions in Food Technology*, CRC Press, 2021, pp. 247–278.

[47]   Saini M, Chaudhary G, Narayan CV, Verma S, Sharma L, Shawney SK. Lipid-based nano carriers for drug delivery. Lipid Based Nanocarriers for Drug Delivery, 2024;97, 43–60. doi: 10.1055/s-0042-1751036.

[48]   Nanda A, Nanda S, Nguyen TA, Rajendran S, Slimani Y (Eds.). *Nanocosmetics: Fundamentals, Applications and Toxicity*, vol. 1, Elsevier publishers, 2020, pp. 59–72.

[49]   Kumar A, Chen F, Mozhi A, Zhang X, Zhao Y, Xue X, Hao Y, Zhang X, Wang PC, Liang XJ. Innovative pharmaceutical development based on unique properties of nanoscale delivery formulation. Nanoscale, 2013;5(18), 8307–8325.

[50]   Patel B, Chakraborty S. Biodegradable polymers: Emerging excipients for the pharmaceutical and medical device industries. Int J Pharm Excipients, 2016 Nov 5;4(4), 126–157.

[51]   Bhairam M, Roy A, Bahadur S, Banafar A, Turkane D. Standardization of herbal medicines–an overview. J Appl Pharma Res, 2013 Dec 23;1(1), 14–21.

[52]   Kar M, Chourasiya Y, Maheshwari R, Tekade RK. Current Developments in Excipient Science: Implication of Quantitative Selection of Each Excipient in Product Development. In *Basic Fundamentals of Drug Delivery*, Academic Press, 2019 Jan 1 pp. 29–83.

[53]   Lopalco A, Iacobazzi RM, Lopedota AA, Denora N. Recent advances in nanodrug delivery systems production, efficacy, safety, and toxicity. CompTox: M&P, 2024 Sep 24; 303–332.

[54]   Mishra N, Singh N, Parashar P. Micro-and Nanoemulsions in Antiviral Treatment. In *Viral Drug Delivery Systems: Advances in Treatment of Infectious Diseases*, Cham: Springer International Publishers, 2023, pp. 119–139.

[55]   Nandal R, Tahlan S, Deep A. Novel approaches of self emulsifying drug delivery systems and recent patents: A comprehensive review. Applied drug research, clinical trials and regulatory affairs: Formerly applied clinical research. Clinical Trials and Regulatory Affairs, 2022 Apr 1;9(1), 42–57.

[56]   Mohapatra P, Gopikrishnan M, Doss CGP, Chandrasekaran N. How precise are nanomedicines in overcoming the blood–brain barrier? A comprehensive review of the literature. Int J Nanomed, 2024 Dec;19(31), 2441–2467.

[57]   Krzyszczyk P, Acevedo A, Davidoff EJ, Timmins LM, Marrero-Berrios I, Patel M, White C, Lowe C, Sherba JJ, Hartmanshenn C, O'Neill KM. The growing role of precision and personalized medicine for cancer treatment. Technology, 2018 Sep 11;6(03n04), 79–100.

[58]   Ezike TC, Okpala US, Onoja UL, Nwike CP, Ezeako EC, Okpara OJ, Okoroafor CC, Eze SC, Kalu OL, Odoh EC, Nwadike UG. Advances in drug delivery systems, challenges and future directions. Heliyon, 2023;9(6), 1–17. doi: 10.1016/j.heliyon.2023.e17488.

[59]   Bhairam M, Pandey RK, Shukla SS, Preparation GB. Optimization, and evaluation of dolutegravir nanosuspension: In vitro and in vivo characterization. J Pharma Innov, 2023 Dec;18(4), 1798–1811.

[60]   Souto EB, Cano A, Martins-Gomes C, Coutinho TE, Zielińska A, Silva AM. Microemulsions and nanoemulsions in skin drug delivery. Bioengineering, 2022 Apr 5;9(4), 158.

[61]   Choudhury A, Deka H, Dey BK, Bhairam M, Sengupta K. Nanomaterials for Skin Repair and Regeneration. In *Nanostructured Materials for Tissue Engineering*, Elsevier, 2023, pp. 497–510.

[62]   Anil L, Kannan K. Microemulsion as drug delivery system for Peptides and Proteins. J Pharma Sci Res, 2018;10(1), 16–25.

[63]   Lustig TA, Weisfeld V (Eds.). Forum on Drug Discovery, Development, and Translation; Board on Health Sciences Policy; Institute of Medicine (IOM). In *International Regulatory Harmonization amid Globalization of Drug Development: Workshop Summary*, 2013, pp. 26–40.

[64]   Ngan CL, Basri M, Lye FF, et al. Comparison of Box–Behnken and central composite designs in optimization of fullerene loaded palm-based nano-emulsions for cosmeceutical application. Ind Crops Prod, 2014;59, 309–317.

Shrikant Dargude, Ajita Khichariya, Anuruddha Chabukswar,
Satish Polshettiwar, and Swati Jagdale*

Chapter 9
Future prospects of micro/nanoemulsions in enhancing drug performance

Abstract: Micro/nanoemulsions offer immense potential in enhancing drug performance through improved solubility, stability, and bioavailability. However, their widespread application is hindered by thermodynamic instability and sensitivity to environmental fluctuations. Long-term stabilization remains a key challenge, particularly for biomedical and pharmaceutical applications. Furthermore, the toxicity and limited biocompatibility of conventional synthetic surfactants underscore the urgent need for safer, sustainable alternatives. Recent advancements emphasize the use of eco-friendly, biodegradable surfactants derived from natural sources, which not only reduce toxicity but also improve compatibility with biological systems.

Innovative approaches are steering the development of environmentally conscious emulsion systems utilizing renewable materials, biodegradable emulsifiers, and low-energy fabrication techniques. These strategies enhance both the structural integrity and functional efficacy of emulsions while enabling targeted and controlled drug delivery. Biopolymer-based stabilizers such as zein, cellulose, starch, and protein–polysaccharide complexes have gained momentum due to their clean-label status and biodegradability. Hybrid systems integrating cellulose nanocrystals, polyphenol crystals, and whey protein microgels have demonstrated superior stability and drug encapsulation efficiency.

Looking ahead, advanced drug delivery platforms combining micro/nanoemulsions with nanocarriers like liposomes, micelles, and dendrimers show promise in achieving precise, stimuli-responsive delivery. Functionalized carriers with peptides, antibodies, or targeting ligands offer the potential for site-specific action and controlled release in response to physiological triggers such as pH, temperature, or CO_2 levels.

Shrikant Dargude, Ajita Khichariya, Both should be considered as the first author.

***Corresponding author: Swati Jagdale,** Department of Pharmaceutical Sciences, School of Health Sciences and Technology, Dr. Vishwanath Karad MIT World Peace University, Survey No. 124, Kothrud 411038, Pune, Maharashtra, India, e-mail: swati.jagdale@mitwpu.edu.in
Shrikant Dargude, Anuruddha Chabukswar, Satish Polshettiwar, Department of Pharmaceutical Sciences, School of Health Sciences and Technology, Dr.Vishwanath Karad MIT World Peace University, Kothrud, Pune, Maharashtra, India
Ajita Khichariya, Kamla Institute of Pharmaceutical Sciences, Shri Shankaracharya Professional University, Bhilai 490020, Chhattisgarh, India

https://doi.org/10.1515/9783111593654-010

This chapter provides a forward-looking perspective on the evolution of micro/nanoemulsions, critically analyzing their current applications, emerging materials, and integration into multidrug delivery systems. It highlights the transformative potential of these systems in addressing future pharmaceutical challenges.

Keywords: Emulsions, smart emulsions, hybrid emulsions, intelligent delivery system, microemulsion, nanoemulsion

9.1 Introduction

Micro- and nanoemulsions have emerged as promising colloidal drug delivery systems, characterized by their thermodynamic stability, small droplet size, and large interfacial surface area. Stabilized by surfactants, these emulsions form finely dispersed systems of oil and water that significantly enhance the solubility, permeability, and bioavailability of poorly water-soluble drugs. Their nanoscale size facilitates improved drug absorption across biological membranes, offering a superior alternative to conventional delivery approaches.

Recent studies by Arredondo-Ochoa et al. and Sharmeen et al. have underscored the potential of micro/nanoemulsion systems as versatile carriers for pharmaceutical compounds. These systems not only improve the pharmacokinetic and pharmacodynamic profiles of drugs but also provide opportunities for targeted and controlled release, thereby minimizing systemic toxicity and enhancing therapeutic efficacy.

This chapter aims to explore the potential of micro/nanoemulsions in advancing drug performance, focusing on novel formulation strategies, technological advancements, regulatory perspectives, and emerging therapeutic applications [1, 2].

9.1.1 Advantages of micro/nanoemulsion

Micro/nanoemulsions facilitate sustained drug release, enhances patient adherence, and maintains therapeutic drug levels over an extended period. They improve stability by shielding drugs from light, oxygen, and enzyme deterioration. Micro/nanoemulsions can be administered via oral, parenteral, topical, and intranasal routes. Nanoemulsions are ideal for topical and transdermal therapies as they enhance drug distribution and reduce toxicity. These novel formulations can include multiple bioactives, making them useful for various pharmaceutical applications [1, 3, 4].

9.1.2 Limitations and challenges in micro/nanoemulsion

Despite their promising potential in enhancing drug delivery, micro/nanoemulsions face several limitations and challenges that must be addressed for their safe and effective therapeutic application. One of the primary concerns is the toxicity and biocompatibility of the formulation components, particularly surfactants and cosurfactants, which can pose risks at higher concentrations or with long-term exposure.

Moreover, maintaining the physical stability of micro/nanoemulsions during storage and upon administration remains a significant hurdle. Issues such as phase separation, droplet coalescence, and Ostwald ripening can compromise the efficacy and shelf life of the formulation. Scalability of production using cost-effective and reproducible methods is another critical challenge, particularly when transitioning from laboratory to industrial-scale manufacturing [5–8].

9.2 Novel formulation strategies

9.2.1 Smart emulsion

Environmental engineering, food science, and drug delivery can greatly benefit from the integration of innovative emulsion systems that respond to environmental stimuli such as heat, pH, and carbon dioxide (CO_2). These stimuli-responsive emulsions – also known as smart emulsions – are formulated using specific emulsifiers, surfactants, and polymers that can adapt their behavior based on changes in pH, temperature, or salt concentration [9–11].

Thermoresponsive emulsions utilize materials that change their hydrophilicity or hydrophobicity with temperature, enabling controlled phase transitions and drug release. CO_2-responsive emulsions can switch reversibly between stable and unstable states by the introduction or removal of CO_2, which alters molecular charge, surface tension, and overall emulsion stability through protonation or deprotonation. Similarly, pH-responsive emulsions exploit functional groups that respond to pH shifts, facilitating site-specific and controlled release of therapeutic agents.

Polymers such as poly(N-isopropylacrylamide) (PNIPAM) and poly(acrylic acid) (PAA), which are sensitive to temperature and pH, respectively, are widely used in these systems to enhance drug targeting, improve therapeutic efficacy, and minimize side effects [12].

The following sections provide a detailed overview of thermo-responsive, pH-responsive, and CO_2-responsive emulsions and their role in smart drug delivery systems.

9.2.1.1 Thermoresponsive emulsion

Thermoresponsive emulsions are a class of smart systems that undergo structural or phase transitions upon temperature variation. These emulsions are typically stabilized by stimuli-responsive polymers or surfactants that change their hydrophilicity/hydrophobicity around a specific transition temperature, such as the lower critical solution temperature. Their ability to offer controlled release and site-specific delivery makes them promising candidates for pharmaceutical, biomedical, and industrial applications.

With an emphasis on surface interaction, Rey et al. had investigated the behavior of stimuli-responsive emulsions stabilized by PNIPAM microgel. Stimuli-responsive emulsions provide a dual benefit, offering both long-term stability and controlled release triggered by external factors like pH or temperature changes. This study demonstrated that the thermo-responsive behavior of emulsions is mainly governed by interactions between interfaces, rather than those occurring within them. Additionally, microgels transferred between emulsion droplets in flocculated emulsions enabled stimuli-responsiveness, irrespective of their internal structure. This highlights the critical importance of microgel morphology and the forces they apply to liquid interfaces in the control and design of stimuli-responsive emulsions and interfaces [13].

Ciarleglio et al. described the synthesis of dual-responsive alginate/PNIPAM microspheres that respond to temperature and pH change. This study details the fabrication of Ozoile-based dual-responsive alginate/poly(N-isopropylacrylamide) (PNIPAM) microspheres, fabricated by the microemulsion-based electrospray method, designed to react to changes in both pH and temperature. Overall, the microspheres developed in this study demonstrated a strong response to pH and temperature changes, validating their stimulus-responsive properties and capability to degrade and release their active ingredients. This research highlighted a promising approach for creating advanced drug delivery systems for hydrophobic compounds like Ozoile, with potential applications in managing oxidative stress and inflammation across various biomedical fields [14].

Zifu et al. formulated a concentrated o/w emulsion of thermo-responsive and pH-responsive PNIPAM-co-methacrylic acid (PNIPAM-co-MAA). Pluronic F-127 hydrogel drug delivery system, polyethylene glycolated hyaluronic acid (PEG-HA) and N,N,N-trimethyl chitosan (TMC) were utilized for the textile-based transdermal treatment that was dual-responsive (pH/temperature). These microgels improved the release of lipophilic molecules. Emulsions above pH 6 remain stable even at 50 °C, suggesting swollen microgels enhance bonding between interconnected microgels and oil droplets, affecting the release of encapsulated molecule [15].

Chatterjee et al. proposed a novel one-step template synthesis method based on the utilization of microfluidic quadruple emulsion to create microcapsules with distinct shells that are responsive to stimuli, allowing flexibility and controlled release. The microcapsules showed resemblance to Trojan horses [16].

In a related advancement, Chuanfeng et al. synthesized a new class of fluorosurfactants using thermo-responsive polymers such as PNIPAM and pEtOx, combined with perfluoropolyethers. These diblock surfactants demonstrated temperature-dependent demulsification: at temperatures below the LCST, the emulsions remained stable due to the hydrophilic nature of the surfactants. However, when the temperature rose above the LCST, the hydrophilic heads became hydrophobic, triggering droplet coalescence and demulsification. This property is particularly valuable in droplet microfluidics, allowing for the production of cell-laden microgels with minimal impact on cell viability. Their potential in microencapsulation and biomedical applications is notable due to the controllable and reversible emulsion behavior [17].

Xiaoxuan et al. developed thermo-responsive Janus silica nanoparticles, with one side functionalized with PNIPAM and the other bearing silanol groups. These Janus particles exhibited enhanced emulsifying efficiency, particularly in oil-water systems such as tetradecane-water. Due to stronger hydrogen bonding and increased hydrophilicity from silanol groups, the Janus particles showed a higher thermal phase transition temperature and improved emulsion stability at elevated temperatures. In crude oil-water systems, they demonstrated superior emulsification performance and temperature-triggered responsiveness, indicating their applicability in oil recovery and industrial separation processes [18].

Feng Huanhuan et al. discussed the development of surfactants using poly(ethylene glycol) and di(ethylene glycol) methacrylate via atom transfer radical polymerization. These surfactants stabilize emulsions for a minimum of four months and have an LCST of 34 °C. The hydrophilic block collapses when the LCST is surpassed, which accelerates the demixing of the emulsion. The study investigated the mechanism of emulsion breaking in a system by observing the surface morphology of individual emulsion droplets during collapse transitions. The results indicated that these well-defined thermoresponsive surfactants provide a promising platform for studying droplet coalescence and temperature-triggered phase inversion in emulsion systems. Additionally, their ability to break highly stable emulsions on demand holds industrial significance for various applications, including film formation in waterborne emulsion paints and product recovery during emulsion-based extraction and reaction processes [19].

9.2.1.2 CO$_2$-responsive emulsions

CO$_2$-responsive emulsions are an innovative class of smart emulsions that can reversibly transition between stable and unstable states upon the introduction or removal of carbon dioxide. This responsiveness enables precise control over emulsification and demulsification processes, making them highly suitable for applications in green chemistry, drug delivery, and biocatalysis. The trigger mechanism typically involves changes in pH or ionic strength caused by CO_2 dissolution in water, leading to a shift in the charge or solubility of emulsion stabilizers.

A notable advancement in this area was presented by Yongkang et al., who developed a CO_2/N_2-responsive oil-in-water (O/W) emulsion stabilized by sodium caseinate (NaCas), a natural protein. Unlike conventional systems that rely on synthetic polymers or surfactants grafted onto solid particles, this approach utilized a protein-based stabilizer, resulting in a simpler and more sustainable emulsion system. This NaCas-stabilized emulsion demonstrated superior reaction efficiency and functional versatility with various oils, setting it apart as the first reported *protein-based* CO_2/N_2-responsive emulsion.

The system belongs to a broader category known as **Pickering emulsions**, which are emulsions stabilized by solid particles rather than traditional surfactants. These particles adsorb irreversibly at the oil-water interface, forming a rigid interfacial layer that enhances stability. In CO_2-responsive Pickering emulsions, responsiveness is often achieved by modifying these particles with CO_2-sensitive polymers or functional groups. However, the use of naturally occurring proteins like NaCas represents a significant shift toward more biocompatible and environmentally friendly alternatives. The work by Yongkang et al. illustrates the potential of such systems in advancing green chemical processes, with promising implications for pharmaceutical, food, and biomedical applications [20].

Pie et al. reported that amine-functionalized hierarchically porous UiO-66-(OH)$_2$ (H-UiO-66-(OH)$_2$) may stabilize emulsion and change the demulsification/re-emulsification process. This occurs by alternating the addition or removal of CO_2. This switching mechanism is caused by the interaction of CO_2 with the metal-organic framework (MOF) aminosilane. This leads to the production of hydrophilic ions that reduce the MOF's wettability [21].

The CO_2-responsive emulsion described by Huaixin et al. demonstrated that Pickering emulsions can be stabilized using the CO_2-switchable surfactant 11-(*N*,*N*-dimethylamino) sodium undecanoate (NCOONa). When CO_2 is bubbled through the emulsion, the surfactant transforms into a more hydrophilic, inactive form, leading to rapid and complete demulsification. This prevents the oil phase from dispersing into the aqueous phase. The method offers an environmentally friendly and efficient approach to emulsion separation and oil product purification, with promising applications in biphasic catalysis, oil transportation, and emulsion extraction [22].

Li et al. investigated an emulsion system responsive to CO_2, focusing on *N*,*N'*-dimethyl-*N*,*N'*-didodecyl butylene diamine (DMDBA), a CO_2-switchable gemini surfactant [23]. The infusion of CO_2 and N_2 triggers a reversible transition between the surfactants active and inactive states, making it particularly promising for drug delivery applications. Two key properties of DMDBA are its cyclic reversibility and CO_2 responsiveness. Upon exposure to CO_2, DMDBA reacts to form a cationic surfactant, DMDBAH, resulting in a gradual increase in conductivity. This change in conductivity quantitatively reflects the reaction rate. The kinetics of the reaction was modeled by fitting the conductivity data during surfactant formation under CO_2 injection at various temperatures. The resulting kinetic equation showed correlation coefficients exceeding 0.99, indicating that the reaction followed a second-order kinetic model. The rate constant (k) increased with temperature, peaking at 55 °C, suggesting that elevated temperatures enhance molecular motion and thus accelerate the reaction. However, higher temperatures also reduce the degree of protonation, likely due to thermal decomposition of the protonated tertiary amine bicarbonate.

9.2.1.3 pH-responsive emulsion

pH-responsive emulsions are a class of smart emulsions that undergo reversible changes in stability, structure, or phase behavior in response to variations in the pH of the surrounding environment. These emulsions are particularly useful in applications such as controlled drug delivery, catalysis, and food formulation, where precise release or reaction control is required. The responsiveness typically arises from pH-sensitive functional groups present in emulsifiers, surfactants, or stabilizing particles that alter their hydrophilic–lipophilic balance or surface charge under different pH conditions.

A notable example of a pH-responsive emulsion system involves the formation of a Pickering emulsion of n-decanol in water using gold-core polymer-metal hybrid micelles – specifically, Au@poly(ethylene oxide)-b-poly(4-vinylpyridine) – as the emulsifying agent. This Pickering emulsion exhibited reversible emulsification and demulsification upon pH cycling in the aqueous phase. Impressively, this pH-induced transition could be repeated several times with minimal loss in emulsion stability or performance. Due to its pH-responsive nature, the emulsion proved highly effective for catalytic reactions at the water–oil interface, demonstrating its potential in interfacial catalysis and other functional applications where on-demand emulsification is desirable [24].

Ruidong et al. reported the development of a novel emulsion system that responds sensitively to pH fluctuations. In this approach, poly(methacrylic acid) (PMAA) brush-coated silica nanoaggregates (P-Si) were synthesized by grafting a diblock copolymer – poly(tert-butyl methacrylate)-b-poly[3-(trimethoxysilyl)propyl methacrylate] (PtBMA-b-PTMSPMA) – onto fumed silica particles. These modified nanoparticles exhibited distinct behavior based on the pH of the aqueous phase. At low pH levels, the P–Si nanoaggregates flocculated, leading to the successful stabilization of a toluene-in-water Pickering emulsion. However, when the pH was raised to neutral or basic levels, the nanoaggregates dispersed more effectively in water, resulting in emulsion breakdown. This reversible flocculation behavior enabled controlled and repeatable cycles of emulsification and demulsification simply by alternating between acid and base conditions. Such a system offers a promising route for designing reusable and tunable emulsions for chemical processing, extraction, and smart delivery applications [25].

Pei et al. reported the development of novel pH-responsive smart surfactants (N +–(n)–N, where n = 14, 16) derived from alkyl trimethylammonium bromides. Depending on the pH of the media, these surfactants may switch between their cationic and hydrophilic Bola forms. They stabilize conventional, Pickering, and oil-in-water emulsions in neutral and alkaline media. The surfactant converts into the hydrophilic Bola form and loses some of its emulsifying power in acidic solutions. Following demulsification, the surfactant returns to the aqueous phase, allowing recycling without contaminating the oil phase. For emulsion polymerization, heterogeneous catalysis, and oil transportation, these intelligent surfactants are crucial for producing short-term stable emulsions. They are environmentally friendly and versatile due to their pH-based transition between stable and unstable states. The recycling of surfactants and nanoparticles during demulsification aligns with green chemistry principles [26].

The formulation of pH-responsive, solid-free nanoemulsions was reported by Ren et al. The formation of a dynamic covalent surfactant (T-DBA) has been shown to alternate between active and inactive states in response to pH variations. The study proved the possibility of controlled stability of nanoemulsions. The study demonstrates the pH-responsive behavior of a nanoemulsion by monitoring the interfacial tension between liquid paraffin and the T-DBA aqueous solution at different pH values. The results show a low interfacial tension of10.3 mN/m after a 10-min equilibration at pH 10, indicating T-DBA is in an active interfacial state. This is due to the introduction of hydrophobic benzene rings and decyl chains through dynamic covalent bonds between hydrophilic taurine and hydrophobic DBA. T-DBA adsorbs to the oil-water interface to stabilize droplets. An emulsifier with an interfacial tension of 10 mN/m was sufficient to stabilize the emulsions. As the pH decreased from 10 to 3, the interfacial tension increased to 34 mN/m, similar to the interfacial tension between liquid paraffin and pure water, DBA in liquid paraffin and pure water, and liquid paraffin and the taurine aqueous solution. The decomposition of interfacial active T-DBA into interfacial inactive taurine and DBA occurred due to the decomposition of the dynamic imine bond. The equilibrium IFT returned to 10.5 mN/m upon increasing the pH to 10, indicating the reformation of interfacial active T-DBA in situ. The reformed interfacial active T-DBA was adsorbed into the oil-water interface again after sonication, resulting in a stable nanoemulsion [27].

9.2.2 Integration with nanotechnology

Nanotechnologies, particularly liposomes and micelles, can enhance treatment by providing drug delivery vehicles at the nano-sized scale. Although the FDA has authorized few micelle-based formulations, several other nanocarrier-based formulations like liposomal medications have been approved for anticancer treatment due to their improved pharmacokinetics and reduced adverse effects [28]. Liposomal systems often face challenges, such as limited controlled release capabilities and low encapsulation efficiency. However, innovatively designed multi-compartment liposomes, acting as microreactors for biochemical synthesis, now offer precise control over cargo loading and release kinetics, effectively overcoming these limitations [29]. This challenge can be addressed by preparing emulsion-like systems that combine the advantages of solid lipid nanoparticles and liposomes. These systems enhance drug loading, stability, and controlled drug release effectiveness due to their solid lipid core and phospholipid bilayer structure [30]. Moreover, integrating liposomes, dendrimers, and quantum dot structures into a single hybrid carrier, such as microemulsion, can combine their advantages and enable simultaneous drug delivery, while also providing a protective function.

eLiposomes represent a novel class of drug delivery systems that integrate emulsions and liposomes to achieve targeted therapeutic delivery. These advanced vesicles encapsulate both a drug and an internal emulsion droplet, allowing for controlled release mechanisms. Upon exposure to external stimuli – particularly ultrasound – the internal emulsion droplet undergoes a phase change from liquid to gas. This expansion generates sufficient force to rupture the surrounding lipid bilayer of the eLiposome, enabling the rapid release of the encapsulated drug or genetic material.

Javadi et al. demonstrated this concept by utilizing perfluoropentane (PFC5) emulsions encapsulated within eLiposomes, further functionalized with a folate ligand for targeted delivery to HeLa cells. Their study evaluated drug transport and plasmid gene transfection efficiency under varying ultrasonic acoustic parameters. Using confocal microscopy, they observed that in the absence of ultrasound, drug delivery and gene transfection remained minimal, even with PFC5 and folate-targeted eLiposomes. However, upon ultrasound activation, there was a marked enhancement in both drug and plasmid delivery, highlighting the potential of this system to disrupt endosomal membranes and release cargo directly into the cytosol. This dual-responsive mechanism – leveraging both pH-sensitivity and ultrasound stimulation – positions eLiposomes as a promising platform for noninvasive, targeted delivery in cancer therapy and gene transfection applications [31].

The complex structure of hybrid multicompartment carriers such as eLiposomes, shown in Figure 9.1 as nanoemulsion droplets encapsulated within a phospholipid bilayer enables the simultaneous delivery of hydrophilic, hydrophobic, and amphiphilic compounds. Owing to this property, these carriers are being actively developed for drug delivery applications. Liposomes are multicompartment hybrid carriers that are intended to deliver drugs effectively. eLiposomes, a type of hybrid multi-compartment carrier, enable the simultaneous delivery of hydrophilic, hydrophobic, and amphiphilic compounds due to their unique structure. These carriers consist of nanoemulsion droplets encapsulated within a phospholipid bilayer, allowing for efficient drug delivery. The stability of the nanoemulsions within the confined liposomal space is crucial for their effectiveness, which is influenced by factors such as ζ-potential and the fraction of the dispersed phase. The efficiency of these nanoemulsions depends on their stability, wherein stable nanoemulsions develop with a ζ-potential of more than 40 mV and a dispersed phase fraction of less than 10%. eLiposomes are stable because of the rapid Brownian motion of their droplets. Stability and biocompatibility must be balanced for eLiposomes to be used in real-world applications. Nanoemulsion stability inside liposomes is predicted using Langevin-dynamics simulations [32].

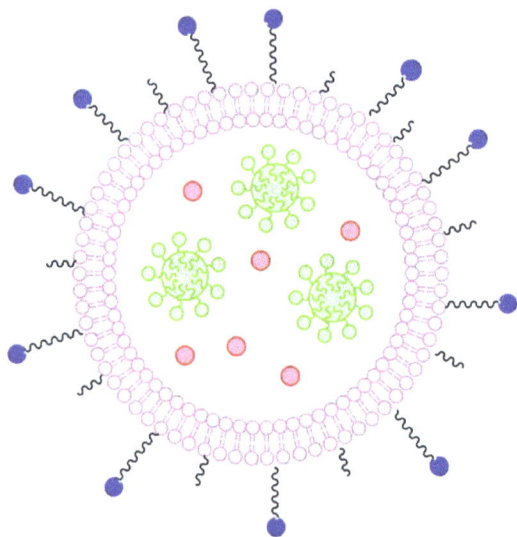

Figure 9.1: Schematic illustration of eLiposome. Emulsion droplets (green spheres) and therapeutic compounds (red spheres) are encapsulated in the lipid bilayer (pink). Stealth polymers and active targeting ligands (blue circles) are found on the exterior of the bilayer. Figure adopted with permission from Elsevier [32].

Husseini et al. showed that drugs may be delivered to tumor tissue by ultrasound utilizing eLiposomes. Nanocarriers are appropriate for in vivo application since they include a perfluorocarbon nanoemulsion droplet, which is stable at body temperature. Experiments were conducted to assess the release of fluorescent calcein molecules from eLiposomes at different temperatures, without the use of ultrasound. The results demonstrated that eLiposomes remain stable at body temperature, and as the temperature exceeded 40 °C, the release of calcein from these innovative nanocarriers increased [33].

Fluorescent quantum dots like CdS and CdTe are stabilized using microgels. This stabilization aids in adjusting the quantum dots' fluorescence characteristics. Unlike isolated CdS or microgels alone, quantum dots implanted in microgels, such as CdS in pNIPAM-co-AA microgels, exhibited photocatalytic activity in dye degradation. 3-Aminophenylboronic acid-functionalized microgels are capable of measuring glucose concentrations. Films with optical detection and adjustable permeability may be produced using these microgels. In response to variations in glucose content, the fluorescence of quantum dots embedded in microgels may be reversibly suppressed and restored, which makes them valuable for optical sensing applications [34].

Rowland Andrew and colleagues studied the production and characteristics of liposome-stabilized all-aqueous emulsions using PEG/dextran, PEG/Ficoll, and PEG/sulfate systems by two different methods: extrusion and vortexing. The data illustrated that although droplet uniformity and stability are slightly better for samples made with extruded vesicles, extrusion is not essential for creating functional microreactors. Emulsions stabilized with vortexed liposomes are equally effective at solute partitioning and allow diffusion across the liposome corona of the droplet. This work broadens the range of compositions available for liposome-stabilized, all-aqueous emulsion droplet bioreactors, making them suitable for a wider variety of reactions. By replacing the liposome extrusion step with vortexing, the production time and cost of bioreactors can be reduced, without significantly lowering emulsion quality [35].

Daniel Dewey and coworkers prepared all-aqueous emulsions stabilized by liposomes for size-controlled, semipermeable bioreactors that preserve ideal interior conditions for biological reactions. The electrostatic stabilization offered by PEGylated, negatively charged liposomes prevents droplet coalescence even at sub monolayer interfacial coatings. The capacity of DNA and RNA to diffuse in and out of droplets allows for localized reactivity, with partitioning capacity to achieve the required concentration. They have demonstrated the potential in biotechnology and medicine using a ribozyme cleavage process within droplets covered with liposomes [36].

9.2.3 Hybrid system

Hybrid nanocarrier systems that combine micro/nanoemulsions with dendrimers, liposomes, and micelle systems offer a multifaceted approach to drug delivery, enhancing solubility, stability, targeting, and therapeutic efficacy. Micro/nanoemulsions are particularly effective in improving the solubility of lipophilic drugs by creating a dispersed system that enhances drug dissolution and absorption. Dendrimers, on the other hand, contribute by providing a highly branched, stable, and protective architecture that shields sensitive drug molecules from degradation [37, 38].

Functionalization of dendrimers with targeting ligands allows for site-specific delivery, directing therapeutic agents to particular cells or tissues and minimizing off-target effects. The micro/nanoemulsion matrix plays a critical role in modulating the release kinetics of the encapsulated drug, ensuring a sustained and controlled release profile that extends the therapeutic effect. Moreover, the synergistic integration of these two nanostructures enhances drug permeation across biological membranes, thereby improving overall bioavailability. This hybrid system represents a promising platform for next-generation drug delivery, particularly in complex conditions requiring precise targeting and prolonged therapeutic action [12].

Dendrimer–micro/nanogels represent an advanced drug delivery technology in which drugs are encapsulated using chemical cross-linking methods. A dendrimer, a highly branched, tree-like macromolecule, is combined with linear polyethylene glycol diacrylate to form a cross-linked gel network. Therapeutic agents such as drugs or genes can be incorporated in situ into the micro/nanoemulsion, which is typically prepared by mixing an oil phase with a surfactant-containing aqueous phase. This emulsion is then subjected to a cross-linking process to form a stable and biocompatible gel matrix.

The efficiency and uniformity of dendrimer-micro/nanogels can be significantly enhanced through advanced manufacturing techniques such as flow chemistry and microfluidics. These continuous production methods offer better control over particle size and morphology compared to conventional batch processes. Additionally, the use of template-guided synthesis allows for the creation of hybrid materials with more complex and functional architectures, enabling precise targeting and controlled release.

By integrating flow chemistry, microfluidic platforms, and template-assisted fabrication, dendrimer-micro/nanogels can be further optimized to achieve superior material properties and targeted therapeutic outcomes [39].

A promising drug delivery approach involves the development of hybrid systems that combine the advantages of both liposomes and micelles. Micelles, formed from amphiphilic block copolymers, offer excellent stability and enhance the solubility of poorly water-soluble drugs. On the other hand, liposomes – spherical microvesicles composed of phospholipid bilayers – provide superior biocompatibility and low immunogenicity, making them ideal carriers for biological applications. In these hybrid systems, researchers have either encapsulated charged micelles within liposomes or linked them to oppositely charged liposomes through electrostatic interactions. This strategic combination integrates the solubilization capacity of micelles with the biocompatibility and targeting potential of liposomes. As a result, the hybrid system exhibits improved drug encapsulation efficiency, enhanced stability, and superior targeting capabilities, offering a robust platform for effective and controlled drug delivery [40].

Innovative drug delivery methods called liposome-micelle-hybrid (LMH) carriers enhance the co-formulation and distribution of chemotherapeutics such as paclitaxel and 5-fluorouracil. Improved permeability and transport across cellular barriers, regulated release, greater drug loading and solubility, and higher cytotoxicity against cancer cells are some of the characteristics of this system. LMH carriers, which provide better therapeutic efficacy and fewer adverse effects for cancer treatment, represent a potential development in nanomedicine [41].

N-Isopropylacrylamide (NIPAM) was used to create ultra-low cross-linked (ULC) nanogels, which efficiently stabilize oil-in-water emulsion at low concentrations. These emulsions remain stable by preventing droplet coalescence and flocculation. ULC nanogels combine the characteristics of cross-linked particles with flexible macromolecules. They improve emulsion stability by providing steric barrier-like particles and stretching uniformly at the interface like polymers. By raising the temperature, the emulsions stabilized by ULC nanogels can be readily ruptured, demonstrating their sensitivity to outside stimuli. Because of this special set of characteristics, ULC nanogels can be used in a variety of fields where enhanced emulsion stability and responsiveness are advantageous, such as food science and heterogeneous catalysis. This hybrid method improves the functionality and performance of emulsions in a variety of domains by utilizing the dual properties of ULC nanogels [42].

Olsen et al. showed the pathway to move barium ions across droplets in a water-in-chloroform emulsion using gold nanoparticles (GNPs) treated with crown ether. Driven by a concentration gradient, GNPs transport barium ions from source droplets to target droplets, causing the precipitation of barium sulfate. By using electron transfer or secondary ion transport, the procedure preserves electroneutrality. The promise of integrating emulsions with nanoparticles for controlled chemical reactions is highlighted by this technique. This imitates biological ion transport and may be used to create nanoscale motors and machinery and potential applications in drug delivery [43].

A novel method for encasing inorganic nanoparticles in block copolymer (BCP) vesicles was developed by Yu et al. The method used 3D co-assembly of BCPs and NPs within organic emulsion droplets to enable spontaneous entry of the aqueous phase. Beyond the critical limits of 0.5, GNPs were encapsulated with high efficiency when the D_N/d_{w0} (ratio of the block layer thickness (d_{w0}) to the NP diameter (D_N)) is 0.66 [44].

9.3 Sustainability and eco-friendly emulsion systems

The growing demand for sustainable solutions in drug delivery, food, and environmental applications has led to the development of eco-friendly emulsion systems. These systems aim to reduce the environmental impact of traditional emulsions by utilizing renewable, biodegradable, and natural materials. Sustainable emulsions, including nanocellulose-based Pickering emulsions, biosurfactant-based emulsions, lecithin-based emulsions, and green chemistry approaches, offer promising alternatives to synthetic surfactants and chemicals.

Nanocellulose-based Pickering emulsions are biodegradable and create stable emulsions without synthetic surfactants. Biosurfactants derived from microorganisms or plants provide a natural, safer alternative to traditional surfactants. Lecithin-based emulsions, sourced from natural materials like soybeans, are widely used for their stability and safety. Green chemistry approaches in micro/nanoemulsion production focus on using renewable resources and minimizing the use of harmful chemicals.

Despite their advantages, challenges remain in optimizing these eco-friendly systems for performance, scalability, and cost-effectiveness, requiring continued innovation for their broader application.

9.3.1 Current applications and challenges

Numerous applications in the food, pharmaceutical, cosmetic, agricultural industries and challenges in developing some sustainable and eco-friendly micro/nanoemulsions have been summarized in Table 9.1.

Table 9.1: Applications and challenges of sustainable and eco-friendly emulsion system.

Emulsion	Applications in food/pharma/cosmetic/agri	Challenges	References
Nanocellulose-based Pickering emulsion	– Sustainable substitutes for synthetic emulsifiers improve smoothness and stability while being more environmentally friendly. – Use for controlled drug delivery, enhancing the bioavailability and stability of active pharmaceutical ingredients. – They provide improved stability and texture in cosmetic formulations, reducing the need for synthetic stabilizers. – Delivery of pesticides and fertilizers, improving efficiency and reducing environmental impact.	– Significant environmental effect of the bleaching process during nanocellulose extraction. – Regulatory framework for Pickering emulsions is required. – More study is required, particularly in low-income nations, on sustainability and eco-design. – Insufficient long-term studies on health effects and cytotoxicity risks for workers.	[45]

Table 9.1 (continued)

Emulsion	Applications in food/pharma/ cosmetic/agri	Challenges	References
Biosurfactant-based emulsion	– Oil and waste frying oil as fat stabilizers, and anti-spattering agents, enhance texture and prolong freshness in products like bread and ice cream. – Antitumor properties; self-assembling potential into nanoparticles of cyclic lipopeptide biosurfactant. – Treatment of acne, dandruff, and body odors, deodorants, nail care products, and toothpaste, and as an anti-wrinkle agent – They enhance soil quality, promote plant growth, and act as biopesticides, contributing to sustainable farming practices.	– Maintaining the precise manufacturing conditions needed for biosurfactants can be difficult. – Compared to synthetic substrates, renewable ones may be more expensive and less effective. – There is minimal information on kinetics and consumption. – Diverse range of applications requires distinct types of biosurfactants, making it difficult to develop a one-size-fits-all solution.	[46]
Lecithin-based emulsion	– Egg yolk and milk lecithins, recommended for early infant nutrition due to high content of sphingolipids as well as arachidonic and docosahexaenoic acids. – Egg yolk lecithins are reported to be the preferred emulsifiers for parenteral oil-in-water emulsions and as a drug carrier for many microemulsions. – Natural phospholipids as cosmetic emulsifiers allows the production of natural cosmetic products, providing skin hydration and enhancing dermal penetration.	– The procedure of choosing a cosurfactant is intricate and necessitates nontoxic substitutes for conventional alcohols. – To ensure that biocompatible oils and surfactants cooperate without sacrificing stability, component compatibility is essential. – There may be toxicity concerns at high quantities. – Intricate formulas are necessary.	[47, 48]

9.3.2 Green chemistry approaches

Biodegradable emulsifiers, renewable resources, and energy-efficient processes are essential for the production of sustainable emulsions. These techniques promote sustainability, reduce the demand for fossil fuels, and use less hazardous materials. Recyclable catalysts also improve overall production processes and cut waste [49].

Lecithin, derived from natural sources like eggs and soybeans, is biodegradable and biocompatible. It enhances the bioavailability of hydrophobic drugs through stable nanoemulsions and is nontoxic, making it ideal for safe, sustainable, and efficient drug delivery systems [50]. There is an increasing need for natural, environmentally friendly emulsifiers. Saccharide-based emulsifiers, including sucrose esters (SEs), sorbitan esters, and alkyl polyglycosides, offer industrial applications with enhanced profitability [51]. Alkyl polyglucosides (APGs) and SEs are emphasized as promising bio-based surfactants. They are biodegradable, biocompatible, and gentle. They perform similarly to conventional surfactants since they are derived from natural sources. The head groups on SEs and APGs can stabilize emulsions lowering surface tension. Their synthesis promotes sustainability and safety with green chemistry principles [52]. Moreover, given the special structural characteristics, cashew nut shell liquid is a sustainable substitute for petrochemical surfactants. It is an environmentally friendly substitute because of its exceptional emulsifying, foaming, and wetting characteristics, which help industries including medicines, cosmetics, and agriculture [53].

Vukašinović et al. designed and developed green emulsifier/emollient-based emulsion vehicles using sustainable components highlighting the use of a natural mixed emulsifier (lauryl glucoside/myristyl glucoside/polyglyceryl-6 laurate) and biodegradable emollients (Emogreen® L15 and L19). The research indicated that these green emulsions are good options to use in pharmaceutical and cosmetic applications since they have a desirable texture, stable and are well-liked by consumers [54]. Additionally, solid particles, such as silica and polymer-based compounds, are ideal for creating stable emulsions because they are more resilient to destabilization than traditional surfactants. Hydrophobic silica and cross-linked polystyrene provides increased stability and utility. By using these solid particles, drug delivery becomes more stable and effective, lowering the requirement for artificial surfactants and adhering to the green chemistry principles [55].

Zhang et al. emphasized the significance of creating biodegradable, nontoxic emulsifiers and solvents to create environmentally friendly and sustainable emulsion systems. The goal of these advancements was to reduce the environmental harm caused by traditional techniques, which may include dangerous chemicals. By applying the principles of green chemistry, scientists can produce solvents and emulsifiers that break down organically and cause minimal harm to ecosystems and human health. This improvement supports environmental sustainability and aligns with increasing regulatory requirements for safer, more responsible chemical processes in industrial application [56].

9.4 Future prospect

Artificial intelligence (AI) is revolutionizing pharmaceutical formulation development by improving accuracy and efficiency. Compared to traditional trial-and-error methods, AI algorithms can analyze large datasets to identify patterns and enhance formulations, saving time and money. AI's ability to predict a drug's solubility, stability, and bioavailability enables early formulation selection. Machine learning algorithms can optimize formulation parameters, including the surfactant content in microemulsion, to achieve desired attributes. Additionally, AI can tailor formulations to individual patients' requirements, increasing treatment efficacy and reducing adverse effects. Future applications include automated formulation development, real-time monitoring, and better drug delivery techniques. While microemulsion offers flexible and efficient drug delivery, AI increases the accuracy and efficiency of formulation development. The integration of these technologies has the potential to revolutionize drug manufacturing and delivery, paving the way for more effective and personalized treatments [57–60].

References

[1]   Arredondo-Ochoa T, Silva-Martínez GA. Microemulsion based nanostructures for drug delivery. Front Nanotechnol. Frontiers Media S.A., 2022;3.

[2]   Rafique S, Das NG, Das SK. Nanoemulsions: An Emerging Technology in Drug Delivery. In *Emerging Technologies for Nanoparticle Manufacturing [Internet]*, JK Patel, YV Pathak, Eds., Cham: Springer International Publishing, 2021, pp. 381–393. Available from: https://doi.org/10.1007/978-3-030-50703-9_17

[3]   Gupta A, Eral HB, Hatton TA, Doyle PS. Nanoemulsions: Formation, properties and applications. Soft Mater. R Soc Chem, 2016;12:2826–2841.

[4]   Pavoni L, Perinelli DR, Bonacucina G, Cespi M, Palmieri GF. An overview of micro- and nanoemulsions as vehicles for essential oils: Formulation, preparation and stability. Nanomaterials. MDPI AG, 2020;10.

[5]   Kumar SS, Natarajan C, George PDC. Micro-nanoemulsion and nanoparticle-assisted drug delivery against drug-resistant tuberculosis: Recent developments. Clin Microbiol Rev [Internet], 2023 Nov 30;36(4):e00088–23. Available from: https://doi.org/10.1128/cmr.00088-23

[6]   Abaszadeh F, Ashoub MH, Amiri M. Nanoemulsions Challenges and Future Prospects as a Drug Delivery System. In *Current Trends in Green Nano-emulsions: Food, Agriculture and Biomedical Sectors [Internet]*, A Husen, RK Bachheti, A Bachheti, Eds., Singapore: Springer Nature Singapore, 2023, pp. 217–243. Available from: https://doi.org/10.1007/978-981-99-5398-1_13

[7]   Anton N, Vandamme TF. Nano-emulsions and micro-emulsions: Clarifications of the critical differences. Pharm Res [Internet], 2011;28(5):978–985. Available from: https://doi.org/10.1007/s11095-010-0309-1

[8]   Suhail N, Alzahrani AK, Basha WJ, Kizilbash N, Zaidi A, Ambreen J, et al. Microemulsions: Unique properties, pharmacological applications, and targeted drug delivery. Front Nanotechnol. Frontiers Media S.A., 2021;3.

[9]   Ciarleglio G, Placido M, Toto E, Santonicola MG. Dual-responsive alginate/PNIPAM microspheres fabricated by microemulsion-based electrospray. Polymers (Basel), 2024 Oct 1;16(19).

[10] Zhang H, Zhang J, Dai W, Zhao Y. Facile synthesis of thermo-, pH-, CO2- and oxidation-responsive poly(amidothioether)s with tunable LCST and UCST behaviors. Polym Chem [Internet], 2017;8 (37):5749–5760. Available from: http://dx.doi.org/10.1039/C7PY01351E

[11] Li Z, Ngai T. Stimuli-responsive gel emulsions stabilized by microgel particles. Colloid Polym Sci [Internet], 2011;289(5):489–496. Available from: https://doi.org/10.1007/s00396-010-2362-z

[12] GhoshMajumdar A, Pany B, Parua SS, Mukherjee D, Panda A, Mohanty M, et al. Stimuli-responsive nanogel/microgel hybrids as targeted drug delivery systems: A comprehensive review. Bionanoscience [Internet], 2024;14(3):3496–3521. Available from: https://doi.org/10.1007/s12668-024-01577-9

[13] Rey M, Kolker J, Richards JA, Malhotra I, Glen TS, Li NYD, et al. Interactions between interfaces dictate stimuli-responsive emulsion behaviour. Nat Commun, 2023 Dec 1;14(1).

[14] Ciarleglio G, Placido M, Toto E, Santonicola MG. Dual-responsive alginate/PNIPAM microspheres fabricated by microemulsion-based electrospray. Polymers (Basel), 2024 Oct 1;16(19).

[15] Li Z, Ngai T. Stimuli-responsive gel emulsions stabilized by microgel particles. Colloid Polym Sci [Internet], 2011;289(5):489–496. Available from: https://doi.org/10.1007/s00396-010-2362-z

[16] Chatterjee S, Hui PCL, Kan CW, Wang W. Dual-responsive (pH/temperature) Pluronic F-127 hydrogel drug delivery system for textile-based transdermal therapy. Sci Rep, 2019 Dec 1;9(1).

[17] An C, Zhang Y, Li H, Zhang H, Zhang Y, Wang J, et al. Thermo-responsive fluorinated surfactant for on-demand demulsification of microfluidic droplets. Lab Chip [Internet], 2021;21(18):3412–3419. Available from: http://dx.doi.org/10.1039/D1LC00450F

[18] Li X, Wang Y, Hou Q, Cai W, Xu Y, Zhao Y. Fabrication of thermo-responsive Janus silica nanoparticles and the structure–performance relationship in Pickering emulsions. Res Chem Intermed [Internet], 2021;47(9):3899–3917. Available from: https://doi.org/10.1007/s11164-021-04486-8

[19] Feng H, Verstappen NAL, Kuehne AJC, Sprakel J. Well-defined temperature-sensitive surfactants for controlled emulsion coalescence. Polym Chem [Internet], 2013;4(6):1842–1847. Available from: http://dx.doi.org/10.1039/C2PY21007J

[20] Xi Y, Liu B, Wang S, Wei S, Yin S, Ngai T, et al. CO2-responsive Pickering emulsions stabilized by soft protein particles for interfacial biocatalysis. Chem Sci [Internet], 2022;13(10):2884–2890. Available from: http://dx.doi.org/10.1039/D1SC06146A

[21] Pei X, Liu J, Song W, Xu D, Wang Z, Xie Y. CO2-switchable hierarchically porous zirconium-based MOF-stabilized Pickering emulsions for recyclable efficient interfacial catalysis. Materials, 2023 Feb 1;16(4).

[22] Li H, Liu Y, Jiang J. CO2-responsive surfactants for switchable Pickering emulsions with a recyclable aqueous phase. Green Chem [Internet], 2022;24(20):8062–8068. Available from: http://dx.doi.org/10.1039/D2GC02630A

[23] Li Y, Tang X, Yang P, Zhang Y, Liu J. Synthesis and kinetics of CO2-responsive Gemini surfactants. Molecules, 2024 Sep 3;29(17):4166.

[24] Fang Z, Yang D, Gao Y, Li H. pH-responsible Pickering emulsion and its catalytic application for reaction at water–oil interface. Colloid Polym Sci, 2015 May 1;293(5):1505–1513.

[25] Luo R, Dong J, Luo Y. pH-Responsive Pickering emulsion stabilized by polymer-coated silica nanoaggregates and applied to recyclable interfacial catalysis. RSC Adv [Internet], 2020;10 (69):42423–42431. Available from: http://dx.doi.org/10.1039/D0RA07957J

[26] Pei X, Zhang S, Zhang W, Liu P, Song B, Jiang J, et al. Behavior of smart surfactants in stabilizing pH-responsive emulsions. Angew Chem Int Ed, 2021 Mar 1;60(10):5235–5239.

[27] Ren G, Li B, Ren L, Lu D, Zhang P, Tian L, et al. PH-responsive nanoemulsions based on a dynamic covalent surfactant. Nanomaterials, 2021;11(6).

[28] Suzuki R, Omata D, Oda Y, Unga J, Negishi Y, Maruyama K. Cancer Therapy with Nanotechnology-Based Drug Delivery Systems: Applications and Challenges of Liposome Technologies for Advanced

Cancer Therapy. In *Nanomaterials in Pharmacology [Internet]*, ZR Lu, S Sakuma, Eds., New York, NY: Springer New York, 2016, pp. 457–482. Available from: https://doi.org/10.1007/978-1-4939-3121-7_23

[29] Shubhra QTH. Multi-compartment liposomes forge new paths in drug delivery. Nat Chem [Internet], 2024;16(10):1578–1579. Available from: https://doi.org/10.1038/s41557-024-01638-2

[30] Singh S, Khurana K, Chauhan SB, Singh I. Emulsomes: New lipidic carriers for drug delivery with special mention to brain drug transport. Futur J Pharm Sci [Internet], 2023;9(1):78. Available from: https://doi.org/10.1186/s43094-023-00530-z

[31] Javadi M, Pitt WG, Tracy CM, Barrow JR, Willardson BM, Hartley JM, et al. Ultrasonic gene and drug delivery using eLiposomes. J Contr Release, 2013 Apr 10;167(1):92–100.

[32] Koroleva MY, Plotniece A. Aggregative stability of nanoemulsions in eLiposomes: Analysis of the results of mathematical simulation. Colloid J, 2022 Apr 1;84(2):162–168.

[33] Husseini GA, Pitt WG, Javadi M. Investigating the stability of eliposomes at elevated temperatures. Technol Cancer Res Treat, 2015 Jan 1;14(4):379–382.

[34] Zhilin DM, Pich A. Nano- and microgels: A review for educators. 2021;3(2):155–167. Available from: https://doi.org/10.1515/cti-2020-0008

[35] Rowland AT, Keating CD. Formation and Properties of Liposome-Stabilized All-Aqueous Emulsions Based on PEG/Dextran, PEG/Ficoll, and PEG/Sulfate Aqueous Biphasic Systems.

[36] Dewey DC, Strulson CA, Cacace DN, Bevilacqua PC, Keating CD. Bioreactor droplets from liposome-stabilized all-aqueous emulsions. Nat Commun, 2014 Aug 20;5.

[37] Kim D, Javius-Jones K, Mamidi N, Hong S. Dendritic nanoparticles for immune modulation: A potential next-generation nanocarrier for cancer immunotherapy. Nanoscale [Internet], 2024;16 (21):10208–10220. Available from: http://dx.doi.org/10.1039/D4NR00635F

[38] Ren S, Xu Y, Dong X, Mu Q, Chen X, Yu Y, et al. Nanotechnology-empowered combination therapy for rheumatoid arthritis: Principles, strategies, and challenges. J Nanobiotechnol. BioMed Central Ltd, 2024;22.

[39] Wang J, Li B, Qiu L, Qiao X, Yang H. Dendrimer-based drug delivery systems: History, challenges, and latest developments. J Biol Eng. BioMed Central Ltd, 2022;16.

[40] Qian J, Guo Y, Xu Y, Wang X, Chen J, Wu X. Combination of micelles and liposomes as a promising drug delivery system: A review. Drug Deliv Transl Res. Springer, 2023;13:2767–2789.

[41] Hassan MM, Romana B, Mao G, Kumar N, Sonvico F, Thordarson P, et al. Liposome-Micelle-Hybrid (LMH) carriers for controlled Co-delivery of 5-FU and paclitaxel as chemotherapeutics. Pharmaceutics, 2023 Jul 1;15(7).

[42] Petrunin AV, Bochenek S, Richtering W, Scotti A. Harnessing the polymer-particle duality of ultra-soft nanogels to stabilise smart emulsions. Phys Chem Chem Phys [Internet], 2023;25(4):2810–2820. Available from: http://dx.doi.org/10.1039/D2CP02700C

[43] Kunstmann-Olsen C, Belić D, Bradley DF, Danks SP, Diaz Fernandez YA, Grzelczak MP, et al. Ion shuttling between emulsion droplets by crown ether modified gold nanoparticles. Nanoscale Adv, 2021 Jun 7;3(11):3136–3144.

[44] Yu Q, Sun N, Hu D, Wang Y, Chang X, Yan N, et al. Encapsulation of inorganic nanoparticles in a block copolymer vesicle wall driven by the interfacial instability of emulsion droplets. Polym Chem [Internet], 2021;12(29):4184–4192. Available from: http://dx.doi.org/10.1039/D1PY00744K

[45] Morais JPS, de Rosa MF, de Brito ES, de Azeredo HMC, de Figueirêdo MCB. Sustainable pickering emulsions with nanocellulose: Innovations and challenges. Foods. Multidisciplinary Digital Publishing Institute (MDPI), 2023;12.

[46] Gayathiri E, Prakash P, Karmegam N, Varjani S, Awasthi MK, Ravindran B. Biosurfactants: Potential and eco-friendly material for sustainable agriculture and environmental safety – A review. Agronomy. MDPI, 2022;12.

[47] Murashova NM. Lecithin microemulsions as drug carriers. Colloid J, 2023 Oct 1;85(5):746–756.

[48] Alhajj MJ, Montero N, Yarce CJ, Salamanca CH. Lecithins from vegetable, land, and marine animal sources and their potential applications for cosmetic, food, and pharmaceutical sectors. Cosmetics. MDPI AG, 2020;7:1–19.

[49] Ratti R. Industrial applications of green chemistry: Status, challenges and prospects. SN Appl Sci. Springer Nature, 2020;2.

[50] Klang V, Valenta C. Lecithin-based nanoemulsions. J Drug Deliv Sci Technol. Editions de Sante, 2011;21:55–76.

[51] Pal A, Mondal MH, Adhikari A, Bhattarai A, Saha B. Scientific information about sugar-based emulsifiers: A comprehensive review. RSC Adv. R Soc Chem, 2021;11:33004–33016.

[52] Stubbs S, Yousaf S, Khan I. A review on the synthesis of bio-based surfactants using green chemistry principles. DARU, J Pharm Sci. Springer Science and Business Media Deutschland GmbH, 2022;30:407–426.

[53] Veeramanoharan A, Kim SC. A comprehensive review on sustainable surfactants from CNSL: Chemistry, key applications and research perspectives. RSC Adv. R Soc Chem, 2024;14:25429–25471.

[54] Vukašinović M, Savić S, Cekić N, Ilić T, Pantelić I, Savić SD. Efficient development of green emulsifier/ emollient-based emulsion vehicles: From RSM optimal experimental design to abridged in vivo assessment. Pharmaceutics, 2023 Feb 1;15(2).

[55] Chang F, Vis CM, Ciptonugroho W, Bruijnincx PCA. Recent developments in catalysis with Pickering emulsions. Green Chem. R Soc Chem, 2021;23:2575–2594.

[56] Zhang Y, Dubé MA. Green Emulsion Polymerization Technology. In *Advances in Polymer Science*, New York LLC: Springer, 2018, pp. 65–100.

[57] Vora LK, Gholap AD, Jetha K, Thakur RRS, Solanki HK, Chavda VP. Artificial intelligence in pharmaceutical technology and drug delivery design. Pharmaceutics. Multidisciplinary Digital Publishing Institute (MDPI), 2023;15.

[58] Ali KA, Mohin S, Mondal P, Goswami S, Ghosh S, Choudhuri S. Influence of artificial intelligence in modern pharmaceutical formulation and drug development. Futur J Pharm Sci, 2024 Mar 29;10(1).

[59] Sampene AK, Nyirenda F. Evaluating the effect of artificial intelligence on pharmaceutical product and drug discovery in China. Futur J Pharm Sci, 2024 Apr 8;10(1).

[60] Gosavi AA, Nandgude TD, Mishra RK, Puri DB. Exploring the potential of artificial intelligence as a facilitating tool for formulation development in fluidized bed processor: A comprehensive review. AAPS PharmSciTech [Internet], 2024;25(5):111. Available from: https://doi.org/10.1208/s12249-024-02816-8

Index

https://doi.org/10.1515/9783111593654-011

www.ingramcontent.com/pod-product-compliance
Lightning Source LLC
Chambersburg PA
CBHW061338210326
41598CB00035B/5811